During 1996–97, the Mathematical Sciences Research Institute held a full academic year program on combinatorics, with special emphasis on the connections with other branches of mathematics, such as algebraic geometry, topology, commutative algebra, representation theory, and convex geometry.

The rich combinatorial problems arising from the study of various algebraic structures are the subject of this book, which represents work done or presented at seminars during the program. It contains contributions on matroid bundles, combinatorial representation theory, lattice points in polyhedra, bilinear forms, combinatorial differential topology and geometry, Macdonald polynomials and geometry, enumeration of matchings, the generalized Baues problem, and Littlewood–Richardson semigroups.

These expository articles, written by some of the most respected researchers in the field, present the state of the art to researchers and graduate students in combinatorics as well as algebra, geometry, and topology.

Mathematical Sciences Research Institute
Publications

38

New Perspectives in Algebraic Combinatorics

Mathematical Sciences Research Institute
Publications

Volumes 1–4 and 6–27 are available from Springer-Verlag

New Perspectives in Algebraic Combinatorics

Edited by

Louis J. Billera
Cornell University

Anders Björner
Royal Institute of Technology, Stockholm

Curtis Greene
Haverford College

Rodica E. Simion
George Washington University

Richard P. Stanley
Massachusetts Institute of Technology

CAMBRIDGE UNIVERSITY PRESS
Cambridge, New York, Melbourne, Madrid, Cape Town, Singapore,
São Paulo, Delhi, Dubai, Tokyo, Mexico City

Cambridge University Press
The Edinburgh Building, Cambridge CB2 8RU, UK

Published in the United States of America by Cambridge University Press, New York

www.cambridge.org
Information on this title: www.cambridge.org/9780521179799

First published 1999
First paperback edition 2010

A catalogue record for this publication is available from the British Library

ISBN 978-0-521-77087-3 Hardback
ISBN 978-0-521-17979-9 Paperback

New Perspectives in Geometric Combinatorics
MSRI Publications
Volume **38**, 1999

Contents

Preface

Algebraic combinatorics involves the use of techniques from algebra, topology and geometry in the solution of combinatorial problems, or the use of combinatorial methods to attack problems in these areas. Problems amenable to the methods of algebraic combinatorics arise in these or other areas of mathematics, or from diverse parts of applied mathematics. Because of this interplay with many fields of mathematics, algebraic combinatorics is an area in which a wide variety of ideas and methods come together.

During 1996–97 MSRI held a full academic year program on Combinatorics, with special emphasis on algebraic combinatorics and its connections with other branches of mathematics, such as algebraic geometry, topology, commutative algebra, representation theory, and convex geometry. Different periods of the year were devoted to research in enumeration, extremal questions, geometric combinatorics and representation theory.

The rich combinatorial problems arising from the study of these various areas are the subject of this book, which represents work done or presented at seminars during the program. It contains contributions on matroid bundles, combinatorial representation theory, lattice points in polyhedra, bilinear forms, combinatorial differential topology and geometry, Macdonald polynomials and geometry, enumeration of matchings, the generalized Baues problem, and Littlewood-Richardson semigroups. These expository articles, written by some of the most respected researchers in the field, present the state-of-the-art to graduate students and researchers in combinatorics as well as algebra, geometry, and topology.

Louis J. Billera
Anders Björner
Curtis Greene
Rodica Simion
Richard P. Stanley

New Perspectives in Geometric Combinatorics
MSRI Publications
Volume **38**, 1999

Matroid Bundles

LAURA ANDERSON

ABSTRACT. Combinatorial vector bundles, or *matroid bundles*, are a combinatorial analog to real vector bundles. Combinatorial objects called *oriented matroids* play the role of real vector spaces. This combinatorial analogy is remarkably strong, and has led to combinatorial results in topology and bundle-theoretic proofs in combinatorics. This paper surveys recent results on matroid bundles, and describes a canonical functor from real vector bundles to matroid bundles.

1. Introduction

Matroid bundles are combinatorial objects that mimic real vector bundles. They were first defined in [MacPherson 1993] in connection with *combinatorial differential manifolds*, or *CD manifolds*. Matroid bundles generalize the notion of the "combinatorial tangent bundle" of a CD manifold. Since the appearance of McPherson's article, the theory has filled out considerably; in particular, matroid bundles have proved to provide a beautiful combinatorial formulation for characteristic classes.

We will recapitulate many of the ideas introduced by McPherson, both for the sake of a self-contained exposition and to describe them in terms more suited to our present context. However, we refer the reader to [MacPherson 1993] for background not given here. We recommend the same paper, as well as [Mnëv and Ziegler 1993] on the combinatorial Grassmannian, for related discussions.

We begin with a key intuitive point of the theory: the notion of an oriented matroid as a combinatorial analog to a vector space. From this we develop matroid bundles as a combinatorial bundle theory with oriented matroids as fibers. Section 2 will describe the category of matroid bundles and its relation to the category of real vector bundles. Section 3 gives examples of matroid bundles arising in both combinatorial and topological contexts, and Section 4 outlines some of the techniques that have been developed to study matroid bundles.

Partially supported by NSF grant DMS 9803615.

1

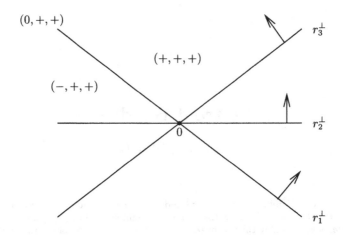

Figure 1. An arrangement of oriented hyperplanes in R^2 and some of the resulting sign vectors.

Acknowledgements. This paper drew great inspiration from Robert MacPherson's 1997 Chern Symposium lecture at Berkeley. Section 4B was written with the help of Eric Babson. The author would like to thank MacPherson, Babson, and James Davis for helpful discussions.

1A. Oriented matroids. We give a brief introduction to oriented matroids, particularly to the idea of oriented matroids as "combinatorial vector spaces". See [Björner et al. 1993] for a more complete introduction to oriented matroids, and [MacPherson 1993, Appendix] for specific notions of importance here.

A rank-n oriented matroid can be considered as a combinatorial analog to an arrangement $\{r_i\}_{i \in E}$ of vectors in $\mathbb{R}^n$, or equivalently, to an arrangement $\{r_i^\perp\}_{i \in E}$ of oriented hyperplanes. (Here we allow the "degenerate hyperplane" $0^\perp = \mathbb{R}^n$.) The idea is as follows. An arrangement $\{r_i^\perp\}_{i \in E}$ of oriented hyperplanes in $\mathbb{R}^n$ partitions $\mathbb{R}^n$ into cones. Each cone C can be identified by a sign vector $v \in \{-, 0, +\}^E$, where v_i indicates on which side of $r_i^\perp$ the cone C lies. (See Figure 1).

The set E together with the collection of sign vectors resulting from this arrangement is called a *realizable oriented matroid*. The sign vectors are called *covectors* of the oriented matroid. Every realizable oriented matroid has 0 as a covector. The hyperplanes describe a cell decomposition of the unit sphere in $\mathbb{R}^n$, with each cell labeled by a nonzero covector.

More generally, an *oriented matroid* M is a finite set E together with a collection $V^*(M)$ of signed sets in $\{-, 0, +\}^E$, satisfying certain combinatorial axioms inspired by the case of realizable oriented matroids. (For a complete definition, see [Björner et al. 1993, Section 4.1].) In this more general context, we still have

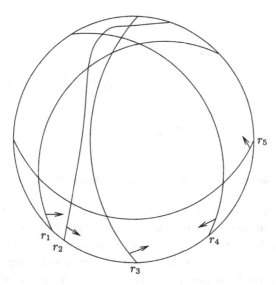

Figure 2. A rank-3 arrangement of five oriented pseudospheres.

a notion of the *rank* of an oriented matroid [MacPherson 1993, Section 5.3], and a beautiful theorem that gives this notion topological meaning.

The *Topological Representation Theorem* of Folkman and Lawrence [Björner et al. 1993, Section 1.4; Folkman and Lawrence 1978] says that the set of nonzero covectors of a rank-n oriented matroid describe a cell decomposition of S^{n-1}. More precisely: a *pseudosphere* in S^{n-1} is a subset S such that some homeomorphism of S^{n-1} takes S to an equator. Thus, a pseudosphere must partition S^{n-1} into two pseudohemispheres. An *oriented pseudosphere* is a pseudosphere together with a choice of positive pseudohemisphere. An *arrangement of oriented pseudospheres* is a set of oriented pseudospheres on S^{n-1} whose intersections behave topologically like intersections of equators. (For a precise definition, see [Björner et al. 1993, Definition 5.1.3].) For an example, see Figure 2.

An arrangement $\{S_i\}_{i \in E}$ of oriented psuedospheres in S^{n-1} determines a collection of signed sets in $\{-, 0, +\}^E$ in the same way that an arrangement of oriented hyperplanes in $\mathbb{R}^n$ does. The Topological Representation Theorem states that any collection of signed sets arising in this way is the set of nonzero covectors of an oriented matroid, and that every oriented matroid arises in this way.

1B. Oriented matroids as "combinatorial vector spaces". A *strong map image* of an oriented matroid M is an oriented matroid N such that $V^*(N) \subseteq V^*(M)$. (Strong maps are called *strong quotients* in [Gelfand and MacPherson 1992]. See [Björner et al. 1993, Section 7.7] for more on strong maps.)

Consider a realizable rank-n oriented matroid M, realized as a set $R = \{r_1^\perp, r_2^\perp, \ldots, r_m^\perp\} \subset \mathbb{R}^n$. If V is a rank-k subspace in $\mathbb{R}^n$, consider the rank-k oriented matroid $\gamma_R(V)$ given by the intersections $\{V \cap r_i^\perp : i \in \{1, \ldots, m\}\}$. In terms of the vector picture of oriented matroids, $\gamma_R(V)$ is given by the orthogonal

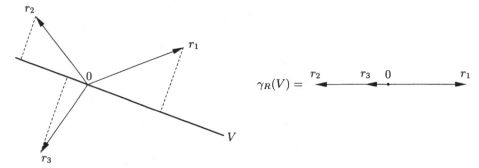

Figure 3. A strong map of realizable oriented matroids.

projections of the elements $\{r_1, \ldots, r_m\}$ onto V. The oriented matroid $\gamma_R(V)$ is a strong map image of M, and encodes considerable geometric data about V. For instance, the loops in $\gamma_R(V)$ are exactly those r_i such that $V \subseteq r_i^\perp$, and the cell decomposition of the unit sphere S_V in V given by the equators $S_V \cap r_i^\perp$ is canonically isomorphic to the cell complex of nonzero covectors of $\gamma_R(V)$. We will think of $\gamma_R(V)$ as a combinatorial model for V, and as a combinatorial "subspace" of M. Figure 3 shows a realization of a rank-2 oriented matroid, a 1-dimensional subspace V of $\mathbb{R}^2$, and the resulting oriented matroid $\gamma_R(V)$.

If M is not realizable, we will still use M as a combinatorial analog to $\mathbb{R}^n$, with the nonzero covectors in $V^*(M)$ playing the role of the unit sphere. Strong map images will be viewed as "pseudosubspaces".

1C. Matroid bundles. Consider a real rank-k vector bundle $\xi : E \to B$ over a compact base space. Choose a collection $\{e_1, \ldots, e_n\}$ of continuous sections of ξ such that at each point b in B, the vectors $\{e_1(b), \ldots, e_n(b)\}$ span the space $\xi^{-1}(b)$. The vectors $\{e_1(b), \ldots, e_n(b)\}$ determine a rank-k oriented matroid $M(b)$ with elements the integers $\{1, \ldots, n\}$. Note that any $b \in B$ has an open neighborhood U_b such that $M(b')$ *weak maps* to $M(b)$ for all $b' \in U_b$. (See [Björner et al. 1993, Section 7.7] for a definition of weak maps. Weak maps are called *specializations* in [MacPherson 1993] and *weak specializations* in [Gelfand and MacPherson 1992].)

PROPOSITION 1.1. *Let $\xi : E \to B$ be a real vector bundle with B finite-dimensional and let $\mu : |T| \to B$ be a triangulation of B. Then there exists a simplicial subdivision T' of T and a spanning collection of sections of ξ such that for every simplex σ of T', the function M is constant on the relative interior of $\mu(|\sigma|)$.*

This is a corollary to the Combinatorialization Theorem in Section 2C.

EXAMPLE. Figure 4 shows the Möbius strip as a line bundle over S^1, and a triangulation of S^1 with vertices a, b, c. The sections $\{\rho_1, \rho_2\}$ associate a single oriented matroid to the interior of each simplex.

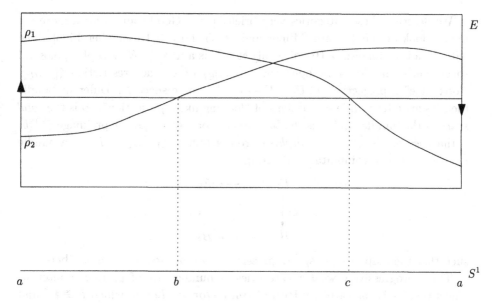

Figure 4. A spanning collection of sections for the Möbius strip.

Such a simplicial complex and the association of an oriented matroid to each cell give the motivating example of a matroid bundle:

DEFINITION 1.2. A rank-k *matroid bundle* is a partially ordered set B (e.g., a simplicial complex with simplices ordered by inclusion) and a rank-k oriented matroid $\mathcal{M}(b)$ associated to each element b, so that $\mathcal{M}(b)$ weak maps to $M(b')$ whenever $b \geq b'$.

This is a simplification of the definition which appears in [MacPherson 1993]. Any matroid bundle in the sense of MacPherson gives a matroid bundle in the present sense. Conversely, given a matroid bundle $(B, \mathcal{M})$ in our current sense, consider the *order complex* ΔB of B, i.e., the simplicial complex of all chains in the partial order. The map associating to each simplex $b_1 \leq \cdots \leq b_m$ in ΔB the oriented matroid $\mathcal{M}(b_m)$ defines a matroid bundle in the sense of MacPherson.

A matroid bundle need not arise from a real vector bundle. For instance, a matroid bundle may include non-realizable oriented matroids as fibers. Section 3 will give examples of matroid bundles arising in combinatorics that do not correspond to any real vector bundles.

1D. What do we want from matroid bundles? The hope is that the category of matroid bundles is closely related to the category of real vector bundles, or perhaps to one of its weaker cousins, such as the category of piecewise-linear microbundles or the category of spherical quasifibrations. (These categories are described below.) Relating bundle theory to oriented matroids promises both combinatorial techniques for bundle theory and bundle-theoretic techniques for combinatorics.

We describe these categories very briefly here. Good sources for a more extended look at bundles are [Milnor and Stasheff 1974; Husemoller 1996]. The loose idea is as follows: a (topological) *bundle* is a map $\xi : E \to B$ of topological spaces such that for some open cover $\{U_i\}_{i \in I}$ of B, each restriction $\xi|_{\xi^{-1}(U_i)}$ "looks like" a projection $p : U_i \times F \to U_i$, for some space F. Different bundle theories arise from different notions of "looking like a projection". E is the *total space* of the bundle, and B is the *base space*. For any $b \in B$, the preimage $\xi^{-1}(b)$ is the *fiber* of ξ over b. A *morphism* from a bundle $\xi_1 : E_1 \to B_1$ to a bundle $\xi_2 : E_2 \to B_2$ is a commutative diagram

$$
\begin{array}{ccc}
E_1 & \longrightarrow & E_2 \\
\xi_1 \downarrow & & \downarrow \xi_2 \\
B_1 & \longrightarrow & B_2
\end{array}
$$

such that the map of total spaces preserves appropriate structure on fibers.

Three progressively weaker categories of bundles are of particular interest. The strongest is the category **Bun** of *real vector bundles*, in which $F \cong \mathbb{R}^k$ and for each U_i we must have a homeomorphism $h : U_i \times \mathbb{R}^k \to \xi^{-1}(U_i)$ such that

$$
\begin{array}{ccc}
U_i \times \mathbb{R}^k & \xrightarrow{\;\;h\;\;} & \xi^{-1}(U_i) \\
& {}_{p}\searrow \quad \swarrow_{\xi} & \\
& U_i &
\end{array}
$$

commutes and h restricts to a linear isomorphism on each fiber. In the weaker category **PL** of *piecewise-linear microbundles*, F is still $\mathbb{R}^k$, but the maps h need only be piecewise-linear homeomorphisms with compatible 0 cross-sections. (See [Milnor 1961] for a precise definition.) A still weaker notion is that of a *quasifibration*, which must only "look like" a projection in that for each $x \in U_i$, $y \in \xi^{-1}(x)$, and $j \in \mathbb{N}$, the map of homotopy groups

$$
p_* : \pi_j(p^{-1}(U_i), p^{-1}(x), y) \longrightarrow \pi_j(U_i, x)
$$

is an isomorphism; see [Dold and Thom 1958, §§ 1.1, and 2.1]. From this condition it follows that each fiber has the same weak homotopy type. We will be interested in the category **Fib** of quasifibrations whose fibers are homotopy spheres. Any real vector bundle or PL microbundle has a canonical associated sphere bundle — essentially by taking a sphere around 0 in each fiber — which is a spherical quasifibration.

Associated to any good bundle theory is a *universal bundle* — that is, a bundle $\Xi : E_\infty \to B_\infty$ such that

1. for any bundle $\xi : E \to B$ there exists a morphism from ξ to Ξ, and
2. if $\xi_1 : E_1 \to B_1$ and $\xi_2 : E_2 \to B_2$ are bundles and $F : \xi_1 \to \xi_2$, $C_1 : \xi_1 \to \Xi$, and $C_2 : \xi_2 \to \Xi$ are morphisms, then there exists a *bundle homotopy* from

C_1 to $C_2 \circ F$, i.e., a morphism H from $\xi_1 \times \mathrm{id} : E_1 \times I \to B_1 \times I$ to Ξ such that $H|_{\xi_1 \times \{0\}} = C_1 \times *$ and $H|_{\xi_1 \times \{1\}} = (C_2 \circ F) \times *$.

In this situation B_∞ is called a *classifying space* for the category. For any bundle ξ and bundle map from ξ to the universal bundle, the map of base spaces is called a *classifying map*. It follows from the properties above that the universal bundle is unique up to bundle homotopy, and that for a fixed universal bundle and fixed real vector bundle, the classifying map is unique up to homotopy. In fact, every vector bundle over a base space B is characterized up to isomorphism by a homotopy class of maps from B to B_∞. Specifically, a bundle ξ over B with classifying map $c(\xi)$is isomorphic to the *pullback* of Ξ by $c(\xi)$, i.e., the bundle $\pi_1 : \{(b,v) : b \in B,\ v \in \Xi^{-1}(c(\xi)(b))\} \to B$. In this way isomorphism classes of bundles over a space B are in bijection with homotopy classes of maps $B \to B_\infty$. Thus if G_1 and G_2 are two categories of bundles with classifying spaces B_∞^1 and B_∞^2 then any map $B_\infty^1 \to B_\infty^2$ gives a functor from isomorphism classes in G_1 to isomorphism classes in G_2.

For rank-k real vector bundles over paracompact base spaces, the classifying space (often called BO_k) is $G(k, \mathbb{R}^\infty)$, the space of all k-dimensional subspaces of $\mathbb{R}^\infty$. The universal bundle is the tautological bundle

$$E_\infty = \{(V, x) : V \in G(k, \mathbb{R}^\infty),\ x \in V\} \longrightarrow G(k, \mathbb{R}^\infty),$$
$$(V, x) \qquad\qquad\qquad \longrightarrow \quad V.$$

(See [Milnor and Stasheff 1974, Chapter 5] for details.) The classifying spaces BPL_k for PL microbundles and BFib_k for spherical quasifibrations are harder to describe explicitly, and we won't attempt it here. (See [Milnor 1961, Chapter 5; Stasheff 1963] for constructions. We note in passing that BFib_k is isomorphic to the classifying space for rank-k spherical fibrations [Stasheff 1963] — see the related discussion in [Anderson and Davis $\geq$ 1999].) Since BO_k has a natural PL microbundle structure and BPL_k has an associated spherical quasifibration, there are canonical (up to homotopy) classifying maps $\mathrm{BO}_k \to \mathrm{BPL}_k \to \mathrm{BFib}_k$, giving canonical functors from real vector bundles to PL bundles to spherical quasifibrations.

How do matroid bundles fit into this picture? In Section 2A we will define morphisms of matroid bundles, leading to a category MB_k of matroid bundles. This category has a universal bundle, whose classifying space is called the *MacPhersonian* $\mathrm{MacP}(k, \infty)$. We can relate matroid bundles to other bundle theories by finding nice maps between $\mathrm{MacP}(k, \infty)$ and other classifying spaces.

Topologically, the category of matroid bundles is awkward in that the fibers are combinatorial objects — oriented matroids — which form no topological total space. In Section 4 we will discuss how the Topological Representation Theorem allows us to associate a spherical quasifibration (easily) and even a PL microbundle (gruelingly) to a matroid bundle, giving maps $\mathrm{MacP}(k, \infty) \to \mathrm{BFib}_k$ and $\mathrm{MacP}(k, \infty) \to \mathrm{BPL}_k$ and hence giving functors of bundle theories. Another key result is the Combinatorialization Theorem described in Section 2C, which implies a map $BO_k \to \mathrm{MacP}(k, \infty)$ and another functor.

Much of the progress on matroid bundles has been in the area of *characteristic classes*. A characteristic class for a bundle theory is a rule assigning to each bundle $\xi : E \to B$ an element $u(\xi)$ of $H^*(B)$ such that if

$$
\begin{array}{ccc}
E_1 & \longrightarrow & E_2 \\
\xi_1 \downarrow & & \downarrow \xi_2 \\
B_1 & \xrightarrow{\ f\ } & B_2
\end{array}
$$

is a bundle map, then $u(\xi_1) = f^*(u(\xi_2))$. (See [Milnor and Stasheff 1974] for much more on characteristic classes.) From the definition of universal bundles, it follows that if B_∞ is the classifying space for a bundle theory, then the characteristic classes are in bijection with the elements of $H^*(B_\infty)$. (Note we have not specified coefficients for cohomology: different coefficients give different interesting characteristic classes.) Thus the maps $BO_k \to \mathrm{MacP}(k, \infty)$, $\mathrm{MacP}(k, \infty) \to \mathrm{BPL}_k$, and $\mathrm{MacP}(k, \infty) \to \mathrm{BFib}_k$ give maps $H^*(\mathrm{BFib}_k) \to H^*(\mathrm{MacP}(k, \infty))$, $H^*(\mathrm{BPL}_k) \to H^*(\mathrm{MacP}(k, \infty))$, and $H^*(\mathrm{MacP}(k, \infty)) \to H^*(\mathrm{BO}_k)$ between the characteristic classes of the respective bundle theories. In various cases (e.g., with $\mathbb{Z}_2$ coefficients) these maps can be shown to be surjective. This gives new results on the topology of $\mathrm{MacP}(k, \infty)$ and connects matroid bundles to the many areas of topology that can be described in terms of characteristic classes.

2. Categories of Matroid Bundles and PL Vector Bundles

2A. The category of matroid bundles. Let B be the poset of cells in a PL cell complex $\mathcal{B}$. Any matroid bundle $(B, \mathcal{M})$ on B induces a canonical matroid bundle structure on the poset of cells of any PL subdivision of $\mathcal{B}$, by associating the oriented matroid $\mathcal{M}(\sigma)$ to each cell in the relative interior of a cell $\sigma \in B$. Two matroid bundles on PL cell complexes are defined to be *equivalent* if there exists a common PL subdivision of the cell complexes such that the resulting matroid bundles on this subdivision are identical.

For B an arbitrary poset, a matroid bundle $(B, \mathcal{M})$ induces a matroid bundle structure $(\Delta B, \mathcal{M}')$ on the cell complex $\|\Delta B\|$ by defining

$$
\mathcal{M}'(\{b_1 < b_2 < \cdots < b_m\}) = \mathcal{M}(b_m).
$$

We extend the above notion of equivalence by defining $(B, \mathcal{M})$ to be equivalent to $(\Delta B, \mathcal{M}')$.

DEFINITION 2.1. If $(B_1, \mathcal{M}_1)$ and $(B_2, \mathcal{M}_2)$ are two matroid bundles, a *morphism* from $(B_1, \mathcal{M}_1)$ to $(B_2, \mathcal{M}_2)$ is a pair $(f, [C_f, \mathcal{M}_f])$, where f is a PL map from ΔB_1 to ΔB_2 and $[C_f, \mathcal{M}_f]$ is an equivalence class of matroid bundle structures on the mapping cylinder of f that restrict to structures equivalent to $(B_1, \mathcal{M}_1)$ and $(B_2, \mathcal{M}_2)$ at either end.

The *composition* of a morphism $(f, [C_f, \mathcal{M}_f])$ from $(B_1, \mathcal{M}_1)$ to $(B_2, \mathcal{M}_2)$ and a morphism $(g, [C_g, \mathcal{M}_g])$ from $(B_2, \mathcal{M}_2)$ to $(B_3, \mathcal{M}_3)$ is $(g \circ f, [C_{g \circ f}, \mathcal{M}_{g \circ f}])$, where $\mathcal{M}_{g \circ f}$ is determined by $\mathcal{M}_3$ on the simplices of B_3 and by $\mathcal{M}_f$ on the rest of the cells of $C_{g \circ f}$.

The set of rank-k matroid bundles and their morphisms form a category.

DEFINITION 2.2. A morphism from $(B_1, \mathcal{M}_1)$ to $(B_2, \mathcal{M}_2)$ is an *isomorphism* if there exists a morphism from $(B_2, \mathcal{M}_2)$ to $(B_1, \mathcal{M}_1)$ such that the composition of these maps is the identity morphism.

We get a better relation to the category of rank-k real vector bundles by considering only isomorphism classes of matroid bundles:

DEFINITION 2.3. MB_k will denote the category of isomorphism classes of rank-k matroid bundles and their morphisms.

The classifying space for matroid bundles. MB_k has a classifying space very similar in spirit (and, as we shall later see, in topology) to the classifying space $G(k, \mathbb{R}^\infty)$ for real rank-k vector bundles. Just as $G(k, \mathbb{R}^\infty)$ is the space of all rank-k subspaces of any $\mathbb{R}^n$, the classifying space for MB_k will be the set of all strong map images of any combinatorial model for any $\mathbb{R}^n$.

DEFINITION 2.4. If M^n is a rank-n oriented matroid then define the *combinatorial Grassmannian* $\Gamma(k, M^n)$ to be the poset of all rank-k strong map images of M^n, with the partial order $M_1 \geq M_2$ if and only if M_1 weak maps to M_2.

In some papers the combinatorial Grassmannian is defined to be the order complex $\Delta\Gamma(k, M^n)$ of $\Gamma(k, M^n)$.

The combinatorial Grassmannian was first introduced in [MacPherson 1993] and was the subject of a previous survey article [Mnëv and Ziegler 1993], to which we refer the reader for further discussion.

A particularly useful case is when M^n is the coordinate oriented matroid:

DEFINITION 2.5. Let M_n be the coordinate oriented matroid with elements $\{1, 2, \ldots, n\}$, i.e., the oriented matroid realized by the coordinate hyperplanes in $\mathbb{R}^n$. Then $\Gamma(k, M_n)$ is a *standard combinatorial Grassmannian*, or *MacPhersonian*, denoted MacP(k, n).

This case is especially important because of a nice alternate description:

PROPOSITION 2.6 [Mnëv and Ziegler 1993]. MacP(k, n) *is the poset of all rank-k oriented matroids with elements* $\{1, 2, \ldots, n\}$, *ordered by weak maps.*

Note that if M_1 strong maps to M_2 then $\Gamma(k, M_2) \subseteq \Gamma(k, M_1)$ (and hence $\Delta\Gamma(k, M_2)$ is a subcomplex of $\Delta\Gamma(k, M_1)$). In particular:

- If $\{1, \ldots, n\}$ is the set of elements of M, $\Gamma(k, M)$ is a subposet of MacP(k, n).
- If M_2 is obtained from M_1 by deleting some elements, there is a natural embedding of $\Gamma(k, M_2)$ into $\Gamma(k, M_1)$. In particular, MacP$(k, n) \hookrightarrow$ MacP$(k, n+1)$ for any k and n.

Thus the direct limit $\lim_{n\to\infty} \Gamma(k, M^n)$ in the category of posets and inclusions is $\bigcup_n \mathrm{MacP}(k, n)$, denoted $\mathrm{MacP}(k, \infty)$.

We can now rephrase the definition of matroid bundles:

DEFINITION 2.7. A rank-k *matroid bundle* is a poset B and a poset map $\mathcal{M}$: $B \to \mathrm{MacP}(k, \infty)$.

Modifying definitions appropriately (to accomodate our combinatorial notion of bundles and bundle morphisms), we see:

PROPOSITION 2.8. *The map* id : $\mathrm{MacP}(k, \infty) \to \mathrm{MacP}(k, \infty)$ *is the universal bundle for* MB_k.

PROOF. A matroid bundle $\mathcal{M} : B \to \mathrm{MacP}(k, \infty)$ determines a simplicial map from ΔB to $\Delta \mathrm{MacP}(k, \infty)$, and $\mathcal{M}$ induces a matroid bundle structure on the mapping cylinder, giving a classifying map. If $(f, [C_f, \mathcal{M}_f])$ is a matroid bundle morphism, then $(C_f, \mathcal{M}_f)$ determines a homotopy between the respective classifying maps.   □

Thus $\mathrm{MacP}(k, \infty)$ is the classifying space for rank-k matroid bundles.

The cohomology of a poset P is defined to be the cohomology of its order complex ΔP. Thus we have:

COROLLARY 2.9. *The characteristic classes for* MB_k *with coefficients in* R *are the elements of the cohomology ring* $H^*(\Delta \mathrm{MacP}(k, \infty); R)$.

The finite combinatorial Grassmannians are of interest in their own right from several perspectives. The spaces $\Delta \Gamma(k, M^n)$ arise as the fibers of a combinatorial Grassmannian bundle in [MacPherson 1993], for instance, and $\Delta \Gamma(n-1, M^n)$ is closely related to the *extension space* $\mathcal{E}(M^n)$ discussed in Section 3.

2B. Relations between the real and combinatorial Grassmannians. We consider more closely the map

$$\gamma_R : G(k, \mathbb{R}^n) \to \Gamma(k, M^n)$$

introduced in Section 1B. The set of preimages of this map give a stratification of $G(k, \mathbb{R}^n)$ which is semialgebraic. This stratification has the property that if the closure of $\gamma_R^{-1}(M_1)$ intersects $\gamma_R^{-1}(M_2)$ then M_1 weak maps to M_2.

By the semi-algebraic triangulation theorem [Hironaka 1975], there exists a triangulation of $G(k, \mathbb{R}^n)$ refining this stratification, giving a simplicial map

$$\tilde{\gamma}_R : G(k, \mathbb{R}^n) \to \Delta \Gamma(k, M^n).$$

(This is described further in [MacPherson 1993] for $\mathrm{MacP}(k, n)$ and in [Anderson and Davis $\geq$ 1999] for more general M^n.) In the direct limit this gives a map $\tilde{\gamma} : G(k, \mathbb{R}^\infty) \to \Delta \mathrm{MacP}(k, \infty)$ of classifying spaces, and hence describes a map from the theory of real vector bundles to the theory of matroid bundles.

The hope is that the map $\tilde{\gamma}_R$ preserves a great deal of the topology of $G(k, \mathbb{R}^n)$. For instance, if the resulting map in cohomology were an isomorphism, then the process of making matroid bundles out of real vector bundles would preserve the theory of characteristic classes. There are numerous grounds for pessimism on this hope. These grounds are detailed in [Björner et al. 1993, Section 2.4]. To give two of the most glaring obstacles, this map is not surjective — it misses all the non-realizable oriented matroids — and the stratification of $G(k, \mathbb{R}^n)$ by preimages of γ_R can have strata with arbitrarily ugly topology [Mnëv 1988]. As an illustration of how bad the topology of combinatorial Grassmannians can be, we note examples by Mnëv and Richter-Gebert [1993] of non-realizable oriented matroids M^n with the property that $\Delta\Gamma(n-1, M^n)$ is disconnected.

This makes the array of positive results on γ_R rather surprising. For small values of k, for the first few homotopy groups, and for realizable oriented matroids we have results relating the real and combinatorial Grassmannians.

THEOREM 2.10 [Folkman and Lawrence 1978]. $\Delta\Gamma(1, M^n)$ *is homeomorphic to* $G(1, \mathbb{R}^n)$. *If* M^n *is realizable,* $\gamma_R : G(1, \mathbb{R}^n) \to \Delta\Gamma(1, M^n)$ *is a homeomorphism.*

THEOREM 2.11 [Babson 1993]. $\Delta\Gamma(2, M^n)$ *is homotopy equivalent to* $G(2, \mathbb{R}^n)$.

We will discuss the proof of Theorem 2.11 in Section 4B.

THEOREM 2.12 (compare [Mnëv and Ziegler 1993]). *Duality holds for the standard combinatorial Grassmannians: if* $|E| = n$, *then* $\Delta\Gamma(k, E) \cong \Delta\Gamma(n-k, E)$.

THEOREM 2.13 (compare [Mnëv and Ziegler 1993; Anderson 1998]).

1. *If* $i = 0$ *or* $i = 1$ *then* $(\tilde{\gamma}_R)_* : \pi_i(G(k, \mathbb{R}^n)) \to \pi_i(\Delta \operatorname{MacP}(k, n))$ *is an isomorphism. Further,*

$$(\tilde{\gamma}_R)_* : \pi_2(G(k, \mathbb{R}^n)) \to \pi_2(\Delta \operatorname{MacP}(k, n))$$

is a surjection.
2. $\eta_* : \pi_i(\Delta \operatorname{MacP}(k, n)) \to \pi_i(\Delta \operatorname{MacP}(k, n+1))$ *is an isomorphism if* $n > k(i + 2)$ *and a surjection for* $n > k(i + 1)$.

THEOREM 2.14 [Anderson and Davis $\geq$ 1999; Anderson et al. $\geq$ 1999].

1. *The maps* $\gamma_R^* : H^*(\Delta\Gamma(k, M^n); \mathbb{Z}_2) \to H^*(G(k, \mathbb{R}^n); \mathbb{Z}_2)$ *for realizable* M^n *and* $\tilde{\gamma}^* : H^*(\Delta \operatorname{MacP}(k, \infty); \mathbb{Z}_2) \to H^*(G(k, \mathbb{R}^\infty); \mathbb{Z}_2)$ *are split surjections.*
2. *The maps* $\gamma_R^* : H^*(\Delta\Gamma(k, M^n); \mathbb{Q}) \to H^*(G(k, \mathbb{R}^n); \mathbb{Q})$ *for realizable* M^n *and* $\tilde{\gamma}^* : H^*(\Delta \operatorname{MacP}(k, \infty); \mathbb{Q}) \to H^*(G(k, \mathbb{R}^\infty); \mathbb{Q})$ *are split surjections.*

Theorem 2.14 follows from the constructions of combinatorial sphere bundles associated to matroid bundles described in Section 4A. In terms of characteristic classes, this theorem implies that matroid bundles have well-defined Stiefel–Whitney and Pontrjagin classes.

Associated to a rank-k oriented real vector bundle $\xi : E \to B$ and its associated sphere bundle $E_0 \to B$ is a cohomology class $u \in H^k(E, E_0, \mathbb{Z})$ whose restriction to each fiber is the orientation class of that fiber. The *Thom isomorphism* of the vector bundle is the isomorphism

$$\phi : H^i(B; \mathbb{Z}) \longrightarrow H^{i+k}(E, E_0; \mathbb{Z})$$
$$x \longrightarrow \xi^*(x) \cup u$$

[Milnor and Stasheff 1974, Chapter 9]. There is a unique class $e(\xi)$ in $H^k(B; \mathbb{Z})$ which is mapped to $u|_E$ under ξ^*. This is a characteristic class of ξ, called the *Euler class*.

The combinatorial sphere bundles associated to matroid bundles admit analogous constructions, giving a further result on characteristic classes.

THEOREM 2.15 [Anderson and Davis $\geq$ 1999]. *There is a well-defined Thom isomorphism and Euler class for matroid bundles.*

2C. The category of PL vector bundles. To compare real vector bundles and matroid bundles, we need to restrict to real vector bundles with triangulable base spaces. Specifically:

DEFINITION 2.16. Define VB_k to be the category whose objects are all rank-k real vector bundles over PL spaces and whose morphisms are the vector bundle maps preserving this PL structure.

This section will give a functor from VB_k to MB_k. Section 1C gave one way to associate a matroid bundle to a real vector bundle with a triangulated base space. Here we will use a related method that is less intuitive but more useful. Consider the map $\gamma_R : G(k, \mathbb{R}^n) \to \Gamma(k, M^n)$ for a realizable M and the direct limit map $\gamma : G(k, \mathbb{R}^\infty) \to \text{MacP}(k, \infty)$. For any $V \in G(k, \mathbb{R}^\infty)$ there is a small neighborhood of V which is mapped by γ to $\text{MacP}(k, \infty)_{\geq \gamma(V)}$.

If a real vector bundle $\xi : E \to B$ over a PL cell complex B has classifying map $c : B \to G(k, \mathbb{R}^\infty)$, the composition $\gamma \circ c$ associates an oriented matroid to each point in B. We call c *tame* if $\gamma \circ c$ is constant on the interior of each cell. Thus a tame classifying map defines a matroid bundle structure on the poset of cells of B.

COMBINATORIALIZATION THEOREM [Anderson and Davis $\geq$ 1999]. *Let* $\xi = (\pi : E \to \|B\|)$ *be a rank-k real vector bundle, where B is a finite-dimensional simplicial complex.*

1. *ξ has a classifying map which is tame with respect to some simplicial subdivision of B.*
2. *For $i = 0, 1$, let $c_i : B \to G(k, V_i)$ be a tame classifying maps for ξ. Then there is a tame classifying map $h : B \times I \to G(k, V_0 \oplus V_1)$, restricting to c_i on $B \times \{i\}$.*

A slightly more complicated form of this theorem holds for bundles over infinite-dimensional spaces: see [Anderson and Davis $\geq$ 1999] for details.

A classifying map to $G(k, \mathbb{R}^n)$ determines a spanning collection of n sections of the bundle, by projecting the unit coordinate vectors in $\mathbb{R}^n$ onto the images of the fibers. Thus Proposition 1.1 is a corollary to the above theorem.

THEOREM 2.17. 1. *Let $\xi : E \to B$ be an element of VB_k and let $\mu_0 : |T_0| \to B$ and $\mu_1 : |T_1| \to B$ be two triangulations of B. Let c_0, c_1 be two classifying maps for ξ such that $\gamma \circ c_i \circ \mu_i^{-1}$ is constant on the interior of each simplex. Then the matroid bundles arising from c_0 and c_1 are isomorphic.*

Thus every element ξ of VB_k gives rise to a unique element $F(\xi)$ of MB_k.

2. *If $\Xi : \xi_1 \to \xi_2$ is a morphism in VB_k, then there exists a simplicial decomposition C_Ξ on the mapping cylinder of the base spaces giving a morphism from $F(\xi_1)$ to $F(\xi_2)$ in MB_k (unique up to equivalence), denoted $F(\Xi)$.*

PROOF. 1. Since c_0 and c_1 are both classifying maps for ordinary vector bundles, we know there exists a homotopy $H : B \times I \to G(k, \mathbb{R}^\infty)$ from c_0 to c_1 in the category of ordinary vector bundles. $B \times I$ has a PL structure induced by B, and μ_0 and μ_1 give PL triangulations of $B \times \{0\}$ and $B \times \{1\}$, respectively. By [Hudson 1969, Corollary 1.6], there exists a triangulation $\mu : |T| \to B \times I$ which restricts to a subdivision of the given triangulations at either end. Note that the composition of $H|_{B \times \{0,1\}}$ with $\tilde{\gamma} : G(k, \mathbb{R}^\infty) \to \Delta \operatorname{MacP}(k, \infty)$ is simplicial with respect to T. By the Simplicial Approximation Theorem, there is a simplicial map homotopic to $\tilde{\gamma} \circ H$. Since the only simplicial approximation to a simplicial map is itself, this simplicial map must restrict to c_i on $B \times \{i\}$.

2. Because the map of base spaces is PL, as above we get a triangulation C_Ξ of the mapping cylinder which restricts to PL triangulations of the base spaces at either end. Any two such triangulations have a common simplicial subdivision. Again applying the simplicial approximation theorem, we get a matroid bundle structure $\mathcal{M}_\Xi$ on C_Ξ which restricts to the appropriate matroid bundle isomorphism classes at either end. Thus C_Ξ and $\mathcal{M}_\Xi$ define a matroid bundle morphism. $\square$

The map F defined in the previous theorem is easily seen to be a covariant functor from VB_k to MB_k.

3. Examples of Matroid Bundles

Matroid bundles arise in various contexts besides that of real vector bundles. We give some examples here.

Combinatorial Grassmannians. For any oriented matroid M^n, the identity map on $\Gamma(k, M^n)$ defines a matroid bundle. As mentioned before, this bundle is of independent combinatorial interest, for instance in relation to the next example.

Extension spaces. Let M be an oriented matroid with elements E. A *nonzero extension* of M by x is an oriented matroid M' with elements $E \cup \{x\}$ such that $M' \backslash x = M$ and M is not a loop in M'. The nonzero extensions of M form a poset $\mathcal{E}(M)$, ordered by weak maps. Since nonzero extensions $M \cup x$ correspond exactly to orientations on corank-1 strong map images M/x of M, there is a canonical double cover $\mathcal{E}(M) \to \Gamma(\operatorname{rank} M - 1, M)$. Thus $\mathcal{E}(M)$ has a natural matroid bundle structure.

If M and all extensions of M are realizable, then the order complex $\Delta\mathcal{E}(M)$ is a $(\operatorname{rank} M - 1)$-sphere, and its matroid bundle structure arises from the standard tangent bundle structure on the sphere. For more general M, life is not nearly so simple. As mentioned before, Mnëv and Richter-Gebert [1993] have found examples for which $\Delta\mathcal{E}(M)$ is not even connected. The topology of general $\Delta\mathcal{E}(M)$ is a mystery of some importance — for instance, due to its connection to the Generalized Baues Conjecture [Mnëv and Ziegler 1993; Reiner 1999].

Combinatorial differential manifolds. The theory of matroid bundles arose out of the theory of *combinatorial differential manifolds*, or *CD manifolds*, the main subject of [MacPherson 1993]. A CD manifold consists of a simplicial complex equipped with an atlas of oriented matroid coordinate charts. Such an atlas yields a matroid bundle structure in a canonical way. All of this is described in detail in [MacPherson 1993]. We review the idea briefly here.

Let N be a differential manifold and $\eta : |T| \to N$ be a piecewise smooth triangulation of N (i.e., smooth on every closed simplex). Consider a point t in the interior of some simplex $|\sigma|$ of $|T|$. For each vertex s of $\overline{\operatorname{star} \sigma}$, the line segment from t to s in $|\sigma \cup \{s\}|$ gives a smooth path $p_s : [0,1] \to N$ from $\eta(t)$ to $\eta(s)$ in N. The tangent vectors $\{p'_s(0) : s \in \overline{\operatorname{star} \sigma}^0\} \subset T_{\eta(t)}(N)$, each defined up to a positive scalar, give an oriented matroid $M(t)$. This oriented matroid can be viewed as the combinatorial remnant of an embedding of $|\operatorname{star} \sigma|$ into $T_{\eta(t)}(N)$, and hence as a combinatorial coordinate chart.

A triangulation is *tame* if there exists a subdivision of T into regular cells so that $M \circ \eta$ is constant on each open cell. Thus associated to a tame triangulation there is a triple $(T, \hat{T}, \mathcal{M})$, with

- T a simplicial complex,
- $\hat{T}$ a regular subdivision of T,
- $\mathcal{M} : \hat{T} \to \operatorname{MacP}(n, |T^0|)$ a map associating to each cell of $\hat{T}$ an oriented matroid.

This triple constitutes a CD manifold.

More generally, a CD manifold is defined to be any triple $(T, \hat{T}, \mathcal{M})$ satisfying certain combinatorial axioms intended to make M look like an atlas of coordinate charts on T (see the complete definition in [MacPherson 1993]). A canonical construction of a matroid tangent bundle associated to any such atlas is given in [MacPherson 1993, Section 3.2].

CD manifolds are the basis for the combinatorial formula for Pontrjagin classes found in [Gelfand and MacPherson 1992]. This formula is discussed in more detail in Section 4C.

Not all CD manifolds arise from differential manifolds (for instance, there exist CD manifolds involving non-realizable oriented matroids). Indeed, nothing in the definition of CD manifolds promises immediately that the base space of a CD manifold is a topological manifold.

CONJECTURE 3.1. *If $(X, \hat{X}, M)$ is a CD manifold, then X is a PL manifold.*

In [Anderson 1999a] this conjecture was proved under the restriction that all oriented matroids involved are *Euclidean*. Euclidean oriented matroids [Björner et al. 1993, Section 10.5] are essentially oriented matroids with the intersection properties of a real vector arrangement. The class of Euclidean oriented matroids includes all realizable oriented matroids and all oriented matroids of rank less than 4.

4. Methods

We outline here some of the most important methods that have developed for studying matroid bundles.

4A. Sphere bundles. One disturbing aspect of the definition of matroid bundles is the absence of a topological total space. Our understanding of real vector bundles $\xi : E \to B$ follows largely from various constructions involving the total space E, or the sphere bundle $\{(b, v) : b \in B, v \in \xi^{-1}(b), v \neq 0\} \to B$. The Topological Representation Theorem leads to combinatorial analogs to these for matroid bundles. As mentioned in Section 1B, the intuition that makes an oriented matroid M a combinatorial model for a vector space also makes $V^*(M)\backslash 0$ a combinatorial model for the unit sphere in that vector space. We will discuss two ways to use these spheres to construct combinatorial sphere bundles over matroid bundles.

A spherical quasifibration associated to a matroid bundle.

DEFINITION 4.1. If $\mathcal{M} : B \to \mathrm{MacP}(k, \infty)$ is a matroid bundle, define

$$E(\mathcal{M}) = \{(b, X) : b \in B, X \in V^*(\mathcal{M}(b))\},$$
$$E_0(\mathcal{M}) = \{(b, X) : b \in B, X \in V^*(\mathcal{M}(b))\backslash 0\}.$$

Each of these sets is partially ordered by $(b_1, X_1) \geq (b_2, X_2)$ if and only if $b_1 \geq b_2$ and $X_1 \geq X_2$.

Note the following properties of $E(\mathcal{M})$ and $E_0(\mathcal{M})$:

- The projections $\pi' : E(\mathcal{M}) \to B$ and $\pi : E_0(\mathcal{M}) \to B$ onto the first component are poset maps. Thus they give simplicial maps $\Delta E(\mathcal{M}) \to \Delta B$ and $\Delta E_0(\mathcal{M}) \to \Delta B$.

Figure 5. Fibers over elements of a poset $1 > 0$.

- For any element b of B, the fiber $\Delta\pi^{-1}(b)$ is the barycentric subdivision of the pseudosphere complex $V^*(M)\backslash 0$, hence is a sphere. The fiber $\Delta(\pi')^{-1}(b)$ is the cone on this sphere.
- If $\xi : \mathcal{E} \to \mathcal{B}$ is a vector bundle with a triangulation $\eta : \|B\| \to \mathcal{B}$ and a tame classifying map giving a matroid bundle $\mathcal{M} : B \to \mathrm{MacP}(k, \infty)$, then for any vertex b of ΔB, the unit sphere in $\xi^{-1}(\eta(b))$ is canonically isomorphic to the fiber over b in $\Delta E_0(\mathcal{M})$.
- The simplicial map $\Delta E_0(M) \to \Delta B$ is *not* necessarily a topological sphere bundle. For example, let B be the poset $1 > 0$ and $\mathcal{M} : B \to \mathrm{MacP}(2,3)$ be as shown in Figure 5.

  The fiber over each vertex of ΔB is an S^1, but the fiber over the 1-simplex $\{1,0\}$ is not an $S^1 \times I$, since it contains a 3-dimensional simplex $\{\{1,0\}, a^-b^+c^+\} > \{\{1,0\}, a^-b^0c^+\} > \{\{0\}, a^-b^0c^+\} > \{\{0\}, a^0b^0c^+\}\}$.

THEOREM 4.2 [Anderson and Davis $\geq$ 1999].

1. $\Delta E_0 \to \Delta B$ *is a spherical quasifibration.*
2. *A morphism of matroid bundles induces a morphism of the corresponding spherical quasifibrations.*
3. *Let $E_G \to G(k, \mathbb{R}^n)$ be the unit sphere bundle over $G(k, \mathbb{R}^n)$, and let M be a realizable rank-n oriented matroid. Then there is a map of spherical quasifibrations*

$$
\begin{array}{ccc}
E_G & \longrightarrow & \Delta E_0 \\
\downarrow & & \downarrow \\
G(k, \mathbb{R}^n) & \xrightarrow{\;c\;} & \Delta\Gamma(k, M)
\end{array}
$$

The $\mathbb{Z}_2$ cohomology of $G(k, \mathbb{R}^n)$ is generated by the *Stiefel–Whitney classes* [Milnor and Stasheff 1974]. The classifying space BFib for spherical quasifibrations also has well-defined Stiefel–Whitney classes [Stasheff 1963]. Given a map of spherical quasifibrations, the induced map on the cohomology rings of the base spaces preserves Stiefel–Whitney classes. Thus as a corollary to Theorem 4.2 we have Theorem 2.14.1. See [Anderson and Davis $\geq$ 1999] for constructions of explicit combinatorial Stiefel–Whitney classes in $H^*(\Delta\Gamma(k, M); \mathbb{Z}_2)$.

In addition, a spherical quasifibration gives a Serre spectral sequence, whose collapsing gives the Thom isomorphism and Theorem 2.15.

A PL microbundle associated to a matroid bundle. A much more complicated construction associates to any matroid bundle a PL microbundle. For the full construction see [Anderson et al. $\geq$ 1999]. Here we will outline some key points of the construction.

THEOREM 4.3 [Anderson 1999b]. *Let* $M_1 \rightsquigarrow M_2$ *be a weak map of oriented matroids.*

1. *If* $X \in V^*(M_1)$ *then there is a unique maximal* $g(X) \in V^*(M_2)$ *such that* $X \geq g(X)$.
2. *The map* $g : V^*(M_1) \to V^*(M_2)$ *thus defined is a poset map.*
3. *If* $Y \in V^*(M_2)$ *then* $g^{-1}(Y)$ *is contractible.*
4. *If* $\operatorname{rank}(M_1) = \operatorname{rank}(M_2)$ *and* $X \in V^*(M_1) \backslash 0$ *then* $g(X) \neq 0$.

Thus a weak map $M_1 \rightsquigarrow M_2$ gives a simplicial map $g : V^*(M_1) \to V^*(M_2)$ of balls, and, if the oriented matroids have the same rank, a simplicial map $g : V^*(M_1) \backslash 0 \to V^*(M_2) \backslash 0$ of spheres.

THEOREM 4.4 [Anderson et al. $\geq$ 1999]. *If* $M_1 \rightsquigarrow M_2 \rightsquigarrow M_3$ *are oriented matroids and* g_{ij} *denotes the poset map* $V^*(M_j) \backslash 0 \to V^*(M_i) \backslash 0$ *as above, then* $g_{32} \circ g_{21}(X) \leq g_{31}(X)$ *for every* $X \in V^*(M_1) \backslash 0$.

If this last inequality were an equality, then g would give a functor from the category Γ_k of rank-k oriented matroids and weak maps to the category of PL spheres and PL maps, and the homotopy colimit of the image of this functor would give a PL sphere bundle over any matroid bundle. With the inequality we have, a much more delicate construction (detailed in [Anderson et al. $\geq$ 1999]) uses g to construct a PL sphere bundle over the second barycentric subdivision of the base space of any matroid bundle. Since PL microbundles have well-defined rational Pontrjagin classes and these classes generate $H^*(G(k, \mathbb{R}^\infty); \mathbb{Q})$, we have Theorem 2.14.2.

4B. Hairs. Babson [1993] generalized combinatorial Grassmannians to *combinatorial flag spaces*, and developed a new tool, *hairs*, to study rank-2 strong map images.

Let $G(v_1, \ldots, v_m, \mathbb{R}^n)$ denote the topological space of all flags $V_1 \subset V_2 \subset \cdots \subset V_m \subseteq \mathbb{R}^n$ of subspaces in $\mathbb{R}^n$ with $\dim V_i = v_i$ for every i. As a combinatorial analog to these flag spaces, we define:

DEFINITION 4.5. *If* M^n *is an oriented matroid and* $\{v_1 < \cdots < v_m \leq n\}$ *is a chain in* $\mathbb{N}$, *let*

$$\Gamma(v_1, \ldots, v_m, M^n)$$
$$= \{(N_1, \ldots, N_m) : N_i \in \Gamma(v_i, N_{i+1}) \text{ if } i < m \text{ and } N_m \in \Gamma(v_m, M)\}.$$

This is a poset, ordered by componentwise weak maps.

THEOREM 4.6 [Babson 1993]. *Let M^n be any oriented matroid of rank greater than 1.*

1. $\Delta\Gamma(1,2,M^n)$ *is homotopy equivalent to* $G(1,2,\mathbb{R}^n)$.
2. $\Delta\Gamma(2,M^n)$ *is homotopy equivalent to* $G(2,\mathbb{R}^n)$.

Let p_1 denote the projection $\Delta\Gamma(1,2,M^n) \to \Delta\Gamma(1,M^n)$ and p_2 the projection $\Delta\Gamma(1,2,M^n) \to \Delta\Gamma(2,M^n)$. The latter projection is easily seen to be a quasifibration. The crucial result in the proof of Theorem 4.6 is the following:

LEMMA 4.7. p_1 *is a quasifibration with fiber a homotopy* $\mathbb{RP}^{n-2}$.

The concept of *hairs* arises in the proof of this lemma. Note that an element of $\Gamma(1,M^n)$ is just a pair $\{X,-X\}$ of antipodal covectors, and an element of $\Gamma(1,2,M^n)$ is such a pair together with an embedded circle in $\Delta V^*(M)$ containing X and $-X$. It is convenient to consider the space $\tilde{\Gamma}(1,2,M^n)$ of such pairs together with an orientation on the circle. Then the projection $p : \tilde{\Gamma}(1,2,M^n) \to \Gamma(1,2,M^n)$ is a double cover, and one shows that $\tilde{p}_1 = p_1 \circ p$ is a quasifibration with fiber a homotopy sphere.

Since each such circle has antipodal symmetry, an element of $\tilde{\Gamma}(1,2,M^n)$ can be represented by an oriented path of covectors from X to $-X$. Babson's proof factors $\tilde{p}_1$ through a series of intermediate spaces which, instead of recording a path from X to $-X$, record only a shorter path which starts at X and is contained in a rank-2 strong image of M^n. These paths are called *hairs*. At each stage of the factorization the ends of the hairs are cut off, until at the last stage they are simply a single covector (an element of $\Gamma(1,M^n)$). All the intermediate projections except the last are homotopy equivalences, while the last is easily seen to be a quasifibration.

To prove Theorem 4.6, fix a basis $\mathcal{B}$ for M^n, and let M_0 be the oriented matroid obtained from M^n by deleting all elements not in $\mathcal{B}$. Then $\Delta\Gamma(1,M_0)$, $\Delta\Gamma(2,M_0)$, and $\Delta\Gamma(1,2,M_0)$ are easily seen to be homotopic to the analogous real flag spaces. The first part of Theorem 4.6 is proved by considering the diagram

$$
\begin{array}{ccc}
\Delta\Gamma(1,2,M^n) & \longrightarrow & \Delta\Gamma(1,M^n) \\
{\scriptstyle d_{12}}\downarrow & & \downarrow{\scriptstyle d_1} \\
\Delta\Gamma(1,2,M_0) & \longrightarrow & \Delta\Gamma(1,M_0)
\end{array}
$$

where the vertical maps are obtained by deleting elements not in $\mathcal{B}$ from each oriented matroid. The map d_1 is easily seen to be a homotopy equivalence: both spaces are homeomorphic to $\mathbb{RP}^{n-1}$. Lemma 4.7 showed that the horizontal maps are quasifibrations with fiber $\mathbb{RP}^{n-2}$. One can also check that the induced maps on fibers are weak homotopy equivalences. A quasifibration has an associated

long exact sequence of homotopy groups [Dold and Thom 1958], so we have a
diagram with exact rows:

$$\cdots \pi_i(p_1^{-1}(x), y) \longrightarrow \pi_i(\Delta\Gamma(1, 2, M^n), y) \longrightarrow \pi_i(\Delta\Gamma(1, M^n), x) \longrightarrow \pi_{i-1}(p_1^{-1}(x), y) \cdots$$

$$\downarrow \qquad\qquad\qquad \downarrow \qquad\qquad\qquad \downarrow \qquad\qquad\qquad \downarrow$$

$$\cdots \pi_i(p_1^{-1}(x'), y') \longrightarrow \pi_i(\Delta\Gamma(1, 2, M_0), y') \longrightarrow \pi_i(\Delta\Gamma(1, M_0), x') \longrightarrow \pi_{i-1}(p_1^{-1}(x'), y') \cdots$$

where x and x' are base points in $\Delta\Gamma(1, M^n)$ and $\Delta\Gamma(1, M_0)$, respectively, and
y and y' are base points in the fibers of p_1 over x and x', respectively. It follows
that $d_{12*} : \pi_i(\Delta\Gamma(1, 2, M^n), y) \to \pi_i(\Delta\Gamma(1, 2, M_0), y')$ is an isomorphism.

   Knowing this, one proves the second statement of Theorem 4.6 by considering
the diagram

$$\begin{array}{ccc} \Delta\Gamma(2, M^n) & \longleftarrow & \Delta\Gamma(1, 2, M^n) \\ d_2 \downarrow & & \downarrow d_{12} \\ \Delta\Gamma(2, M_0) & \longleftarrow & \Delta\Gamma(1, 2, M_0). \end{array}$$

   The horizontal maps are quasifibrations, and a diagram as before proves d_2 is
a homotopy equivalence.

4C. Pontrjagin classes. The theorem of the previous section is the first step in
an approach of Gelfand and MacPherson to finding a combinatorial formula for
the rational Pontrjagin classes of a differential manifold ([Gelfand and MacPherson 1992]). As described in Section 3, a triangulation of a differential manifold
yields a CD manifold, which in turn yields a matroid bundle. Associated to any
matroid bundle $(B, \mathcal{M})$ there is a quasifibration $Y \to B$, where the fiber over a
vertex b is $\Delta\Gamma(2, \mathcal{M}(b))$. In turn, there is a quasifibration $Z \to Y$ in which the
fiber over a vertex (b, N) is $\Delta\Gamma(1, N)$. This is analogous to the association to a
tangent bundle $TM \to M$ the Grassmannian 2-plane bundle $\mathcal{G}_2(TM) \to M$ and
the circle bundle $\mathcal{G}_1(\mathcal{G}_2(TM)) \to \mathcal{G}_2(TM)$. If the matroid bundle arose from a
differential manifold, we also get a *fixing cycle*, analogous to the orientation class
of $\mathcal{G}_2(TM)$. These analogies allow one to reproduce in a combinatorial context
a formula for Pontrjagin classes arising in Chern–Weil theory.

5. Areas for Further Research

1. We do not know the kernel of the map

$$\tilde{\gamma}^* : H^*(\Delta\Gamma(k, M^n), R) \to H^*(G(k, \mathbb{R}^n), R)$$

for any coefficients R. Any nontrivial elements would give exotic characteristic
classes for matroid bundles which may have interesting combinatorial interpretations.

2. Is $\tilde{\gamma}^*$ surjective in integer coefficients? One motivation for this question is that
integer characteristic classes are used to distinguish exotic differential structures

on spheres. Thus a positive answer to this question would suggest that CD manifolds can distinguish exotic differential structures.

3. Any smooth manifold has a tame triangulation, giving a CD manifold. Thus if Conjecture 3.1 is true in general, then CD manifolds lie somewhere between the differential and PL categories. The question is where. Does every PL manifold have a CD structure? Is every CD manifold smoothable?

4. The computational question of calculating the homotopy groups of some of the finite MacPhersonians becomes more enticing in light of the stability result, Theorem 2.2. Theorem 1.1 gives π_0 and π_1, but beyond this the question is open.

5. Under the right notion of "complex oriented matroid", one should get a useful theory of complex matroid bundles. There are two likely candidates for the fibers. Ziegler's notion of complex matroids [Ziegler 1993] has already proved to encode nontrivial aspects of complex structure, and gives a "combinatorial complex Grassmannian" by exactly the same construction as $\Gamma(k, M^n)$. Alternatively, one could define a complex oriented matroid to be a direct sum of two real oriented matroids. Neither avenue to complex matroid bundles has yet been explored.

6. To reiterate a question from [MacPherson 1993], what are the combinatorial analogs to such topological theories as transversality, cobordism, surgery, and so on?

References

[Anderson 1998] L. Anderson, "Homotopy groups of the combinatorial Grassmannian", *Discrete Comput. Geom.* **20**:4 (1998), 549–560.

[Anderson 1999a] L. Anderson, "Representing weak maps of oriented matroids", preprint, Texas A&M University, 1999. Available at http://www.math.tamu.edu/~laura.anderson/.

[Anderson 1999b] L. Anderson, "Topology of combinatorial differential manifolds", *Topology* **38**:1 (1999), 197–221.

[Anderson and Davis ≥ 1999] L. Anderson and J. Davis, "Mod 2 cohomology of the combinatorial Grassmannian". In preparation.

[Anderson et al. ≥ 1999] L. Anderson, E. Babson, and J. Davis, "Rational cohomology of the combinatorial Grassmannian". In preparation.

[Babson 1993] E. Babson, *A combinatorial flag space*, Ph.D. thesis, Mass. Inst. Technology, Cambridge, 1993.

[Björner et al. 1993] A. Björner, M. Las Vergnas, B. Sturmfels, N. White, and G. M. Ziegler, *Oriented matroids*, Encyclopedia of Mathematics and its Applications **46**, Cambridge University Press, New York, 1993.

[Dold and Thom 1958] A. Dold and R. Thom, "Quasifaserungen und unendliche symmetrische Produkte", *Ann. Math.* (2) **67** (1958), 239–281.

[Folkman and Lawrence 1978]   J. Folkman and J. Lawrence, "Oriented matroids", *J. Combin. Theory Ser. B* **25** (1978), 199–236.

[Gelfand and MacPherson 1992]   I. M. Gelfand and R. D. MacPherson, "A combinatorial formula for the Pontrjagin classes", *Bull. Amer. Math. Soc. (N.S.)* **26**:2 (1992), 304–309.

[Hironaka 1975]   H. Hironaka, "Triangulations of algebraic sets", pp. 165–185 in *Algebraic geometry* (Arcata, CA, 1974), edited by R. Hartshorne, Proc. Sympos. Pure Math. **29**, Amer. Math. Soc., Providence, 1975.

[Hudson 1969]   J. F. P. Hudson, *Piecewise linear topology*, W. A. Benjamin, New York, 1969.

[Husemoller 1996]   D. Husemoller, *Fibre bundles*, Springer, New York, 1996.

[MacPherson 1993]   R. D. MacPherson, "Combinatorial differential manifolds: a symposium in honor of John Milnor's sixtieth birthday", pp. 203–221 in *Topological methods in modern mathematics* (Stony Brook, NY, 1991), edited by L. R. Goldberg and A. Phillips, Publish or Perish, Houston, 1993.

[Milnor 1961]   J. Milnor, "Microbundles and differentiable structures", 1961. Mimeographed lecture notes.

[Milnor and Stasheff 1974]   J. W. Milnor and J. D. Stasheff, *Characteristic classes*, Annals of Mathematics Studies **76**, Princeton University Press, Princeton, NJ, 1974.

[Mnëv 1988]   N. Mnëv, "The universality theorems on the classification problem of configuration varieties and convex polytope varieties", pp. 527–544 in *Topology and Geometry: Rohlin Seminar*, edited by O. Y. Viro, Lecture Notes in Mathematics **1346**, Springer, Berlin, 1988.

[Mnëv and Richter-Gebert 1993]   N. E. Mnëv and J. Richter-Gebert, "Two constructions of oriented matroids with disconnected extension space", *Discrete Comput. Geom.* **10**:3 (1993), 271–285.

[Mnëv and Ziegler 1993]   N. E. Mnëv and G. M. Ziegler, "Combinatorial models for the finite-dimensional Grassmannians", *Discrete Comput. Geom.* **10**:3 (1993), 241–250.

[Reiner 1999]   V. Reiner, "The generalized Baues problem", in *New perspectives in algebraic combinatorics*, edited by L. Billera et al., Math. Sci. Res. Inst. Publications **37**, Cambridge University Press, New York, 1999.

[Stasheff 1963]   J. Stasheff, "A classification theorem for fibre spaces", *Topology* **2** (1963), 239–246.

[Ziegler 1993]   G. Ziegler, "What is a complex matroid?", *Discrete Comput. Geom.* **10**:3 (1993), 313–348.

LAURA ANDERSON
DEPARMENT OF MATHEMATICS
TEXAS A&M UNIVERSITY
COLLEGE STATION, TX 77843-6028
UNITED STATES
Laura.Anderson@math.tamu.edu

New Perspectives in Geometric Combinatorics
MSRI Publications
Volume **38**, 1999

Combinatorial Representation Theory

HÉLÈNE BARCELO AND ARUN RAM

ABSTRACT. We survey the field of combinatorial representation theory, describe the main results and main questions and give an update of its current
status. Answers to the main questions are given in Part I for the fundamen-
tal structures, S_n and $\mathrm{GL}(n, \mathbb{C})$, and later for certain generalizations, when
known. Background material and more specialized results are given in a
series of appendices. We give a personal view of the field while remaining
aware that there is much important and beautiful work that we have been
unable to mention.

CONTENTS

Key words and phrases. Algebraic combinatorics, representations.

Barcelo was supported in part by National Science Foundation grant DMS-9510655.

Ram was supported in part by National Science Foundation grant DMS-9622985.

This paper was written while both authors were in residence at MSRI. We are grateful for the
hospitality and financial support of MSRI.

Appendices

Introduction

In January 1997, during the special year in combinatorics at MSRI, at a dessert party at Hélène's house, Gil Kalai, in his usual fashion, began asking very pointed questions about exactly what all the combinatorial representation theorists were investigating. After several unsuccessful attempts at giving answers that Gil would find satisfactory, it was decided that some talks should be given in order to explain to other combinatorialists what the specialty is about and what its main questions are.

In the end, Arun gave two talks at MSRI in which he tried to clear up the situation. After the talks several people suggested that it would be helpful if someone would write a survey article containing what had been covered in the two talks and including further interesting details. After some arm twisting it was agreed that Arun and Hélène would write such a paper on combinatorial representation theory. What follows is our attempt to define the field of combinatorial representation theory, describe the main results and main questions and give an update of its current status.

Of course this is wholly impossible. Everybody in the field has their own point of view and their own preferences of questions and answers. Furthermore, there is much too much material in the field to possibly collect it all in a single article (even conceptually). We therefore feel that we must stress the obvious: in this article we give a personal viewpoint on the field while remaining aware that there is much important and beautiful work that we have not been able to mention.

On the other hand, we have tried very hard to give a focused approach and to make something that will be useful to both specialists and non specialists, for understanding what we do, for learning the concepts of the field, and for tracking down history and references. We have chosen to write in an informal style in the hope that this way we can better convey the conceptual aspects of the field. Readers should keep this in mind and refer to the notes and references and the appendices when there are questions about the precision in definitions and statements of results. We have included a table of contents at the beginning of the paper which should help with navigation. The survey articles [Hanlon 1984; Howe 1995; Stanley 1983] are excellent complements to this one. Having made these points, and put in a lot of work, we leave it to you the reader, with the earnest hope that you find it useful.

We would like to thank the many people in residence at the special year 1996–97 in combinatorics at MSRI, for their interest, their suggestions, and for continually encouraging us to explain and write about the things that we enjoy doing. We both are extremely indebted to our graduate advisors, A. Garsia and H. Wenzl, who (already many years ago) introduced us to and taught us this wonderful field.

PART I

1. What is Combinatorial Representation Theory?

What do we mean by "combinatorial representation theory"? First and foremost, combinatorial representation theory is representation theory. The adjective "combinatorial" will refer to the way in which we answer representation theoretic questions; we will discuss this more fully later. For the moment we begin with:

What is Representation Theory?

If representation theory is a black box, or a machine, then the input is an algebra A. The output of the machine is information about the modules for A.

An **algebra** is a vector space A over $\mathbb{C}$ with a multiplication.

An important example: define the **group algebra** of a group G to be

$$A = \mathbb{C}G = \mathbb{C}\text{-span } \{g \in G\},$$

so that the elements of G form a basis of A. The multiplication in the group algebra is inherited from the multiplication in the group.

We want to study the algebra A via its actions on vector spaces.

An A-**module** is a finite-dimensional vector space M over $\mathbb{C}$ with an A action.

See Appendix A1 for a complete definition. We shall use the words module and **representation** interchangeably. Representation theorists are always trying to break up modules into pieces.

An A-module M is **indecomposable** if $M \not\cong M_1 \oplus M_2$ where M_1 and M_2 are nonzero A-modules.

An A-module M is **irreducible** or **simple** if it has no submodules.

In reference to modules the words "irreducible" and "simple" are used completely interchangeably.

The algebra A is **semisimple** if

$$\textbf{indecomposable} = \textbf{irreducible}$$

for A-modules.

The nonsemisimple case, where indecomposable is not the same as irreducible, is called modular representation theory. We will not consider this case much in these notes. However, before we banish it completely we describe the flavor of modular representation theory.

A **composition series** for M is a sequence

$$M = M_0 \supseteq M_1 \supseteq \cdots \supseteq M_k = 0$$

of submodules of M such that each M_i/M_{i+1} is simple.

The Jordan–Hölder theorem says that two different composition series of M will always produce the same multiset $\{M_i/M_{i+1}\}$ of simple modules. Modular representation theorists are always trying to determine this multiset of *composition factors of M*.

Remarks

(1) We shall not make life difficult in this article but one should note that it is common to work over general fields rather than just using the field $\mathbb{C}$.

(2) If one is bold one can relax things in the definition of module and let M be infinite-dimensional.

(3) Of course the definition of irreducible modules is not correct since 0 and M are always submodules of M. So we are ignoring these two submodules in this definition. But conceptually the definition is the right one, we want a simple module to be something that has no submodules.

(4) The definition of semisimple above is not technically correct; see Appendix A1 for the proper definition. However, the power of semisimplicity is exactly that it makes all indecomposable modules irreducible. So "indecomposable = irreducible" is really the right way to think of semisimplicity.

(5) A good reference for the basics of representation theory is [Curtis and Reiner 1962]. The book [Bourbaki 1958] contains a completely general and comprehensive treatment of the theory of semisimple algebras. Appendix A1 to this article also contains a brief (and technically correct) introduction with more specific references.

Main Questions in Representation Theory

I. What are the irreducible A-modules?

What do we mean by this question? We would like to be able to give some kind of answer to the following more specific questions.

(a) How do we index/count them?

(b) What are their dimensions?

The dimension of a module is just its dimension as a vector space.

(c) What are their characters?

The character of a module M is the function $\chi_M : A \to \mathbb{C}$, where $\chi_M(a)$ is the trace of the linear transformation determined by the action of a on M. More precisely,

$$\chi_M(a) = \sum_{b_i \in B} ab_i\big|_{b_i},$$

where the sum is over a basis B of the module M and $ab_i\big|_{b_i}$ denotes the coefficient of b_i in ab_i when we expand in terms of the basis B.

C. How do we construct the irreducible modules?

S. Special/Interesting representations M

(a) How does M decompose into irreducibles?

If we are in the semisimple case then M will always be a direct sum of irreducible modules. If we group the irreducibles of the same type together we can write

$$M \cong \bigoplus_\lambda (V^\lambda)^{\oplus c_\lambda},$$

where the modules V^λ are the irreducible A-modules and c_λ is the number of times an irreducible of type V^λ appears as a summand in M. It is common to abuse notation and write

$$M = \sum_\lambda c_\lambda V^\lambda.$$

(b) What is the character of M?

Special modules often have particularly nice formulas describing their characters. It is important to note that having a nice character formula for M does not necessarily mean that it is easy to see how M decomposes into irreducibles. Thus this question really is different from the previous one.

(c) How do we find interesting representations?

Sometimes special representations turn up by themselves and other times one has to work hard to construct the right representation with the right properties. Often very interesting representations come from other fields.

(d) Are they useful?

A representation may be particularly interesting just because of its structure while other times it is a special representation that helps to prove some particularly elusive theorem. Sometimes these representations lead to a completely new understanding of previously known facts. A famous example (which unfortunately we won't have space to discuss; see [Humphreys 1972]) is the Verma module, which was discovered in the mid 1960s and completely changed representation theory.

M. The modular case

In the modular case we have the following important question in addition to those above.

(a) What are the indecomposable representations?

(b) What are the structures of their composition series?

For each indecomposable module M there is a multiset of irreducibles $\{M_i/M_{i+1}\}$ determined by a composition series of M. One would like to determine this multiset.

Even better (especially for combinatorialists), the submodules of M form a lattice under inclusion of submodules and one would like to understand this lattice. This lattice is always a modular lattice and we may imagine that each edge of the Hasse diagram is labeled by the simple module N_1/N_2 where N_1 and N_2 are the modules on the ends of the edge. With this point of view the various composition series of M are the maximal chains in this lattice of modules. The Jordan–Hölder theorem says that every maximal chain in the lattice of submodules of M has the same multiset of labels on its edges. What modular representation theorists try to do is determine the set of labels on a maximal chain.

Remark. The abuse of notation that allows us to write $M = \sum_\lambda c_\lambda V^\lambda$ has been given a formal setting called the *Grothendieck ring*. In other words, the formal object that allows us to write such identities has been defined carefully. See [Serre 1971] for precise definitions of the Grothendieck ring.

Answers Should Be of the Form...

Now we come to the adjective "Combinatorial." It refers to the way in which we give the answers to the main questions of representation theory.

I. What are the irreducible A-modules?

(a) How do we index/count them?

We want to answer with a **bijection**

$$\text{nice combinatorial objects } \lambda \xleftrightarrow{\text{1-1}} \text{irreducible representations } V^\lambda.$$

(b) What are their dimensions?

We should answer with a **formula** of the form

$$\dim(V^\lambda) = \# \text{ of nice combinatorial objects.}$$

(c) What are their characters?

We want a **character formula** of the type

$$\chi(a) = \sum_T \text{wt}^a(T),$$

where the sum runs over all T in a set of nice combinatorial objects and wt^a is a weight on these objects which depends on the element $a \in A$ where we are evaluating the character.

C. How do we construct the irreducible modules?

We want to give constructions that have a **very explicit and very combinatorial flavor**. What we mean by this will be clearer from the examples; see C(i), C(ii) in Section 2 (page 31).

S. Special/Interesting representations M

(a) How does M decompose into irreducibles?

If M is an interesting representation we want to determine **the positive integers c_λ** in the decomposition

$$M \cong \bigoplus_\lambda (V^\lambda)^{\oplus c_\lambda}$$

in the form

$$c_\lambda = \# \text{ of nice combinatorial objects.}$$

In the formula for the decomposition of M the sum is over all λ which are objects indexing the irreducible representations of A.

(b) What is the character of M?

As in case **I(c)**, we want a **character formula** of the type

$$\chi_M(a) = \sum_T \mathrm{wt}^a(T),$$

where the sum runs over all T in some set of nice combinatorial objects and wt^a is a weight on these objects which depends on the element $a \in A$ where we are evaluating the character.

(c) How do we find interesting representations?

It is particularly pleasing when interesting representations arise in **other parts of combinatorics!** One such example is a representation on the homology of the partition lattice, which also, miraculously, appears as a representation on the free Lie algebra. We won't have space to discuss this here, see the original references [Hanlon 1981; Joyal 1986; Klyachko 1974; Stanley 1982], the article [Garsia 1990] for some further basics, and [Barcelo 1990] for a study of how it can be that this representation appears in two completely different places. Another fascinating example can be found in Sundaram's work [1994]. In this paper she study the homology representations of the symmetric group on Cohen–Macaulay subposets of the partition lattice. The coefficients discussed in **(S2)** of Section 2 (page 32) also appear in her work.

(d) Are they useful?

Are they useful for **solving combinatorial problems?** How about for **making new ones?** Sometimes a representation is exactly what is most helpful for solving a combinatorial problem. One example of this is in the recent solution of the last few plane partition conjectures. See [Stanley 1971; Macdonald 1995, Chap. I, § 5, Ex. 13–18] for the statement of the problem and [Kuperberg 1994c; 1994b; 1996a; Stembridge 1994a; 1994b; 1995] for the solutions. These solutions were motivated by the method of Proctor [1984].

The main point of all this is that a combinatorialist thinks in a special way (nice objects, bijections, weighted objects, etc.) and this method of thinking should be an integral part of the form of the solution to the problem.

2. Answers for S_n, the Symmetric Group

Most people in the field of combinatorial representation theory agree that the field begins with the fundamental results for the symmetric group S_n. We give the answers to the main questions for the case of S_n. The precise definitions of all the objects used below can be found in Appendix A2. As always, by a representation of the symmetric group we mean a representation of its group algebra $A = \mathbb{C}S_n$.

I. What are the irreducible S_n-modules?

(a) How do we index/count them?

There is a **bijection**

$$\text{partitions } \lambda \text{ of } n \overset{1-1}{\longleftrightarrow} \text{irreducible representations } S^\lambda.$$

(b) What are their dimensions?

The **dimension** of the irreducible representation S^λ is

$$\dim(S^\lambda) = \text{\# of standard tableaux of shape } \lambda = \frac{n!}{\prod_{x \in \lambda} h_x},$$

where h_x is the hook length at the box x in λ (see Appendix A2).

(c) What are their characters?

Let $\chi^\lambda(\mu)$ be the character of the irreducible representation S^λ evaluated at a permutation of cycle type $\mu = (\mu_1, \mu_2, \ldots, \mu_l)$. Then the **character** $\chi^\lambda(\mu)$ is given by

$$\chi^\lambda(\mu) = \sum_T \text{wt}^\mu(T),$$

where the sum is over all standard tableaux T of shape λ and

$$\text{wt}^\mu(T) = \prod_{i=1}^{n} f(i, T),$$

with

$$f(i,T) = \begin{cases} -1 & \text{if } i \notin B(\mu) \text{ and } i+1 \text{ is SW of } i, \\ 0 & \text{if } i, i+1 \notin B(\mu), i+1 \text{ is NE of } i, \text{ and } i+2 \text{ is SW of } i+1, \\ 1, & \text{otherwise,} \end{cases}$$

and $B(\mu) = \{\mu_1 + \mu_2 + \cdots + \mu_k \mid 1 \le k \le l\}$. In the formula for $f(i,T)$, SW means strictly south and weakly west and NE means strictly north and weakly east.

C. How do we construct the irreducible modules?

There are several interesting constructions of the irreducible S^λ.

(i) Via Young symmetrizers. Let T be a tableau. Set

$$R(T) = \text{permutations that fix the rows of } T, \text{ as sets};$$
$$C(T) = \text{permutations that fix the columns of } T, \text{ as sets};$$
$$P(T) = \sum_{w \in R(T)} w, \quad \text{and} \quad N(T) = \sum_{w \in C(T)} \varepsilon(w) w,$$

where $\varepsilon(w)$ is the sign of the permutation w. Then

$$S^\lambda \cong \mathbb{C}S_n P(T) N(T),$$

where the action of the symmetric group is by left multiplication.

(ii) Young's seminormal construction. Let

$$S^\lambda = \mathbb{C}\text{-span-}\{v_T \mid T \text{ is a standard tableau of shape } \lambda\},$$

so that the vectors v_T are a basis of S^λ. The **action** of S_n on S^λ is given by

$$s_i v_T = (s_i)_{TT} v_T + (1 + (s_i)_{TT}) v_{s_i T},$$

where s_i is the transposition $(i, i+1)$ and

$$(s_i)_{TT} = \frac{1}{c(T(i+1)) - c(T(i))},$$

with the following notation:

$T(i)$ denotes the box containing i in T;

$c(b)$ is the **content** of the box b, that is, $j - i$ if b has position (i, j) in λ;

$s_i T$ is the same as T except that the entries i and $i+1$ are switched;

$v_{s_i T} = 0$ if $s_i T$ is not a standard tableau.

There are other important constructions of the irreducible representations S^λ. We do not have room to discuss these constructions here, see Remarks (10)–(12) on page 34 and Appendix A3. The main ones are:

(iii) Young's orthonormal construction,

(iv) the Kazhdan–Lusztig construction, and

(v) the Springer construction.

S. Particularly interesting representations

(S1) Let $k + l = n$. The module $S^\lambda \downarrow^{S_n}_{S_k \times S_l}$ is the same as S^λ except that we only look at the action of the subgroup $S_k \times S_l$. Then

$$S^\lambda \downarrow^{S_n}_{S_k \times S_l} = \bigoplus_{\mu \vdash k, \nu \vdash l} (S^\mu \otimes S^\nu)^{\oplus c^\lambda_{\mu\nu}} = \sum_{\mu,\nu} c^\lambda_{\mu\nu}(S^\mu \otimes S^\nu),$$

where $\mu \vdash k$ means that μ is a partition of k, as usual, and the positive integers $c^\lambda_{\mu\nu}$ are the **Littlewood–Richardson coefficients**: $c^\lambda_{\mu\nu}$ is the number of column strict fillings of λ/μ of content ν such that the word of the filling is a lattice permutation. See Appendix A2.

(S2) Let $\mu = (\mu_1, \ldots, \mu_l)$ be a partition of n. Let $S_\mu = S_{\mu_1} \times \cdots \times S_{\mu_l}$. The module $1\uparrow^{S_n}_{S_\mu}$ is the vector space

$$1\uparrow^{S_n}_{S_\mu} = \mathbb{C}(S_n/S_\mu) = \mathbb{C}\text{-span}\{wS_\mu \mid w \in S_n\}$$

where the action of S_n on the cosets is by left multiplication. Then

$$1\uparrow^{S_n}_{S_\mu} = \sum_\lambda K_{\lambda\mu} S^\lambda,$$

where

$$K_{\lambda\mu} = \# \text{ of column strict tableaux of shape } \lambda \text{ and weight } \mu.$$

This representation also occurs in the following context:

$$1\uparrow_{S_\mu}^{S_n} \cong H^*(\mathcal{B}_u),$$

where u is a unipotent element of $GL(n, \mathbb{C})$ with Jordan decomposition μ and $\mathcal{B}_u$ is the variety of Borel sugroups in $GL(n, \mathbb{C})$ containing u. This representation is related to the Springer construction mentioned as item **C(v)** on the preceding page. See Appendix A3 for further details.

(S3) If $\mu, \nu \vdash n$, the tensor product S_n-module $S^\mu \otimes S^\nu$ is defined by $w(m \otimes n) = wm \otimes wn$, for all $w \in S_n$, $m \in S^\mu$ and $n \in S^\nu$. There are **positive integers** $\gamma_{\mu\nu\lambda}$ such that

$$S^\mu \otimes S^\nu = \sum_{\lambda \vdash n} \gamma_{\mu\nu\lambda} S^\lambda.$$

Except for a few special cases the $\gamma_{\mu\nu\lambda}$ are still unknown. See [Remmel 1992] for a combinatorial description of the cases for which the coefficients $\gamma_{\mu\nu\lambda}$ are known.

Remarks

(1) The bijection in **(Ia)**, page 31, between irreducible representations and partitions, is due to Frobenius [1900]. Frobenius is the founder of representation theory and the symmetric group was one of the first examples that he worked out.

(2) The formula in **(Ib)**, page 31, for the dimension of S^λ as the number of standard tableaux is immediate from the work of Frobenius, but it really came into the fore with the work of Young [1901; 1902; 1928; 1930a; 1930b; 1931; 1934a; 1934b]. The "hook formula" for $\dim(S^\lambda)$ is due to Frame, Robinson, and Thrall [Frame et al. 1954].

(3) The formula for the characters of the symmetric group given in **(Ic)**, page 31, is due to Fomin and Greene [1998]. For them, this formula arose by application of their theory of noncommutative symmetric functions. Roichman [1997] discovered this formula independently in the more general case of the Iwahori–Hecke algebra. The formula for the Iwahori–Hecke algebra is exactly the same as the one for the S_n case except that the 1 appearing in case 3 of the definition of $f(i, T)$ should be changed to a q.

(4) There is a different and more classical formula for the characters than the one given in **(Ic)**, called the Murnaghan–Nakayama rule [Murnaghan 1937; Nakayama 1941]. We describe it in Theorem A2.2 (page 57). Once the formula in **(Ic)** is given it is not hard to show combinatorially that it is equivalent to the Murnaghan–Nakayama rule, but if one does not know the formula it is nontrivial to guess it from the Murnaghan–Nakayama rule.

(5) We do not know if anyone has compared the algorithmic complexity of the formula given in **(Ic)** with that of the Murnaghan–Nakayama rule. One would expect that they have the same complexity: the formula above is a sum over more objects than the sum in the Murnaghan–Nakayama rule but these objects are easier to create and many of them have zero weight.

(6) One of the beautiful things about the formula for the character of S^λ given in **(Ic)** is that it is a sum over the same set that we have used to describe the dimension of S^λ.

(7) The construction of S^λ by Young symmetrizers is due to Young [1901; 1902]. It is used so often and has so many applications that it is considered classical. A review and generalization of this construction to skew shapes appears in [Garsia and Wachs 1989].

(8) The seminormal form construction of S^λ is also due to Young [1931; 1934b], although it was discovered some thirty years after the Young symmetrizer construction.

(9) Young's orthonormal construction differs from the seminormal construction only by multiplication of the basis vectors by certain constants. A comprehensive treatment of all three constructions of Young is given in [Rutherford 1948].

(10) The Kazhdan–Lusztig construction uses the Iwahori–Hecke algebra in a crucial way. It is combinatorial but relies crucially on certain polynomials which seem to be impossible to compute in practice except for very small n; see [Brenti 1994] for further information. This construction has important connections to geometry and other parts of representation theory. The paper [Garsia and McLarnan 1988] and the book [Humphreys 1990] give elementary treatments of the Kazhdan–Lusztig construction.

(11) Springer's construction is a geometric construction. In this construction the irreducible module S^λ is realized as the top cohomology group of a certain variety; see [Springer 1978; Chriss and Ginzburg 1997], and Appendix A3.

(12) There are many ways of constructing new representations from old ones. Among the common techniques are *restriction*, *induction*, and *tensoring*. The special representations **(S1)**, **(S2)**, and **(S3)** given on pages 32–33 are particularly nice examples of these constructions. One should note that tensoring of representations works for group algebras (and Hopf algebras) but not for general algebras.

3. Answers for $\mathrm{GL}(n, \mathbb{C})$, the General Linear Group

The results for the general linear group are just as beautiful and just as fundamental as those for the symmetric group. The results are surprisingly similar and yet different in many crucial ways. We shall see that the results for $\mathrm{GL}(n, \mathbb{C})$ have been generalized to a very wide class of groups whereas the results for S_n have only been generalized successfully to groups that look very similar to symmetric groups. The representation theory of $GL(n, \mathbb{C})$ was put on a very firm footing from the fundamental work of Schur [1901; 1927].

I. What are the irreducible $GL(n, \mathbb{C})$-modules?

(a) How do we index/count them?

There is a **bijection**

partitions λ with at most n rows
$$\overset{1-1}{\longleftrightarrow}\ \textbf{irreducible polynomial representations } V^\lambda.$$

See Appendix A4 for a definition and discussion of what it means to be a *polynomial* representation.

(b) What are their dimensions?

The **dimension** of the irreducible representation V^λ is

$$\dim(V^\lambda) = \#\ \textbf{of column strict tableaux of shape } \lambda$$

$$\qquad\qquad\qquad \textbf{filled with entries from } \{1, 2, \ldots, n\}$$

$$= \prod_{x \in \lambda} \frac{n + c(x)}{h_x},$$

where $c(x)$ is the content of the box x and h_x is the hook length at the box x.

(c) What are their characters?

Let $\chi^\lambda(g)$ be the character of the irreducible representation V^λ evaluated at an element $g \in GL(n, \mathbb{C})$. The **character** $\chi^\lambda(g)$ is given by

$$\chi^\lambda(g) = \sum_T x^T$$

$$= \frac{\sum_{w \in S_n} \varepsilon(w)\, wx^{\lambda + \delta}}{\sum_{w \in S_n} \varepsilon(w)\, wx^\delta} = \frac{\det(x_i^{\lambda_j + n - j})}{\det(x_i^{n-j})},$$

where the sum is over all column strict tableaux T of shape λ filled with entries from $\{1, 2, \ldots, n\}$ and

$$x^T = x_1^{\mu_1} x_2^{\mu_2} \cdots x_n^{\mu_n},$$

where μ_i is the number of i's in T and $x_1, x_2, \ldots, x_n$ are the eigenvalues of the matrix g. (Let us not worry at the moment about the quotient of sums on the second line of the display above. It is routine to rewrite it as the second expression on that line, which is one of the standard expressions for the Schur function; see [Macdonald 1995, Chap. I, § 3].)

C. How do we construct the irreducible modules?

There are several interesting constructions of the irreducible V^λ.

(C1) Via Young symmetrizers. Recall that the irreducible S^λ of the symmetric group S_k was constructed via Young symmetrizers in the form

$$S^\lambda \cong \mathbb{C}S_n P(T)N(T).$$

We can construct the irreducible $\mathrm{GL}(n, \mathbb{C})$-module in a similar form. If λ is a **partition** of k then

$$V^\lambda \cong V^{\otimes k} P(T)N(T).$$

This important construction is detailed in Appendix A5.

(C2) Gelfand–Tsetlin bases. This construction of the irreducible $\mathrm{GL}(n, \mathbb{C})$ representations V^λ is analogous to Young's seminormal construction of the irreducible representations S^λ of the symmetric group. Let

$$V^\lambda = \mathrm{span}\text{-}\left\{ v_T \;\middle|\; \begin{array}{l} T \text{ is a column strict tableau of shape } \lambda \\ \text{filled with elements of } \{1, 2, \ldots, n\} \end{array} \right\},$$

so that the vectors v_T are a basis of V^λ. Define an action of symbols $E_{k-1,k}$, for $2 \le k \le n$, on the basis vectors v_T by

$$E_{k-1,k} v_T = \sum_{T^-} a_{T^- T}(k) v_{T^-},$$

where the sum is over all column strict tableaux T^- which are obtained from T by changing a k to a $k-1$ and the coefficients $a_{T^- T}(k)$ are given by

$$a_{T^- T}(k) = -\frac{\prod_{i=1}^{k}(T_{ik} - T_{j,k-1} + j - k)}{\prod_{\substack{i=1 \\ i \ne j}}^{k-1}(T_{i,k-1} - T_{j,k-1} + j - k)},$$

where j is the row number of the entry where T^- and T differ and T_{ik} is the position of the rightmost entry $\le k$ in row i of T. Similarly, define an action of symbols $E_{k,k-1}$, for $2 \le k \le n$, on the basis vectors v_T by

$$E_{k,k-1} v_T = \sum_{T^+} b_{T^+ T}(k) v_{T^+},$$

where the sum is over all column strict tableaux T^+ which are obtained from T by changing a $k-1$ to a k and the coefficients $b_{T^+ T}(k)$ are given by

$$b_{T^+ T}(k) = \frac{\prod_{i=1}^{k-2}(T_{i,k-2} - T_{j,k-1} + j - k)}{\prod_{\substack{i=1 \\ i \ne j}}^{k-1}(T_{i,k-1} - T_{j,k-1} + j - k)},$$

where j is the row number of the entry where T^+ and T differ and T_{ik} is the position of the rightmost entry $\le k$ in row i.

Since

$$g_i(z) = \begin{pmatrix} 1 & 0 & \cdots & & & & 0 \\ 0 & \ddots & & & & & \\ & & 1 & & & & \vdots \\ \vdots & & & z & & & \\ & & & & 1 & & \\ & & & & & \ddots & 0 \\ 0 & & & \cdots & & 0 & 1 \end{pmatrix}, \quad \text{for } z \in \mathbb{C}^*,$$

$$g_{i-1,i}(z) = \begin{pmatrix} 1 & 0 & \cdots & & & & 0 \\ 0 & \ddots & & & & & \\ & & 1 & z & & & \vdots \\ \vdots & & 0 & 1 & & & \\ & & & & \ddots & 0 & \\ 0 & & & \cdots & & 0 & 1 \end{pmatrix}, \quad \text{for } z \in \mathbb{C},$$

$$g_{i,i-1}(z) = \begin{pmatrix} 1 & 0 & \cdots & & & & 0 \\ 0 & \ddots & & & & & \\ & & 1 & 0 & & & \vdots \\ \vdots & & z & 1 & & & \\ & & & & \ddots & 0 & \\ 0 & & & \cdots & & 0 & 1 \end{pmatrix}, \quad \text{for } z \in \mathbb{C},$$

generate $\mathrm{GL}(n, \mathbb{C})$, the action of these matrices on the basis vectors v_T will determine the action of all of $\mathrm{GL}(n, \mathbb{C})$ on the space V^λ. The action of these generators is given by

$$g_i(z) v_T = z^{(\# \text{ of } i\text{'s in } T)} v_T,$$
$$g_{i-1,i}(z) v_T = e^{z E_{i-1,i}} v_T = (1 + z E_{i-1,i} + \tfrac{1}{2!} z^2 E_{i-1,i}^2 + \cdots) v_T,$$
$$g_{i,i-1}(z) v_T = e^{z E_{i,i-1}} v_T = (1 + z E_{i,i-1} + \tfrac{1}{2!} z^2 E_{i,i-1}^2 + \cdots) v_T.$$

(C3) The Borel–Weil–Bott construction. Let λ be a partition of n. Then λ defines a character (one-dimensional representation) of the group T_n of diagonal matrices in $G = \mathrm{GL}(n, \mathbb{C})$. This character can be extended to the group $B = B_n$ of upper triangular matrices in $G = \mathrm{GL}(n, \mathbb{C})$ by letting it act trivially on U_n the group of upper unitriangular matrices in $G = \mathrm{GL}(n, \mathbb{C})$. Then the fiber product

$$\mathcal{L}_\lambda = G \times_B \lambda$$

is a line bundle on G/B. Finally,

$$V^\lambda \cong H^0(G/B, \mathcal{L}_\lambda),$$

where $H^0(G/B, \mathcal{L}_\lambda)$ is the **space of global sections** of the line bundle $\mathcal{L}_\lambda$. More details on the construction of the character λ and the line bundle $\mathcal{L}_\lambda$ are given in Appendix A6.

S. Special/Interesting representations

(S1) Let

$$
\mathrm{GL}(k) \times \mathrm{GL}(l) = \left(\begin{pmatrix} \mathrm{GL}(k, \mathbb{C}) \end{pmatrix} \quad 0 \\ 0 \quad \begin{pmatrix} \mathrm{GL}(l, \mathbb{C}) \end{pmatrix} \right) \subseteq \mathrm{GL}(n),
$$

where $k + l = n$. Then

$$
V^\lambda \Big\downarrow^{\mathrm{GL}(n)}_{\mathrm{GL}(k) \times \mathrm{GL}(l)} = \sum_{\mu, \nu} c^\lambda_{\mu\nu} (V^\mu \otimes V^\nu),
$$

where the $c^\lambda_{\mu\nu}$ are the **Littlewood–Richardson coefficients** that appeared in the decomposition of $S^\lambda \big\downarrow^{S_n}_{S_k \times S_l}$ in terms of $S^\mu \otimes S^\nu$ (page 32). We may write this expansion in the form

$$
V^\lambda \Big\downarrow^{\mathrm{GL}(n)}_{\mathrm{GL}(k) \times \mathrm{GL}(l)} = \sum_{F \text{ fillings}} V^{\mu(F)} \otimes V^{\nu(F)}.
$$

We could do this precisely if we wanted. We won't do it now, but the point is that it may be nice to write this expansion as a sum over combinatorial objects. This will be the form in which this will be generalized later.

(S2) Let V^μ and V^ν be irreducible polynomial representations of $\mathrm{GL}(n)$. Then

$$
V^\mu \otimes V^\nu = \sum_\lambda c^\lambda_{\mu\nu} V^\lambda,
$$

where $\mathrm{GL}(n)$ acts on $V^\mu \otimes V^\nu$ by $g(m \otimes n) = gm \otimes gn$, for $g \in \mathrm{GL}(n, \mathbb{C})$, $m \in V^\mu$ and $n \in V^\nu$. Amazingly, the coefficients $c^\lambda_{\mu\nu}$ are the **Littlewood–Richardson coefficients** again, the same that appeared in the **(S1)** case immediately above and in the **(S1)** case for the symmetric group (page 32).

Remarks

(1) There is a strong similarity between the results for the symmetric group and the results for $\mathrm{GL}(n, \mathbb{C})$. One might wonder whether there is any connection between these two pictures.

There are **two distinct** ways of making concrete connections between the representation theories of $\mathrm{GL}(n, \mathbb{C})$ and the symmetric group. In fact these two are so different that **different symmetric groups** are involved.

(a) *If λ is a partition of n* then the "zero weight space", or $(1, 1, \ldots, 1)$ weight space, of the irreducible $\mathrm{GL}(n, \mathbb{C})$-module V^λ is isomorphic to the irreducible module S^λ for the group S_n, where the action of S_n is determined by the fact that S_n is the subgroup of permutation matrices in $GL(n, \mathbb{C})$. This relationship is reflected in the combinatorics: the standard tableaux of shape λ are exactly the column strict tableaux of shape λ which are of weight $\nu = (1, 1, \ldots, 1)$.

(b) Schur–Weyl duality (see Appendix A5) says that the action of the symmetric group S_k on $V^{\otimes k}$ by permutation of the tensor factors generates the full centralizer of the $GL(n, \mathbb{C})$-action on $V^{\otimes k}$ where V is the standard n-dimensional representation of $GL(n, \mathbb{C})$. By double centralizer theory, this duality induces a correspondence between the irreducible representations of $GL(n, \mathbb{C})$ which appear in $V^{\otimes k}$ and the irreducible representations of S_k which appear in $V^{\otimes k}$. These representations are indexed by *partitions* λ *of* k.

(2) Note that the word *character* has two different and commonly used meanings and the use of the word character in (C3) (page 37) is different from that in Section 1. In (C3) above the word character means *one-dimensional representation*. This terminology is used particularly (but not exclusively) in reference to representations of abelian groups, like the group T_n in (C3). In general one has to infer from the context which meaning is intended.

(3) The indexing and the formula for the characters of the irreducible representations is due to Schur [1901].

(4) The formula for the dimensions of the irreducibles as the number of column strict tableaux follows from the work of Kostka [1882] and Schur [1901]. The "hook-content" formula appears in [Macdonald 1995, Chap. I, § 3, Ex. 4], where the book [Littlewood 1940] is quoted.

(5) The construction of the irreducibles by Young symmetrizers appeared in 1939 in the influential book [Weyl 1946]. It was generalized to the symplectic and orthogonal groups by H. Weyl in the same book. Further important information about this construction in the symplectic and orthogonal cases is found in [Berele 1986a] and [King and Welsh 1993]. It is not known how to generalize this construction to arbitrary complex semisimple Lie groups.

(6) The Gelfand–Tsetlin basis construction originates in [Gel'fand and Tsetlin 1950a]. A similar construction was given for the orthogonal group at the same time [Gel'fand and Tsetlin 1950b] and was generalized to the symplectic group by Zhelobenko [1987; 1970]. This construction does not generalize well to other complex semisimple groups since it depends crucially on a tower $G \supseteq G_1 \supseteq \cdots \supseteq G_k \supseteq \{1\}$ of "nice" Lie groups such that all the combinatorics is controllable.

(7) The Borel–Weil–Bott construction is not a combinatorial construction of the irreducible module V^λ. It is very important because it is a construction that generalizes well to all other compact connected real Lie groups.

(8) The facts about the special representations which we have given above are found in [Littlewood 1940].

4. Answers for Finite-Dimensional Complex Semisimple Lie Algebras $\mathfrak{g}$

Although the foundations for generalizing the $\mathrm{GL}(n, \mathbb{C})$ results to all complex semisimple Lie groups and Lie algebras were laid in the fundamental work of Weyl [1925; 1926], it was only recently that a complete generalization of the tableaux results for $\mathrm{GL}(n, \mathbb{C})$ was obtained [Littelmann 1995]. The results which we state below are generalizations of those given for $\mathrm{GL}(n, \mathbb{C})$ in the last section; partitions get replaced by points in a lattice called P^+, and column strict tableaux get replaced by paths. See Appendix A7 for some basics on complex semisimple Lie algebras.

I. What are the irreducible $\mathfrak{g}$-modules?

(a) How do we index/count them?

There is a **bijection**

$$\lambda \in P^+ \overset{1-1}{\longleftrightarrow} \text{ irreducible representations } V^\lambda,$$

where P^+ is the cone of dominant integral weights for $\mathfrak{g}$ (see Appendix A8).

(b) What are their dimensions?

The **dimension** of the irreducible representation V^λ is

$$\dim(V^\lambda) = \# \text{ of paths in } \mathcal{P}\pi_\lambda = \prod_{\alpha > 0} \frac{\langle \lambda + \rho, \alpha \rangle}{\langle \rho, \alpha \rangle},$$

where

$\rho = \frac{1}{2} \sum_{\alpha > 0} \alpha$ is the half sum of the positive roots,

π_λ is the straight line path from 0 to λ, and

$\mathcal{P}\pi_\lambda = \{f_{i_1} \cdots f_{i_k} \pi_\lambda | 1 \le i_1, \ldots, i_k \le n\}$, where $f_1, \ldots, f_n$ are the path operators introduced in [Littelmann 1995].

We shall not define the operators f_i; we will just say that they act on paths and are partial permutations in the sense that if f_i acts on a path π then the result is either 0 or another path. See Appendix A8 (page 65) for a few more details.

(c) What are their characters?

The **character** of the irreducible module V^λ is

$$\mathrm{char}(V^\lambda) = \sum_{\eta \in \mathcal{P}\pi_\lambda} e^{\eta(1)} = \frac{\sum_{w \in W} \varepsilon(w) e^{w(\lambda + \rho)}}{\sum_{w \in W} \varepsilon(w) e^{w\rho}},$$

where $\eta(1)$ is the **endpoint** of the path η. These expressions live in the group algebra of the weight lattice P, $\mathbb{C}[P] = \mathrm{span}\{e^\mu \mid \mu \in P\}$, where e^μ is a formal variable indexed by μ and the multiplication is given by $e^\mu e^\nu = e^{\mu + \nu}$, for $\mu, \nu \in P$. See Appendix A7 for more details.

S. Special/Interesting representations

(S1) Let $\mathfrak{l} \subseteq \mathfrak{g}$ be a Levi subalgebra of $\mathfrak{g}$ (this is a Lie algebra corresponding to a subgraph of the Dynkin diagram associated with $\mathfrak{g}$). The subalgebra $\mathfrak{l}$ corresponds to a subset J of the set $\{\alpha_1, \ldots, \alpha_n\}$ of simple roots. The **restriction rule** from $\mathfrak{g}$ to $\mathfrak{l}$ is

$$V^\lambda \!\downarrow_{\mathfrak{l}}^{\mathfrak{g}} = \sum_\eta V^{\eta(1)},$$

where

the sum is over all paths $\eta \in \mathcal{P}\pi_\lambda$ such that $\eta \in \bar{C}_{\mathfrak{l}}$, and

$\eta \in \bar{C}_{\mathfrak{l}}$ means that $\langle \eta(t), \alpha_i \rangle \geq 0$, for all $t \in [0,1]$ and all $\alpha_i \in J$.

(S2) The **tensor product** of two irreducible modules is given by

$$V^\mu \otimes V^\nu = \sum_\eta V^{\mu+\eta(1)},$$

where the sum is over all paths $\eta \in \mathcal{P}\pi_\nu$ such that $\pi_\mu * \eta \in \bar{C}$,

π_μ and π_ν are straight line paths from 0 to μ and 0 to ν, respectively,

$\mathcal{P}\pi_\nu$ is as in (Ib) on page 40,

$\pi_\mu * \eta$ is the path obtained by attaching η to the end of π_μ, and

$(\pi_\mu * \eta) \in \bar{C}$ means that $\langle (\pi_\mu * \eta)(t), \alpha_i \rangle \geq 0$, for all $t \in [0,1]$ and all simple roots α_i.

Remarks

(1) The indexing of irreducible representations given in **(Ia)**, page 40, is due to Cartan and Killing, the founders of the theory, from around the turn of the century. Introductory treatments of this result can be found in [Fulton and Harris 1991; Humphreys 1972].

(2) The first equality in **(Ib)**, page 40, is due to Littelmann [1994], but his later article [1995] has some improvements and can be read independently, so we recommend the later article. This formula for the dimension of the irreducible representation, the number of paths in a certain set, is exactly analogous to the formula in the $\mathrm{GL}(n, \mathbb{C})$ case, the number of tableaux which satisfy a certain condition. The second equality is the Weyl dimension formula, originally proved in [Weyl 1925; 1926]. It can be proved easily from the Weyl character formula given in **(Ic)** on page 40; see [Humphreys 1972; Stembridge 1994a, Lemma 2.5]. This product formula is an analogue of the "hook-content" formula given in the $\mathrm{GL}(n, \mathbb{C})$ case.

(3) A priori, it might be possible that the set $\mathcal{P}\pi_\lambda$ is an infinite set, at least the way that we have defined it. In fact, this set is always finite and there is a description of the paths that are contained in it. The paths in this set are called *Lakshmibai–Seshadri paths*; see [Littelmann 1995]. The explicit

description of these paths is a generalization of the types of indexings that were used in the "standard monomial theory" of Lakshmibai and Seshadri [1991].

(4) The first equality in **(Ic)** is due to Littelmann [1995]. This formula, a weighted sum over paths, is an analogue of the formula for the irreducible character of $GL(n, \mathbb{C})$ as a weighted sum of column strict tableaux. The second equality in **(Ic)** is the celebrated Weyl character formula, originally proved in [Weyl 1925; 1926]. Modern treatments can be found in [Bröcker and tom Dieck 1985; Humphreys 1972; Varadarajan 1984].

(5) The general restriction formula **(S1)** on the preceding page is due to Littelmann [1995]. This is an analogue of the rule given in **(S1)** of the $GL(n, \mathbb{C})$ results, page 38. In this case the formula is as a sum over paths which satisfy certain conditions whereas in the $GL(n, \mathbb{C})$ case the formula is a sum over column strict fillings which satisfy a certain condition.

(6) The general tensor product formula in **(S2)** on the preceding page is due to Littelmann [1995]. This formula is an analogue of the formula given in **(S2)** of the $GL(n, \mathbb{C})$ results, page 38.

(7) The results of Littelmann given above are some of the most exciting results of combinatorial representation theory in recent years. They were very much inspired by some very explicit conjectures of Lakshmibai that arose out of the "standard monomial theory" developed by Lakshmibai and Seshadri [1991]. Although Littelmann's theory is actually much more general than we have stated above, the special set of paths $\mathcal{P}\pi_\lambda$ used in **(Ib)** and **(Ic)** is a modified description of the same set which appeared in Lakshmibai's conjecture. Another important influence on Littelmann in his work was Kashiwara's work [1990] on crystal bases.

PART II

5. Generalizing the S_n Results

Having the above results for the symmetric group in hand we would like to try to generalize as many of the S_n results as we can to other similar groups and algebras. Work along this line began almost immediately after the discovery of the S_n results and it continues today. In the current state of results this has been largely

1. **successful** for the complex reflection groups $G(r, p, n)$ and their "Hecke algebras",
2. **successful** for tensor power centralizer algebras and their q-analogues, and
3. **unsuccessful** for general Weyl groups and finite Coxeter groups.

Some partial results giving answers to the main questions for the complex reflection groups $G(r, p, n)$, their "Hecke algebras", and some tensor power centralizer algebras can be found in Appendices B1–B8. Here we limit ourselves to a brief description of these objects and some notes and references.

5.1. The Reflection Groups $G(r, p, n)$

A **finite Coxeter group** is a finite group generated by reflections in $\mathbb{R}^n$. In other words, take a bunch of linear transformations of $\mathbb{R}^n$ which are reflections (in the sense of reflections and rotations in the orthogonal group) and see what group they generate. If the group is finite then it is a finite Coxeter group. Actually, this definition of finite Coxeter group is not the usual one (which is given in Appendix B1), but since we have the following theorem we are not too far astray.

THEOREM 5.1. *A group is a finite group generated by reflections if and only if it is a finite Coxeter group.*

The finite Coxeter groups have been classified completely and there is one group of each of the following "types"

$$A_n, \ B_n, \ D_n, \ E_6, \ E_7, \ E_8, \ F_4, \ H_3, \ H_4, \ \text{or } I_2(m).$$

The finite *crystallographic* reflection groups are called *Weyl groups* because of their connection with Lie theory. These are the finite Coxeter groups of types

$$A_n, \ B_n, \ D_n, \ E_6, \ E_7, \ E_8, \ F_4, \ \text{and } G_2 = I_2(6).$$

A **complex reflection group** is a group generated by complex reflections, that is, invertible linear transformations of $\mathbb{C}^n$ which have finite order and which have exactly one eigenvalue that is not 1. Every finite Coxeter group is also a finite complex reflection group. The finite complex reflection groups have been classified by Shephard and Todd [1954] and each such group is one of the groups

(a) $G(r, p, n)$, where r, p, n are positive integers such that p divides r, or
(b) one of 34 "exceptional" finite complex reflection groups.

The groups $G(r, p, n)$ are very similar to the symmetric group S_n in many ways and this is probably why generalizing the S_n theory has been so successful for these groups. The symmetric groups and the finite Coxeter groups of types B_n, and D_n are all special cases of the groups $G(r, p, n)$.

I. What are the irreducible modules?

The indexing, dimension formulas and character formulas for the representations of the groups $G(r, p, n)$ are originally due to

– Young [1901] for finite Coxeter groups of types B_n and D_n;
– Specht [1932] for the group $G(r, 1, n)$.

We do not know who first did the general $G(r, p, n)$ case but it is easy to generalize Young and Specht's results to this case. See [Ariki 1995; Halverson and Ram 1998] for recent accounts.

Essentially what one does to determine the indexing, dimensions and the characters of the irreducible modules is to use Clifford theory to reduce the $G(r, 1, n)$ case to the case of the symmetric group S_n. Then one can use Clifford theory again to reduce the $G(r, p, n)$ case to the $G(r, 1, n)$ case. The original reference for Clifford theory is [Clifford 1937]; the book by Curtis and Reiner [1981; 1987] has a modern treatment. The articles [Stembridge 1989b; Halverson and Ram 1998] explain how the reduction from $G(r, p, n)$ to $G(r, 1, n)$ is done. The dimension and character theory for the case $G(r, 1, n)$ has an excellent modern treatment in [Macdonald 1995, Chap. I, App. B].

C. How do we construct the irreducible modules?

The construction of the irreducible representations by Young symmetrizers was extended to the finite Coxeter groups of types B_n and D_n by Young himself [1930b]. The authors don't know when the general case was first treated in the literature, but it is not difficult to extend Young's results to the general case $G(r, p, n)$. The $G(r, p, n)$ case does appear periodically in the literature, see [Allen 1997], for example.

Young's seminormal construction was generalized to the "Hecke algebras" of $G(r, p, n)$ in [Ariki and Koike 1994; Ariki 1995]. One can easily set $q = 1$ in the constructions of Ariki and Koike and obtain the appropriate analogues for the groups $G(r, p, n)$. We do not know if the analogue of Young's seminormal construction for the groups $G(r, p, n)$ appeared in the literature previous to the work of Ariki and Koike on the "Hecke algebra" case.

S. Special/Interesting representations

We do not know if the analogues of the S_n results, (S1)–(S3) of Section 2 (page 32), have explicitly appeared in the literature. It is easy to use symmetric functions and the character formulas of Specht (see [Macdonald 1995, Chap. I, App. B]) to derive formulas for the $G(r, 1, n)$ case in terms of the symmetric group results. Then one proceeds as described above to compute the necessary formulas for $G(r, p, n)$ in terms of the $G(r, 1, n)$ results. See [Stembridge 1989b] for how this is done.

5.2. The "Hecke algebras" of reflection groups

In group theory, a Hecke algebra is a specific kind of double coset algebra associated with a finite group G and a subgroup B of G; see Appendix B3 for the definition. In connection with finite Coxeter groups and the groups $G(r, p, n)$ certain algebras can be defined that are analogous in some sense to Hecke algebras.

The **Iwahori–Hecke algebra** of a finite Coxeter group W is a certain algebra that is a q-analogue, or q-deformation, of the group algebra W. (See Appendix B3 for a proper definition.) It has a basis T_w, for $w \in W$ (so it is the same dimension as the group algebra of W), but multiplication depends on a particular number $q \in \mathbb{C}$, which can be chosen arbitrarily. These algebras are true Hecke algebras only when W is a finite Weyl group.

The **"Hecke algebras" of the groups** $G(r, p, n)$ are q-analogues of the group algebras of the groups $G(r, p, n)$; see Appendix B4 for a discusson of how they are defined. They were defined only recently, as follows:

- Ariki and Koike [1994] for the case $G(r, 1, n)$;
- Broué and Malle [1993] and Ariki [1995] for the general case.

It is important to note that these algebras are not true Hecke algebras. In group theory a Hecke algebra is very specific kind of double coset algebra and the "Hecke algebras" of the groups $G(r, p, n)$ do not fit this mold.

Appendix B4 gives some partial answers to the main questions for these algebras; here we give references to the literature for these answers.

I. What are the irreducibles?

Results of [Ariki and Koike 1994; Ariki 1995] say that the "Hecke algebras" of $G(r, p, n)$ are q-deformations of the group algebras of the groups $G(r, p, n)$. Thus, it follows from the Tits deformation theorem [Carter 1985, Chap. 10, 11.2; Curtis and Reiner 1987, § 68.17] that the indexings and dimension formulas for the irreducible representations of these algebras must be the same as the indexings and dimension formulas for the groups $G(r, p, n)$. Finding analogues of the character formulas requires a bit more work and a Murnaghan–Nakayama type rule for the "Hecke algebras" of $G(r, p, n)$ was given by Halverson and Ram [1998]. As far as we know, the formula for the irreducible characters of S_n as a weighted sum of standard tableaux which we gave in the symmetric group section has not yet been generalized to the case of $G(r, p, n)$ and its "Hecke algebras".

C. How do we construct the irreducible modules?

Analogues of Young's seminormal representations have been given by

- Hoefsmit [1974] and Wenzl [1988a], independently, for Iwahori–Hecke algebras of type A_{n-1};
- Hoefsmit [1974], for Iwahori–Hecke algebras of types B_n and D_n;
- Ariki and Koike [1994] for the "Hecke algebras" of $G(r, 1, n)$;
- Ariki [1995] for the general "Hecke algebras" of $G(r, p, n)$.

There seems to be more than one appropriate choice for the analogue of Young symmetrizers for Hecke algebras. The definitions in the literature are due to

- Gyoja [1986], for the Iwahori–Hecke algebras of type A_{n-1};
- Dipper and James [1986] and Murphy [1992; 1995] for the Iwahori–Hecke algebras of type A_{n-1};
- King and Wybourne [1992] and Duchamp et al. [1995] for the Iwahori–Hecke algebras of type A_{n-1};

- Dipper, James, and Murphy [1992; 1995] for the Iwahori–Hecke algebras of type B_n;
- Pallikaros [1994], for the Iwahori–Hecke algebras of type D_n;
- Mathas [1997] and Murphy [≥ 1999], for the "Hecke algebras" of $G(r, p, n)$.

The paper [Graham and Lehrer 1996] also contains important ideas in this direction.

S. Special/Interesting representations

It follows from the Tits deformation theorem (or rather, an extension of it) that the results for the "Hecke algebras" of $G(r, p, n)$ must be the same as for the case of the groups $G(r, p, n)$.

5.3. Tensor Power Centralizer Algebras

A **tensor power centralizer algebra** is an algebra that is isomorphic to $\mathrm{End}_G(V^{\otimes k})$ for some group (or Hopf algebra) G and some representation V of G. In this definition

$$\mathrm{End}_G(V^{\otimes k}) = \{T \in \mathrm{End}(V^{\otimes k}) \mid Tgv = gTv, \text{ for all } g \in G \text{ and all } v \in V^{\otimes k}\}.$$

Some examples of tensor power centralizer algebras have been particularly important:

(a) The **group algebras**, $\mathbb{C}S_k$, **of the symmetric groups** S_k.

(b) **Iwahori–Hecke algebras**, $H_k(q)$, **of type** A_{k-1} were introduced by Iwahori [1964] in connection with $\mathrm{GL}(n, \mathbb{F}_q)$. Jimbo [1986] realized that they arise as tensor power centralizer algebras for quantum groups.

(c) **Temperley–Lieb algebras**, $\mathrm{TL}_k(x)$, were introduced independently by several people. Some of the discoverers were Rumer, Teller, and Weyl [Rumer et al. 1932], Penrose [1969; 1971], Temperley and Lieb [1971], Kaufmann [1987] and Jones [1983]. The work of V. Jones was crucial in making them so important in combinatorial representation theory today.

(d) **Brauer algebras**, $B_k(x)$, were defined in [Brauer 1937]. Brauer also proved that they are tensor power centralizers.

(e) **Birman–Murakami–Wenzl algebras**, $\mathrm{BMW}_k(r, q)$, are due to Birman and Wenzl [1989] and Murakami [1987]. It was realized early on [Reshetikhin 1987; Wenzl 1990] that they arise as tensor power centralizers but there was no proof in the literature for some time. See the references in [Chari and Pressley 1994, § 10.2].

(f) **Spider algebras** were written down combinatorially and studied by G. Kuperberg [1994a; 1996b].

(g) **Rook monoid algebras** are the "group algebras" of some very natural monoids, and L. Solomon (unpublished work) realized these as tensor power centralizers. Their combinatorics has also been studied in [Garsia and Remmel 1986].

(h) **Solomon–Iwahori algebras** were introduced in [Solomon 1990]. The fact that they are tensor power centralizer algebras is an unpublished result of Solomon [1995].

(i) **Rational Brauer algebras** were introduced in a nice combinatorial form in [Benkart et al. 1994] and in other forms in [Koike 1989; Procesi 1976] and other older invariant theory works [Weyl 1946]. All of these works were related to tensor power centralizers and/or fundamental theorems of invariant theory.

(j) The **q-rational Brauer algebras** were introduced by Kosuda and Murakami [1992; 1993] and subsequently studied in [Leduc 1994; Halverson 1996; 1995].

(k) **partition algebras** were introduced by V. Jones [1994] and have been studied subsequently by P. Martin [1996].

The following paragraphs give references on the results associated with objects (b)–(e); see also Appendices B5 to B8 for the results themselves and missing definitions. We will not discuss objects (f)–(k) further; see the references given in the individual entries above.

I. What are the irreducibles?

Indexing of the representations of tensor power centralizer algebras follows from double centralizer theory [Weyl 1946] and a good understanding of the indexings and tensor product rules for the group or algebra that it is centralizing: $GL(n, \mathbb{C})$, $O(n, \mathbb{C})$, $U_q\mathfrak{sl}(n)$, etc. The references for resulting indexings and dimension formulas for the irreducible representations are as follows:

Temperley–Lieb algebras. These results are classical and can be found in [Goodman et al. 1989].

Brauer algebras. These results were known to Brauer [1937] and Weyl [1946]. An important combinatorial point of view was given by Berele [1986b; 1986a] and further developed by Sundaram [1990c; 1990b; 1990a].

Iwahori–Hecke algebras of type A_{n-1}. These results follow from the Tits deformation theorem and the corresponding results for the symmetric group.

Birman–Murakami–Wenzl algebras. These results follow from the Tits deformation theorem and the corresponding results for the Brauer algebra.

The indexings and dimension formulas for the Temperley–Lieb and Brauer algebras also follow easily by using the techniques of the Jones basic construction [Wenzl 1988b; Halverson and Ram 1995].

The references for the irreducible characters of the various tensor power centralizer algebras are as follows:

Temperley–Lieb algebras. Character formulas can be derived easily by using Jones Basic Construction techniques [Halverson and Ram 1995].

Iwahori–Hecke algebras of type A_{n-1}. The analogue of the formula for the irreducible characters of S_n as a weighted sum of standard tableaux was found by

Roichman [1997]. Murnaghan–Nakayama type formulas were found by several authors [King and Wybourne 1992; Van der Jeugt 1991; Vershik and Kerov 1988; Ueno and Shibukawa 1992; Ram 1991].

Brauer algebras and Birman–Murakami–Wenzl algebras. Murnaghan–Nakayama type formulas were derived in [Ram 1995] and [Halverson and Ram 1995], respectively.

Brauer algebra and Birman–Murakami–Wenzl algebra analogues of the formula for the irreducible characters of the symmetric groups as a weighted sum of standard tableaux have not appeared in the literature.

C. How do we construct the irreducibles?

Temperley–Lieb algebras. An application of the Jones Basic Construction [Wenzl 1988b; Halverson and Ram 1995] gives a construction of the irreducible representations of the Temperley–Lieb algebras. This construction is classical and has been rediscovered by many people. In this case the construction is an analogue of the Young symmetrizer construction. The analogue of the seminormal construction appears in [Goodman et al. 1989].

Iwahori–Hecke algebras of type A_{n-1}. The analogue of Young's seminormal construction for this case is due, independently, to Hoefsmit [1974] and Wenzl [1988a]. Various analogues of Young symmetrizers have been given by Gyoja [1986], Dipper and James [1986], Murphy [1992; 1995], King and Wybourne [1992] and Duchamp et al. [1995].

Brauer algebras. Analogues of Young's seminormal representations have been given, independently, by Nazarov [1996] and Leduc and Ram [1997]. An analogue of the Young symmetrizer construction can be obtained by applying the Jones Basic Construction to the classical Young symmetrizer construction and this is the one that has been used by many authors [Benkart et al. 1990; Hanlon and Wales 1989; Kerov 1992; Graham and Lehrer 1996]. The actual element of the algebra that is the analogue of the Young symmetrizer involves a central idempotent for which there is no known explicit formula and this is the reason that most authors work with a quotient formulation of the appropriate module.

Birman–Murakami–Wenzl algebras. Analogues of Young's seminormal representations have been given by Murakami [1990] and Leduc and Ram [1997], using different methods: Murakami uses the physical theory of Boltzmann weights, whereas Leduc and Ram use the theory of ribbon Hopf algebras and quantum groups. Exactly in the same way as for the Brauer algebra, an analogue of the Young symmetrizer construction can be obtained by applying the Jones Basic Construction to the Young symmetrizer constructions for the Iwahori–Hecke algebra of type A_{n-1}. As in the Brauer algebra case one should work with a quotient formulation of the module to avoid using a central idempotent for which there is no known explicit formula.

5.4. Reflection Groups of Exceptional Type

Generalizing the S_n theory to finite Coxeter groups of exceptional type, finite complex reflection groups of exceptional type and the corresponding Iwahori–Hecke algebras, has been largely unsuccessful. This is not to say that there haven't been some very nice partial results, only that at the moment nobody has any understanding of how to make a good combinatorial theory to encompass all the classical and exceptional types at once. Two amazing partial results along these lines are the **Springer construction** and the **Kazhdan–Lusztig construction**.

The Springer construction is a construction of the irreducible representations of the crystallographic reflection groups on cohomology of unipotent varieties [Springer 1978]. It is a geometric construction and not a combinatorial construction. See Appendix A3 for more information in the symmetric group case. It is possible that this construction may be combinatorialized in the future, but to date no one has done this.

The Kazhdan–Lusztig construction [1979] is a construction of certain representations called *cell representations* and it works for all finite Coxeter groups. The cell representations are almost irreducible but unfortunately not irreducible in general, and nobody understands how to break them up into irreducibles, except in a case by case fashion. The other problem with these representations is that they depend crucially on certain polynomials, the Kazhdan–Lusztig polynomials, which seem to be exceedingly difficult to compute or understand well except in very small cases; see [Brenti 1994] for more information. See [Carter 1985] for a summary and tables of the known facts about representation of finite Coxeter groups of exceptional type.

Remarks

(1) As mentioned in Section 5.1.2, a Hecke algebra is a specific double coset algebra that depends on a group G and a subgroup B. Iwahori [1964] studied these algebras in the case that G is a finite Chevalley group and B is a Borel subgroup of G, and defined what are now called *Iwahori–Hecke algebras*. These are q-analogues of the group algebras of finite Weyl groups. The work of Iwahori yields a presentation for these algebras which can easily be extended to define Iwahori–Hecke algebras for all Coxeter groups but, except for the original Weyl group case, these have never been realized as true Hecke algebras, i.e., double coset algebras corresponding to an appropriate G and B. The "Hecke algebras" corresponding to the groups $G(r, p, n)$ are q-analogues of the group algebras of $G(r, p, n)$. Although these algebras are not true Hecke algebras either, Broué and Malle [1993] have shown that many of these algebras arise in connection with nondefining characteristic representations of finite Chevalley groups and Deligne–Lusztig varieties.

(2) There is much current research on generalizing symmetric group results to affine Coxeter groups and affine Hecke algebras. The case of affine Coxeter groups was done by Kato [1983] using ideas from Clifford theory. The case of affine Hecke algebras has been intensely studied [Lusztig 1987a; 1987b; 1987b; 1989a; 1988; 1989b; 1995b; 1995a, Kazhdan and Lusztig 1987; Ginsburg 1987; Chriss and Ginzburg 1997] Most of this work is very geometric and relies on methods involving intersection cohomology and K-theory. Hopefully some of this work will be made combinatorial in the near future.

(3) Wouldn't it be great if we had a nice combinatorial representation theory for finite simple groups!

6. Generalizations of $\mathrm{GL}(n, \mathbb{C})$ Results

There have been successful generalizations of the $\mathrm{GL}(n, \mathbb{C})$ results **(Ia)–(Ic)** and **(S1)–(S2)** of Section 3 (pages 35–38) to the following classes of groups and algebras.

(1) **Connected complex semisimple Lie groups**. Examples: $\mathrm{SL}(n, \mathbb{C})$, $\mathrm{SO}(n, \mathbb{C})$, $\mathrm{Sp}(2n, \mathbb{C})$, $\mathrm{PGL}(n, \mathbb{C})$, $\mathrm{PSO}(2n, \mathbb{C})$, $\mathrm{PSp}(2n, \mathbb{C})$.

(2) **Compact connected real Lie groups**. Examples: $\mathrm{SU}(n, \mathbb{C})$, $\mathrm{SO}(n, \mathbb{R})$, $\mathrm{Sp}(n)$, where $\mathrm{Sp}(n) = Sp(2n, \mathbb{C}) \cap U(2n, \mathbb{C})$.

(3) **Finite-dimensional complex semisimple Lie algebras**. Examples: $\mathfrak{sl}(n, \mathbb{C})$, $\mathfrak{so}(n, \mathbb{C})$, $\mathfrak{sp}(2n, \mathbb{C})$. See Appendix A7 for the complete list of the finite-dimensional complex semisimple Lie algebras.

(4) **Quantum groups corresponding to complex semisimple Lie algebras.**

The method of generalizing the $\mathrm{GL}(n, \mathbb{C})$ results to these objects is to reduce them all to case (3) and then solve case (3). The results for case (3) are given in Section 4. The reduction of cases (1) and (2) to case (3) are outlined in [Serre 1987], and given in more detail in [Varadarajan 1984] and [Bröcker and tom Dieck 1985]. The reduction of (4) to (3) is given in [Chari and Pressley 1994] and in [Jantzen 1996].

Partial Results for Further Generalizations

Some partial results along the lines of the results **(Ia)–(Ic)** and **(S1)–(S2)** of Section 3 have been obtained for the following groups and algebras.

(1) **Kac–Moody Lie algebras and groups**
(2) **Yangians**
(3) **Simple Lie superalgebras**

Other groups and algebras, for which the combinatorial representation theory is not understood very well, are

(4) **Finite Chevalley groups**
(5) **p-adic Chevalley groups**
(6) **Real reductive Lie groups**
(7) **The Virasoro algebra**

There are many many possible ways that we could extend this list but probably these four cases are the most fundamental cases where the combinatorial representation theory has not been formulated. There has been intense work on all of these cases, but hardly any by combinatorialists. Thus many beautiful results are known, but very few of them have been stated or interpreted through a combinatorialist's eyes. They are gold veins yet to be mined!

Remarks

(1) An introductory reference to Kac–Moody Lie algebras is [Kac 1990]. This book contains a good description of the basic representation theory of these algebras. We don't know of a good introductory reference for the Kac–Moody groups case. We would suggest beginning with [Kostant and Kumar 1986] and following the references given there.

(2) The basic introductory reference for Yangians and their basic representation theory is [Chari and Pressley 1994, Chapter 12]. See also the references given there.

(3) The best introductory reference for Lie superalgebras is [Scheunert 1979]. For an update on the combinatorial representation theory of these cases see [Sergeev 1984; Berele and Regev 1987; Benkart et al. 1998; Serganova 1996].

(4) Finding a general combinatorial representation theory for finite Chevalley groups has been elusive for many years. After the fundamental work of J. A. Green [1955], which established a combinatorial representation theory for $GL(n, \mathbb{F}_q)$, there has been a concerted effort to extend these results to other finite Chevalley groups. G. Lusztig [1978; 1977; 1984; 1974] has made important contributions to this field; in particular, the results of Deligne and Lusztig [1976] are fundamental. However, this is a geometric approach rather than a combinatorial one and there is much work to be done for combinatorialists, even in interpreting the known results from a combinatorial viewpoint. A good introductory treatment of this theory is the book by Digne and Michel [1991]. The original work of Green is treated in [Macdonald 1995, Chap. IV].

(5) The representation theory of p-adic Lie groups has been studied intensely by representation theorists but essentially not at all by combinatorialists. It is clear that there is a beautiful (although possibly very difficult) combinatorial representation theory lurking here. The best introductory reference to this work is the paper of R. Howe [1994] on p-adic $GL(n)$. Recent results of G.

Lusztig [1995a] are a very important step in providing a general combinatorial representation theory for p-adic groups.

(6) The best places to read about the representation theory of real reductive groups are the books [Knapp and Vogan 1995; Vogan 1987; 1981; Wallach 1988; 1992].

(7) The Virasoro algebra is a Lie algebra that seems to turn up in every back alley of representation theory. One can only surmise that it must have a beautiful combinatorial representation theory that is waiting to be clarified. A good place to read about the Virasoro algebra is in [Feĭgin and Fuchs 1990].

APPENDICES

A1. Basic Representation Theory

An *algebra* A is a vector space over $\mathbb{C}$ with a multiplication that is associative, distributive, has an identity and satisfies the equation

$$(ca_1)a_2 = a_1(ca_2) = c(a_1a_2), \quad \text{for all } a_1, a_2 \in A \text{ and } c \in \mathbb{C}.$$

An *A-module* is a vector space M over $\mathbb{C}$ with an A-action $A \times M \to M$, $(a, m) \mapsto am$, that satisfies

$$1m = m,$$
$$a_1(a_2m) = (a_1a_2)m,$$
$$(a_1 + a_2)m = a_1m + a_2m,$$
$$a(c_1m_1 + c_2m_2) = c_1(am_1) + c_2(am_2).$$

for all $a, a_1, a_2 \in A$, $m, m_1, m_2 \in M$ and $c_1, c_2 \in \mathbb{C}$. We shall use the words module and *representation* interchangeably.

A module M is *indecomposable* if there do not exist non zero A-modules M_1 and M_2 such that

$$M \cong M_1 \oplus M_2.$$

A module M is *irreducible* or *simple* if the only submodules of M are the zero module 0 and M itself. A module M is *semisimple* if it is the direct sum of simple submodules.

An algebra is *simple* if the only ideals of A are the zero ideal 0 and A itself. The *radical* rad(A) of an algebra A is the intersection of all the maximal left ideals of A. An algebra A is *semisimple* if all its modules are semisimple. An algebra A is *Artinian* if every decreasing sequence of left ideals of A stabilizes, that is for every chain

$$A_1 \supseteq A_2 \supseteq A_3 \supseteq \cdots$$

of left ideals of A there is an integer m such that $A_i = A_m$ for all $i \geq m$.

The following statements follow directly from the definitions.

Let A be an algebra.

(a) *Every irreducible A-module is indecomposable.*

(b) *The algebra A is semisimple if and only if every indecomposable A-module is irreducible.*

The proofs of the following statements are more involved and can be found in [Bourbaki 1958, Chap. VIII, § 6 n°4 and § 5 n°3].

THEOREM A1.1. (a) *If A is an Artinian algebra, the radical of A is the largest nilpotent ideal of A.*

(b) *An algebra A is semisimple if and only if A is Artinian and* $\mathrm{rad}(A) = 0$.

(c) *Every semisimple algebra is a direct sum of simple algebras.*

The case when A is not necessarily semisimple is often called *modular representation theory.* Let M be an A-module. A *composition series* of M is a chain

$$M = M_k \supseteq M_{k-1} \supseteq \cdots \supseteq M_1 \supseteq M_0 = 0,$$

such that, for each $1 \leq i \leq k$, the modules M_i/M_{i-1} are irreducible. The irreducible modules M_i/M_{i-1} are the *factors* of the composition series. The following theorem is proved in [Curtis and Reiner 1962, (13.7)].

THEOREM A1.2 (JORDAN–HÖLDER). *If there exists a composition series for M then any two composition series must have the same multiset of factors (up to module isomorphism).*

An important combinatorial point of view is as follows: The analogue of the subgroup lattice of a group can be studied for any A-module M. More precisely, the *submodule lattice* $L(M)$ of M is the lattice defined by the submodules of M with the order relations given by inclusions of submodules. The composition series are maximal chains in this lattice.

References. All these results can be found in [Bourbaki 1958, Chap. VIII; Curtis and Reiner 1962].

A2. Partitions and Tableaux

Partitions. A *partition* is a sequence $\lambda = (\lambda_1, \ldots, \lambda_n)$ of integers such that

$$\lambda_1 \geq \cdots \geq \lambda_n \geq 0.$$

We write $\lambda \vdash N$ if $N = \lambda_1 + \cdots + \lambda_n$. It is conventional to identify a partition with its Ferrers diagram, which has λ_i boxes in the i-th row. For example, the partition $\lambda = (55422211)$ has the Ferrers diagram shown at the top of the next page.

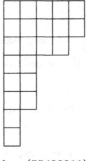

$$\lambda = (55422211)$$

We number the rows and columns of the Ferrers diagram as is conventionally done for matrices. If x is the box in λ in position (i,j), the *content* of x is $c(x) = j - i$ and the *hook length* of x is $h_x = \lambda_i - j + \lambda'_j - i + 1$, where λ'_j is the length of the j-th column of λ.

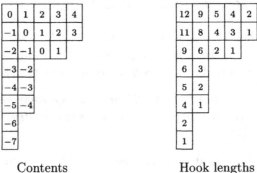

Contents   Hook lengths

If μ and λ are partitions such that the Ferrers diagram of μ is contained in that of λ, we write $\mu \subseteq \lambda$ and we denote the difference of the Ferrers diagrams by λ/μ. We refer to λ/μ as a *shape* or, more specifically, a *skew shape*.

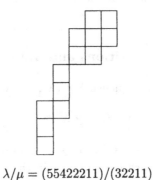

$$\lambda/\mu = (55422211)/(32211)$$

Tableaux. Suppose that λ has k boxes. A *standard tableau* of shape λ is a filling of the Ferrers diagram of λ with $1, 2, \ldots, k$ such that the rows and columns are increasing from left to right and from top to bottom respectively.

1	2	5	9	13
3	6	10	14	16
4	8	15	17	
7	12			
11	20			
18	21			
19				
22				

Let λ/μ be a shape. A *column strict tableau* of shape λ/μ filled with $1, 2, \ldots n$ is a filling of the Ferrers diagram of λ/μ with elements of the set $\{1, 2, \ldots, n\}$ such that the rows are weakly increasing from left to right and the columns are strictly increasing from top to bottom. The *weight* of a column strict tableau T is the sequence of positive integers $\nu = (\nu_1, \ldots, \nu_n)$, where ν_i is the number of i's in T.

1	1	1	2	3
2	2	4	4	5
3	3	5	5	
6	7			
7	8			
8	9			
10				
11				

Shape $\lambda = (55422211)$

Weight $\nu = (33323122111)$

The *word* of a column strict tableau T is the sequence

$$w = w_1 w_2 \cdots w_p$$

obtained by reading the entries of T from right to left in successive rows, starting with the top row. A word $w = w_1 \cdots w_p$ is a *lattice permutation* if for each $1 \le r \le p$ and each $1 \le i \le n-1$ the number of occurrences of the symbol i in $w_1 \cdots w_r$ is not less than the number of occurrences of $i+1$ in $w_1 \cdots w_r$.

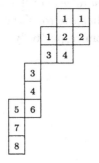

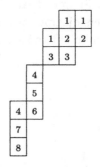

$w = 1122143346578$ $w = 1122133456478$

Not a lattice permutation   Lattice permutation

A *border strip* is a skew shape λ/μ that

(a) is connected (two boxes are connected if they share an edge), and
(b) does not contain a 2×2 block of boxes.

The weight of a border strip λ/μ is given by

$$\mathrm{wt}(\lambda/\mu) = (-1)^{r(\lambda/\mu)-1},$$

where $r(\lambda/\mu)$ is the number of rows in λ/μ.

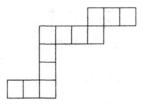

$$\lambda/\mu = (86333)/(5222)$$
$$\mathrm{wt}(\lambda/\mu) = (-1)^{5-1}$$

Let λ and $\mu = (\mu_1, \ldots, \mu_l)$ be partitions of n. A μ-*border strip tableau* of shape λ is a sequence of partitions

$$T = (\varnothing = \lambda^{(0)} \subseteq \lambda^{(1)} \subseteq \cdots \subseteq \lambda^{(l-1)} \subseteq \lambda^{(l)} = \lambda)$$

such that, for each $1 \leq i \leq l$,

(a) $\lambda^{(i)}/\lambda^{(i-1)}$ is a border strip, and
(b) $|\lambda^{(i)}/\lambda^{(i-1)}| = \mu_i$.

The weight of a μ-border strip tableau T of shape λ is

$$\mathrm{wt}(T) = \prod_{i=1}^{l-1} \mathrm{wt}(\lambda^{(i)}/\lambda^{(i-1)}). \qquad (A2\text{--}1)$$

THEOREM A2.1 (MURNAGHAN–NAKAYAMA RULE). *Let λ and μ be partitions of n and let $\chi^\lambda(\mu)$ denote the irreducible character of the symmetric group S_n indexed by λ evaluated at a permutation of cycle type μ. Then*

$$\chi^\lambda(\mu) = \sum_T \mathrm{wt}(T),$$

where the sum is over all μ-border strip tableaux T of shape λ and $\mathrm{wt}(T)$ is as given in (A2–1).

References. All these facts can be found in [Macdonald 1995, Chap. I]. The proof of theorem A2.2 is given in [Macdonald 1995, Chap. I, §7, Ex. 5].

A3. The Flag Variety, Unipotent Varieties, and Springer Theory for $\mathrm{GL}(n, \mathbb{C})$

Borel subgroups, Cartan subgroups, and unipotent elements. Define

$$B_n = \left\{ \begin{pmatrix} * & * & \cdots & * \\ 0 & * & & \vdots \\ \vdots & & \ddots & * \\ 0 & \cdots & 0 & * \end{pmatrix} \right\}, \quad T_n = \left\{ \begin{pmatrix} * & 0 & \cdots & 0 \\ 0 & * & & \vdots \\ \vdots & & \ddots & 0 \\ 0 & \cdots & 0 & * \end{pmatrix} \right\}, \quad U_n = \left\{ \begin{pmatrix} 1 & * & \cdots & * \\ 0 & 1 & & \vdots \\ \vdots & & \ddots & * \\ 0 & \cdots & 0 & 1 \end{pmatrix} \right\}$$

as the subgroups of $\mathrm{GL}(n, \mathbb{C})$ consisting of upper triangular, diagonal, and upper unitriangular matrices, respectively.

A *Borel subgroup* of $\mathrm{GL}(n, \mathbb{C})$ is a subgroup that is conjugate to B_n.

A *Cartan subgroup* of $\mathrm{GL}(n, \mathbb{C})$ is a subgroup that is conjugate to T_n.

A matrix $u \in \mathrm{GL}(n, \mathbb{C})$ is *unipotent* if it is conjugate to an upper unitriangular matrix.

The flag variety. There are one-to-one correspondences among the following sets:

(1) $\mathcal{B} = \{$Borel subgroups of $\mathrm{GL}(n, \mathbb{C})\}$,
(2) G/B, where $G = \mathrm{GL}(n, \mathbb{C})$ and $B = B_n$,
(3) $\{$flags $0 \subseteq V_1 \subseteq V_2 \subseteq \cdots \subseteq V_n = \mathbb{C}^n$ such that $\dim(V_i) = i\}$.

Each of these sets naturally has the structure of a complex algebraic variety, that is called the *flag variety*.

The unipotent varieties. Given a unipotent element $u \in \mathrm{GL}(n, \mathbb{C})$ with Jordan blocks given by the partition $\mu = (\mu_1, \ldots, \mu_l)$ of n, define an algebraic variety

$$\mathcal{B}_\mu = \mathcal{B}_u = \{$Borel subgroups of $\mathrm{GL}(n, \mathbb{C})$ that contain $u\}.$$

By conjugation, the structure of the subvariety $\mathcal{B}_u$ of the flag variety depends only on the partition μ. Thus $\mathcal{B}_\mu$ is well defined, as an algebraic variety.

Springer theory. It is a deep theorem of Springer [1978] (which holds in the generality of semisimple algebraic groups and their corresponding Weyl groups) that there is an action of the symmetric group S_n on the cohomology $H^*(\mathcal{B}_u)$ of the variety $\mathcal{B}_u$. This action can be interpreted nicely as follows. The imbedding

$$\mathcal{B}_u \subseteq \mathcal{B} \quad \text{induces a surjective map} \quad H^*(\mathcal{B}) \longrightarrow H^*(\mathcal{B}_u).$$

It is a famous theorem of Borel that there is a ring isomorphism

$$H^*(\mathcal{B}) \cong \mathbb{C}[x_1, \ldots, x_n]/I^+, \tag{A3-1}$$

where I^+ is the ideal generated by symmetric functions without constant term. It follows that $H^*(\mathcal{B}_u)$ is also a quotient of $\mathbb{C}[x_1, \ldots, x_n]$. From the work of Kraft [1981], DeConcini and Procesi [1981] and Tanisaki [1982], one has that the ideal $\mathcal{J}_u$ which it is necessary to quotient by in order to obtain an isomorphism

$$H^*(\mathcal{B}_u) \cong \mathbb{C}[x_1, \ldots, x_n]/\mathcal{J}_u,$$

can be described explicitly.

The symmetric group S_n acts on the polynomial ring $\mathbb{C}[x_1, \ldots, x_n]$ by permuting the variables. It turns out that the ideal $\mathcal{J}_u$ remains invariant under this action, thus yielding a well defined action of S_n on $\mathbb{C}[x_1, \ldots, x_n]/\mathcal{J}_u$. This action coincides with the Springer action on $H^*(\mathcal{B}_u)$. Hotta and Springer [1977] have established that, if u is a unipotent element of shape μ then, for every permutation $w \in S_n$,

$$\sum_i q^i \varepsilon(w) \operatorname{trace}(w^{-1}, H^{2i}(\mathcal{B}_u)) = \sum_{\lambda \vdash n} \tilde{K}_{\lambda\mu}(q) \chi^\lambda(w),$$

where

$\varepsilon(w)$ is the sign of the permutation w,

$\operatorname{trace}(w^{-1}, H^{2i}(\mathcal{B}_u))$ is the trace of the action of w^{-1} on $H^{2i}(\mathcal{B}_u)$,

$\chi^\lambda(w)$ is the irreducible character of the symmetric group evaluated at w, and

$\tilde{K}_{\lambda\mu}(q)$ is a variant of the Kostka–Foulkes polynomial; see [Macdonald 1995, Chap. III, § 7, Ex. 8, and § 6].

It follows from this discussion and some basic facts about the polynomials $\tilde{K}_{\lambda\mu}(q)$ that the top degree cohomology group in $H^*(\mathcal{B}_\mu)$ is a realization of the irreducible representation of S_n indexed by μ,

$$S^\mu \cong H^{\mathrm{top}}(\mathcal{B}_\mu).$$

This construction of the irreducible modules of S_n is the *Springer construction*.

References. See [Macdonald 1995, Chap. II, § 3, Ex. 1] for a description of the variety $\mathcal{B}_u$ and its structure. The theorem of Borel stated in (A3–1) is given in [Borel 1953; Bernšteĭn et al. 1973]. The references quoted in the text above will provide a good introduction to the Springer theory. The beautiful combinatorics of Springer theory has been studied by Barcelo [1993], Garsia and Procesi [1992], Lascoux [1991], Lusztig and Spaltenstein [1985], Shoji [1979], Spaltenstein [1976], Weyman [1989], and others.

A4. Polynomial and Rational Representations of $GL(n, \mathbb{C})$

If V is a $GL(n, \mathbb{C})$-module of dimension d then, by choosing a basis of V, we can define a map

$$\rho_V : GL(n, \mathbb{C}) \longrightarrow GL(d, \mathbb{C})$$
$$g \longmapsto \rho(g),$$

where $\rho(g)$ is the transformation of V that is induced by the action of g on V. Let

g_{ij} denote the (i, j) entry of the matrix g, and
$\rho(g)_{kl}$ denote the (k, l) entry of the matrix $\rho(g)$.

The map ρ depends on the choice of the basis of V, but the following definitions do not.

The module V is a *polynomial representation* if there are polynomials $p_{kl}(x_{ij})$, for $1 \leq k, l \leq d$, such that

$$\rho(g)_{kl} = p_{kl}(g_{ij}), \quad \text{for all } 1 \leq k, l \leq d.$$

In other words $\rho(g)_{jk}$ is the same as the polynomial p_{kl} evaluated at the entries g_{ij} of the matrix g.

The module V is a *rational representation* if there are rational functions (quotients of two polynomials) $p_{kl}(x_{ij})/q_{kl}(x_{ij})$, for $1 \leq k, l \leq d$, such that

$$\rho(g)_{kl} = p_{kl}(g_{ij})/q_{kl}(g_{ij}), \quad \text{for all } 1 \leq k, l \leq n.$$

Clearly, every polynomial representation is a rational one.

The theory of rational representations of $GL(n, \mathbb{C})$ can be reduced to the theory of polynomial representations of $GL(n, \mathbb{C})$. This is accomplished as follows. The determinant $\det : GL(n, \mathbb{C}) \to \mathbb{C}$ defines a 1-dimensional (polynomial) representation of $GL(n, \mathbb{C})$. Any integral power

$$\det^k : \quad GL(n, \mathbb{C}) \longrightarrow \quad \mathbb{C}$$
$$g \longmapsto \quad \det(g)^k$$

of the determinant also determines a 1-dimensional representation of $GL(n, \mathbb{C})$. **All irreducible rational representations $GL(n, \mathbb{C})$ can be constructed in the form**

$$\det^k \otimes V^\lambda,$$

for some $k \in \mathbb{Z}$ and some irreducible polynomial representation V^λ of $\mathrm{GL}(n, \mathbb{C})$.

There exist representations of $\mathrm{GL}(n, \mathbb{C})$ that are not rational representations, for example

$$g \mapsto \begin{pmatrix} 1 & \ln|\det(g)| \\ 0 & 1 \end{pmatrix}.$$

There is no known classification of representations of $\mathrm{GL}(n, \mathbb{C})$ that are not rational.

References. See [Stembridge 1989a; 1987] for a study of the combinatorics of the rational representations of $\mathrm{GL}(n, \mathbb{C})$.

A5. Schur–Weyl Duality and Young Symmetrizers

Let V be the usual n-dimensional representation of $\mathrm{GL}(n, \mathbb{C})$ on column vectors of length n, that is

$$V = \mathrm{span}\{b_1, \ldots, b_n\} \quad \text{where} \quad b_i = (0, \ldots, 0, 1, 0, \ldots, 0)^t,$$

and the 1 in b_i appears in the i-th entry. Then

$$V^{\otimes k} = \mathrm{span}\{b_{i_1} \otimes \cdots \otimes b_{i_k} \mid 1 \le i_1, \ldots, i_k \le n\}$$

is the span of the words of length k in the letters b_i (except that the letters are separated by tensor symbols). The general linear group $\mathrm{GL}(n, \mathbb{C})$ and the symmetric group S_k act on $V^{\otimes k}$ by

$$g(v_1 \otimes \cdots \otimes v_k) = gv_1 \otimes \cdots \otimes gv_k, \quad \text{and} \quad (v_1 \otimes \cdots \otimes v_k)\sigma = v_{\sigma(1)} \otimes \cdots \otimes v_{\sigma(k)},$$

where $g \in \mathrm{GL}(n, \mathbb{C})$, $\sigma \in S_k$, and $v_1, \ldots, v_k \in V$. We have chosen to make the S_k-action a right action here; one could equally well choose the action of S_k to be a left action, but then the formula would be

$$\sigma(v_1 \otimes \cdots \otimes v_k) = v_{\sigma^{-1}(1)} \otimes \cdots \otimes v_{\sigma^{-1}(k)}.$$

The following theorem is the amazing relationship between the group S_k and the group $\mathrm{GL}(n, \mathbb{C})$ discovered by Schur [1901] and exploited with such success by Weyl [1946].

THEOREM A5.1 (SCHUR–WEYL DUALITY).

(a) *The action of S_k on $V^{\otimes k}$ generates* $\mathrm{End}_{Gl(n,\mathbb{C})}(V^{\otimes k})$.
(b) *The action of $\mathrm{GL}(n, \mathbb{C})$ on $V^{\otimes k}$ generates* $\mathrm{End}_{S_k}(V^{\otimes k})$.

This theorem has an corollary that provides an intimate correspondence between the representation theory of S_k and *some* of the representations of $\mathrm{GL}(n, \mathbb{C})$ (the ones indexed by partitions of k).

COROLLARY A5.2. *As* $GL(n, \mathbb{C}) \times S_k$ *bimodules*

$$V^{\otimes k} \cong \bigoplus_{\lambda \vdash k} V^\lambda \otimes S^\lambda,$$

where V^λ is the irreducible $GL(n, \mathbb{C})$-module and S^λ is the irreducible S_k-module indexed by λ.

If λ is a partition of k, the irreducible $GL(n, \mathbb{C})$-representation V^λ is given by

$$V^\lambda \cong V^{\otimes k} P(T) N(T),$$

where T is a tableau of shape λ and $P(T)$ and $N(T)$ are as defined in Section 2, Question **C** (page 31).

A6. The Borel–Weil–Bott Construction

Let $G = GL(n, \mathbb{C})$ and let $B = B_n$ be the subgroup of upper triangular matrices in $GL(n, \mathbb{C})$. A *line bundle on G/B* is a pair $(\mathcal{L}, p)$ where $\mathcal{L}$ is an algebraic variety and p is a map (morphism of algebraic varieties)

$$p : \mathcal{L} \longrightarrow G/B,$$

such that the fibers of p are lines and such that $\mathcal{L}$ is a locally trivial family of lines. In this definition, *fibers* means the sets $p^{-1}(x)$ for $x \in G/B$ and *lines* means one-dimensional vector spaces. For the definition of *locally trivial family of lines* see [Shafarevich 1994, Chap. VI, § 1.2]. By abuse of language, a line bundle $(\mathcal{L}, p)$ is simply denoted by $\mathcal{L}$. Conceptually, a line bundle on G/B means that we are putting a one-dimensional vector space over each point in G/B.

A *global section* of the line bundle $\mathcal{L}$ is a map (morphism of algebraic varieties)

$$s : G/B \to \mathcal{L}$$

such that $p \circ s$ is the identity map on G/B. In other words a global section is any possible "right inverse map" to the line bundle.

Each partition $\lambda = (\lambda_1, \ldots, \lambda_n)$ determines a character (one-dimensional representation) of the group T_n of diagonal matrices in $GL(n, \mathbb{C})$ via

$$\lambda \left(\begin{pmatrix} t_1 & 0 & \cdots & 0 \\ 0 & t_2 & & \vdots \\ \vdots & & \ddots & 0 \\ 0 & \cdots & 0 & t_n \end{pmatrix} \right) = t_1^{\lambda_1} t_2^{\lambda_2} \cdots t_n^{\lambda_n}.$$

Extend this character to be a character of $B = B_n$ by letting λ ignore the strictly upper triangular part of the matrix, that is $\lambda(u) = 1$, for all $u \in U_n$. Let $\mathcal{L}_\lambda$ be the fiber product $G \times_B \lambda$, that is, the set of equivalence classes of pairs (g, c), with $g \in G$ and $c \in \mathbb{C}^*$, under the equivalence relation

$$(gb, c) \sim (g, \lambda(b^{-1})c), \quad \text{for all } b \in B.$$

Then $\mathcal{L}_\lambda = G \times_B \lambda$ with the map

$$
\begin{array}{rcl}
p: \quad G \times_B \lambda & \longrightarrow & G/B \\
(g, c) & \longmapsto & gB
\end{array}
$$

is a line bundle on G/B.

The Borel–Weil–Bott theorem says that the irreducible representation V^λ of $GL_n(\mathbb{C})$ is

$$V^\lambda \cong H^0(G/B, \mathcal{L}_\lambda),$$

where $H^0(G/B, \mathcal{L}_\lambda)$ is the space of global sections of the line bundle $\mathcal{L}_\lambda$.

References. See [Fulton and Harris 1991] and G. Segal's article in [Carter et al. 1995] for further information and references on this very important construction.

A7. Complex Semisimple Lie Algebras

A *finite-dimensional complex semisimple Lie algebra* is a finite-dimensional Lie algebra $\mathfrak{g}$ over $\mathbb{C}$ such that $\mathrm{rad}(\mathfrak{g}) = 0$. The following theorem classifies all finite-dimensional complex semisimple Lie algebras.

THEOREM A7.1. (a) *Every finite-dimensional complex semisimple Lie algebra*
  $\mathfrak{g}$ *is a direct sum of complex simple Lie algebras.*
(b) *There is one complex simple Lie algebra corresponding to each of the following types*

$$A_{n-1}, \quad B_n, \quad C_n, \quad D_n, \quad E_6, \quad E_7, \quad E_8, \quad F_4, \quad G_2.$$

The complex simple Lie algebras of types A_n, B_n, C_n and D_n are the ones of *classical type* and they are:

$$
\begin{array}{ll}
\text{Type } A_{n-1}: & \mathfrak{sl}(n, \mathbb{C}) = \{A \in M_n(\mathbb{C}) \mid \mathrm{Tr}(A) = 0\}, \\
\text{Type } B_n: & \mathfrak{so}(2n+1, \mathbb{C}) = \{A \in M_{2n+1}(\mathbb{C}) \mid A + A^t = 0\}, \\
\text{Type } C_n: & \mathfrak{sp}(2n, \mathbb{C}) = \{A \in M_{2n}(\mathbb{C}) \mid AJ + JA^t = 0\}, \\
\text{Type } D_n: & \mathfrak{so}(2n) = \{A \in M_{2n}(\mathbb{C}) \mid A + A^t = 0\},
\end{array}
$$

where J is the matrix of a skew-symmetric form on a $2n$-dimensional space.

Let $\mathfrak{g}$ be a complex semisimple Lie algebra. A *Cartan subalgebra* of $\mathfrak{g}$ is a maximal abelian subalgebra $\mathfrak{h}$ of $\mathfrak{g}$. Fix a Cartan subalgebra $\mathfrak{h}$ of $\mathfrak{g}$. If V is a finite-dimensional $\mathfrak{g}$-module and $\mu : \mathfrak{h} \to \mathbb{C}$ is any linear function, define

$$V_\mu = \{v \in V \mid hv = \mu(h)v \quad \text{for all } h \in \mathfrak{h}\}.$$

The space V_μ is the *μ-weight space* of V. It is a nontrivial theorem [Serre 1987] that

$$V = \bigoplus_{\mu \in P} V_\mu,$$

where P is a $\mathbb{Z}$-lattice in $\mathfrak{h}^*$ which can be identified with the $\mathbb{Z}$-lattice P defined in Appendix A8. The vector space $\mathfrak{h}^*$ is the space of linear functions from $\mathfrak{h}$ to $\mathbb{C}$.

Let $\mathbb{C}[P]$ be the group algebra of P. It can be given explicitly as

$$\mathbb{C}[P] = \mathbb{C}\text{-span}\{e^\mu \mid \mu \in P\}, \quad \text{with multiplication } e^\mu e^\nu = e^{\mu+\nu}, \text{ for } \mu, \nu \in P,$$

where the e^μ are formal variables indexed by the elements of P. The *character* of a $\mathfrak{g}$-module is

$$\text{char}(V) = \sum_{\mu \in P} \dim(V_\mu) e^\mu.$$

References. Theorem (A7.1) is due to the founders of the theory, Cartan and Killing, from the late 1800's. The beautiful text of Serre [1987] gives a review of the definitions and theory of complex semisimple Lie algebras. See [Humphreys 1972] for further details.

A8. Roots, Weights and Paths

To each of the "types", A_n, B_n, etc., there is an associated hyperplane arrangement $\mathcal{A}$ in $\mathbb{R}^n$ (Figure 1). The space $\mathbb{R}^n$ has the usual Euclidean inner product $\langle \cdot, \cdot \rangle$. For each hyperplane in the arrangement $\mathcal{A}$ we choose two vectors orthogonal to the hyperplane and pointing in opposite directions. This set of chosen vectors is called the *root system* R associated to $\mathcal{A}$ (Figure 2, left). There is a convention for choosing the lengths of these vectors but we shall not worry about that here.

Choose a chamber (connected component) C of $\mathbb{R}^n \setminus \bigcup_{H \in \mathcal{A}} H$ (Figure 2, right). For each root $\alpha \in R$ we say that α is *positive* if it points toward the same side of the hyperplane as C is and *negative* if it points toward the opposite side. It is standard notation to write

$$\begin{cases} \alpha > 0, & \text{if } \alpha \text{ is a positive root,} \\ \alpha < 0, & \text{if } \alpha \text{ is a negative root.} \end{cases}$$

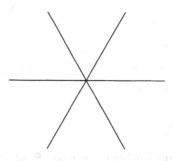

Figure 1. Hyperplane arrangement for A_2.

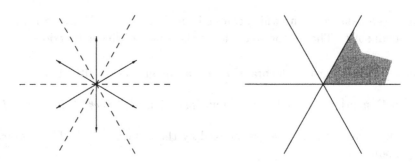

Figure 2. Left: Root system for A_2. Right: A chamber for A_2.

The positive roots associated to hyperplanes that form the walls of C are the *simple roots* $\{\alpha_1, \ldots, \alpha_n\}$. The *fundamental weights* are the vectors $\{\omega_1, \ldots, \omega_n\}$ in $\mathbb{R}^n$ such that

$$\langle \omega_i, \alpha_j^\vee \rangle = \delta_{ij}, \quad \text{where} \quad \alpha_j^\vee = \frac{2\alpha_j}{\langle \alpha_j, \alpha_j \rangle}.$$

Then

$$P = \sum_{i=1}^{r} \mathbb{Z}\omega_i \quad \text{and} \quad P^+ = \sum_{i=1}^{n} \mathbb{N}\,\omega_i, \quad \text{where} \quad \mathbb{N} = \mathbb{Z}_{\geq 0},$$

are the lattice of *integral weights* and the cone of *dominant integral weights*, respectively (Figure 3). There is a one-to-one correspondence between the irreducible representations of $\mathfrak{g}$ and the elements of the cone P^+ in the lattice P.

If λ is a point in P^+, the *straight line path* from 0 to λ is the map

$$\begin{aligned} \pi_\lambda : \quad [0,1] &\longrightarrow \mathbb{R}^n \\ t &\longmapsto t\lambda. \end{aligned}$$

The set $\mathcal{P}\pi_\lambda$ is given by

$$\mathcal{P}\pi_\lambda = \{f_{i_1} \cdots f_{i_k} \pi_\lambda \mid 1 \leq i_1, \ldots, i_k \leq n\}$$

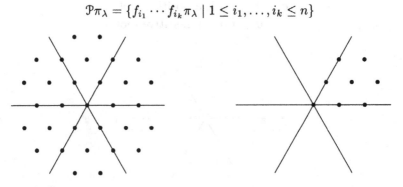

Figure 3. Left: The lattice of integral weights. Right: The cone of dominant integral weights.

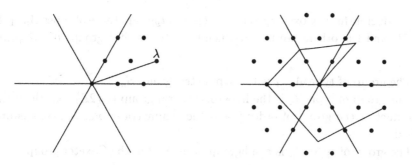

Figure 4. Left: Straight line path from 0 to λ. Right: A path in $\mathcal{P}\pi_\lambda$.

where $f_1, \ldots, f_n$ are the path operators introduced in [Littelmann 1995]. These paths are always piecewise linear and end at a point in P. An example is shown in Figure 4, right.

References. The basics of root systems can be found in [Humphreys 1972]. The reference for the path model is [Littelmann 1995].

B1. Coxeter Groups, Groups Generated by Reflections, and Weyl Groups

A *Coxeter group* is a group W presented by generators $S = \{s_1, \ldots, s_n\}$ and relations

$$s_i^2 = 1, \qquad \text{for } 1 \le i \le n,$$
$$(s_i s_j)^{m_{ij}} = 1, \quad \text{for } 1 \le i \ne j \le n,$$

where each m_{ij} is either ∞ or a positive integer greater than 1.

A *reflection* is a linear transformation of $\mathbb{R}^n$ that is a reflection in some hyperplane.

A *finite group generated by reflections* is a finite subgroup of $\mathrm{GL}(n, \mathbb{R})$ generated by reflections.

THEOREM B1.1. *The finite Coxeter groups are exactly the finite groups generated by reflections.*

A finite Coxeter group is *irreducible* if it cannot be written as a direct product of finite Coxeter groups.

THEOREM B1.2 (CLASSIFICATION OF FINITE COXETER GROUPS).

(a) *Every finite Coxeter group can be written as a direct product of irreducible finite Coxeter groups.*

(b) *There is one irreducible finite Coxeter group corresponding to each of the following "types"*

$$A_{n-1}, \ B_n, \ D_n, \ E_6, \ E_7, \ E_8, \ F_4, \ H_3, \ H_4, \ I_2(m).$$

The irreducible finite Coxeter groups of *classical type* are those of types A_{n-1}, B_n, and D_n and the others are the irreducible finite Coxeter groups of *exceptional type*.

(a) The group of type A_{n-1} is the symmetric group S_n.
(b) The group of type B_n is the hyperoctahedral group $(\mathbb{Z}/2\mathbb{Z}) \wr S_n$, the wreath product of the group of order 2 and the symmetric group S_n. It has order $2^n n!$.
(c) The group of type D_n is a subgroup of index 2 in the Coxeter group of type B_n.
(d) The group of type $I_2(m)$ is a dihedral group of order $2m$.

A finite group W generated by reflections in $\mathbb{R}^n$ is *crystallographic* if there is a lattice in $\mathbb{R}^n$ stable under the action of W. The crystallographic finite Coxeter groups are also called *Weyl groups*. The irreducible Weyl groups are the irreducible finite Coxeter groups of types

$$A_{n-1}, \quad B_n, \quad D_n, \quad E_6, \quad E_7, \quad E_8, \quad F_4, \quad G_2 = I_2(6).$$

References. The most comprehensive reference for finite groups generated by reflections is [Bourbaki 1968]. See also the [Humphreys 1990].

B2. Complex Reflection Groups

A *complex reflection* is an invertible linear transformation of $\mathbb{C}^n$ of finite order which has exactly one eigenvalue that is not 1. A *complex reflection group* is a group generated by complex reflections in $\mathbb{C}^n$. The finite complex reflection groups have been classified by Shepard and Todd [1954]. Each finite complex reflection group is either

(a) $G(r, p, n)$ for some positive integers r, p, n such that p divides r, or
(b) one of 34 other "exceptional" finite complex reflection groups.

Let r, p, d and n be positive integers such that $pd = r$. The complex reflection group $G(r, p, n)$ is the set of $n \times n$ matrices such that

(a) The entries are either 0 or r-th roots of unity,
(b) There is exactly one nonzero entry in each row and each column,
(c) The d-th power of the product for the nonzero entries is 1.

The group $G(r, p, n)$ is a normal subgroup of $G(r, 1, n)$ of index p and

$$|G(r, p, n)| = dr^{n-1} n!.$$

In addition:

(a) $G(1, 1, n) \cong S_n$ the symmetric group or *Weyl group of type A_{n-1}*,
(b) $G(2, 1, n)$ is the *hyperoctahedral group* or *Weyl group of type B_n*,

(c) $G(r, 1, n) \cong (\mathbb{Z}/r\mathbb{Z}) \wr S_n$, the wreath product of the cyclic group of order r with S_n,

(d) $G(2, 2, n)$ is the *Weyl group of type* D_n.

Partial Results for $G(r, 1, n)$

Here are the answers to the main questions **(Ia)–(Ic)** (page 27) for the groups $G(r, 1, n) \cong (\mathbb{Z}/r\mathbb{Z}) \wr S_n$. For the general $G(r, p, n)$ case see [Halverson and Ram 1998].

I. What are the irreducible $G(r, 1, n)$-modules?

(a) How do we index/count them?

There is a **bijection**

$$r\text{-tuples } \lambda = (\lambda^{(1)}, \ldots, \lambda^{(r)}) \text{ of partitions such that } \sum_{i=1}^{r} |\lambda^{(i)}| = n$$
$$\longleftrightarrow \quad \textbf{irreducible representations } C^\lambda.$$

(b) What are their dimensions?

The **dimension** of the irreducible representation C^λ is

$$\dim(C^\lambda) = \# \text{ of standard tableaux of shape } \lambda = n! \prod_{i=1}^{r} \prod_{x \in \lambda^{(i)}} \frac{1}{h_x},$$

where h_x is the hook length at the box x. A standard tableau of shape $\lambda = (\lambda^{(1)}, \ldots, \lambda^{(r)})$ is any filling of the boxes of the $\lambda^{(i)}$ with the numbers $1, 2, \ldots, n$ such that the rows and the columns of each $\lambda^{(i)}$ are increasing.

(c) What are their characters?

A Murnaghan–Nakayama type rule for the characters of the groups $G(r, 1, n)$ was originally given by Specht [1932]. See also [Osima 1954; Halverson and Ram 1998].

References. The original paper of Shepard and Todd [1954] remains a basic reference. Further information about these groups can be found in [Halverson and Ram 1998]. The articles [Orlik and Solomon 1980; Lehrer 1995; Stembridge 1989b; Malle 1995] contain other recent work on the combinatorics of these groups.

B3. Hecke Algebras and "Hecke Algebras" of Coxeter Groups

Let G be a finite group and let B be a subgroup of G. The *Hecke algebra* of the pair (G, B) is the subalgebra

$$\mathcal{H}(G, B) = \left\{ \sum_{g \in G} a_g g \;\middle|\; a_g \in \mathbb{C}, \text{ and } a_g = a_h \text{ if } BgB = BhB. \right\}$$

of the group algebra of G. The elements

$$T_w = \frac{1}{|B|} \sum_{g \in BwB} g,$$

as w runs over a set of representatives of the double cosets $B\backslash G/B$, form a basis of $\mathcal{H}(G, B)$.

Let G be a finite Chevalley group over the field $\mathbb{F}_q$ with q elements and fix a Borel subgroup B of G. The pair (G, B) determines a pair (W, S) where W is the Weyl group of G and S is a set of simple reflections in W (with respect to B). The *Iwahori–Hecke algebra* corresponding to G is the Hecke algebra $\mathcal{H}(G, B)$. In this case the basis elements T_w are indexed by the elements w of the Weyl group W corresponding to the pair (G, B) and the multiplication is given by

$$T_s T_w = \begin{cases} T_{sw}, & \text{if } l(sw) > l(w), \\ (q-1)T_w + qT_{sw}, & \text{if } l(sw) < l(w), \end{cases}$$

if s is a simple reflection in W. In this formula $l(w)$ is the *length* of w, i.e., the minimum number of factors needed to write w as a product of simple reflections.

A particular example of the Iwahori–Hecke algebra occurs when $G = \mathrm{GL}(n, \mathbb{F}_q)$ and B is the subgroup of upper triangular matrices. Then the Weyl group W, is the symmetric group S_n, and the simple reflections in the set S are the transpositions $s_i = (i, i+1)$, for $1 \le i \le n-1$. In this case the algebra $\mathcal{H}(G, B)$ is the *Iwahori–Hecke algebra of type A_{n-1}* and (as we will see in Theorems B5.1 and B5.2) can be presented by generators $T_1, \ldots, T_{n-1}$ and relations

$$\begin{aligned} T_i T_j &= T_j T_i, & &\text{for } |i - j| > 1, \\ T_i T_{i+1} T_i &= T_{i+1} T_i T_{i+1}, & &\text{for } 1 \le i \le n-2, \\ T_i^2 &= (q-1)T_i + q, & &\text{for } 2 \le i \le n. \end{aligned}$$

See Appendix B5 for more facts about the Iwahori–Hecke algebras of type A. In particular, these Iwahori–Hecke algebras also appear as tensor power centralizer algebras; see Theorem B5.3. This is rather miraculous: the Iwahori–Hecke algebras of type A are the only Iwahori–Hecke algebras which arise naturally as tensor power centralizers.

In view of the multiplication rules for the Iwahori–Hecke algebras of Weyl groups it is easy to define a "Hecke algebra" for all Coxeter groups (W, S), just by defining it to be the algebra with basis T_w, for $w \in W$, and multiplication

$$T_s T_w = \begin{cases} T_{sw}, & \text{if } l(sw) > l(w), \\ (q-1)T_w + qT_{sw}, & \text{if } l(sw) < l(w), \end{cases}$$

if $s \in S$. These algebras are not true Hecke algebras except when W is a Weyl group.

References. For references on Hecke algebras see [Curtis and Reiner 1981, §11]. For references on Iwahori–Hecke algebras see [Bourbaki 1968, Chap. IV, §2, Ex. 23–25; Curtis and Reiner 1987, §67; Humphreys 1990, Chap. 7]. The article [Curtis 1988] is also very informative.

B4. "Hecke algebras" of the Groups $G(r, p, n)$

Let q and $u_0, u_1, \ldots, u_{r-1}$ be indeterminates. Let $H_{r,1,n}$ be the algebra over the field $\mathbb{C}(u_0, u_1, \ldots, u_{r-1}, q)$ given by generators $T_1, T_2, \ldots, T_n$ and relations

(1) $T_i T_j = T_j T_i$,   for $|i - j| > 1$,

(2) $T_i T_{i+1} T_i = T_{i+1} T_i T_{i+1}$,   for $2 \leq i \leq n - 1$,

(3) $T_1 T_2 T_1 T_2 = T_2 T_1 T_2 T_1$,

(4) $(T_1 - u_0)(T_1 - u_1) \cdots (T_1 - u_{r-1}) = 0$,

(5) $(T_i - q)(T_i + q^{-1}) = 0$,   for $2 \leq i \leq n$.

Upon setting $q = 1$ and $u_{i-1} = \xi^{i-1}$, where ξ is a primitive r-th root of unity, one obtains the group algebra $\mathbb{C}G(r, 1, n)$. In the special case where $r = 1$ and $u_0 = 1$, we have $T_1 = 1$, and $H_{1,1,n}$ is isomorphic to an Iwahori–Hecke algebra of type A_{n-1}. The case $H_{2,1,n}$ when $r = 2$, $u_0 = p$, and $u_1 = p^{-1}$, is isomorphic to an Iwahori–Hecke algebra of type B_n.

Now suppose that p and d are positive integers such that $pd = r$. Let $x_0^{1/p}, \ldots, x_{d-1}^{1/p}$ be indeterminates, let $\varepsilon = e^{2\pi i/p}$ be a primitive p-th root of unity and specialize the variables $u_0, \ldots, u_{r-1}$ according to the relation

$$u_{ld+kp+1} = \varepsilon^l x_k^{1/p},$$

where the subscripts on the u_i are taken mod r. The *"Hecke algebra" $H_{r,p,n}$ corresponding to the group $G(r, p, n)$* is the subalgebra of $H_{r,1,n}$ generated by the elements

$$a_0 = T_1^p, \quad a_1 = T_1^{-1} T_2 T_1, \quad \text{and} \quad a_i = T_i, \quad \text{for } 2 \leq i \leq n.$$

Upon specializing $x_k^{1/p} = \xi^{kp}$, where ξ is a primitive r-th root of unity, $H_{r,p,n}$ becomes the group algebra $\mathbb{C}G(r, p, n)$. Thus $H_{r,p,n}$ is a "q-analogue" of the group algebra of the group $G(r, p, n)$.

References. The algebras $H_{r,1,n}$ were first constructed by Ariki and Koike [1994]. They were classified as cyclotomic Hecke algebras of type B_n by Broué and Malle [1993] and the representation theory of $H_{r,p,n}$ was studied by Ariki [1995]. See [Halverson and Ram 1998] for information about the characters of these algebras.

B5. The Iwahori–Hecke Algebras $H_k(q)$ of Type A

A *k-braid* is viewed as two rows of k vertices, one above the other, and k strands that connect top vertices to bottom vertices in such a way that each vertex is incident to precisely one strand. Strands cross over and under one another in three-space as they pass from one vertex to the next.

$$t_1 = \quad , \quad t_2 = \quad .$$

We multiply k-braids t_1 and t_2 using the concatenation product given by identifying the vertices in the top row of t_2 with the corresponding vertices in the bottom row of t_1 to obtain the product $t_1 t_2$.

$$t_1 t_2 = $$

Given a permutation $w \in S_k$ we will make a k-braid T_w by tracing the edges in order from left to right across the top row. Any time an edge that we are tracing crosses an already traced edge we raise the pen briefly so that the edge being traced goes under the edge that is already there. Applying this process to all of the permutations in S_k produces a set of $k!$ braids.

$$w = \qquad\qquad T_w = $$

Fix $q \in \mathbb{C}$. The *Iwahori–Hecke algebra* $H_k(q)$ of type A_{k-1} is the span of the $k!$ braids produced by tracing permutations in S_k with multiplication determined by the braid multiplication and the identity

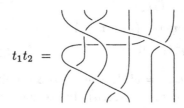

This identity can be applied in any local portion of the braid.

THEOREM B5.1. *The algebra $H_k(q)$ is the associative algebra over $\mathbb{C}$ presented by generators $T_1, \ldots, T_{k-1}$ and relations*

$$
\begin{aligned}
T_i T_j &= T_j T_i, & &\text{for } |i - j| > 1, \\
T_i T_{i+1} T_i &= T_{i+1} T_i T_{i+1}, & &\text{for } 1 \le i \le n - 2, \\
T_i^2 &= (q - 1) T_i + q, & &\text{for } 2 \le i \le n.
\end{aligned}
$$

The Iwahori–Hecke algebra of type A is a q-analogue of the group algebra of the symmetric group. If we allow ourselves to be imprecise (about the limit) we can write

$$\lim_{q \to 1} H_k(q) = \mathbb{C}S_k.$$

Let q be a power of a prime and let $G = \mathrm{GL}(n, \mathbb{F}_q)$ where $\mathbb{F}_q$ is the finite field with q elements. Let B be the subgroup of upper triangular matrices in G and let $\mathbf{1}_B^G$ be the trivial representation of B induced to G, i.e., the G-module given by

$$\mathbf{1}_B^G = \mathbb{C}\text{-span}\{gB \mid g \in G\},$$

where G acts on the cosets by left multiplication. Using the description of $H_n(q)$ as a double coset algebra (Appendix B3), one gets an action of $H_n(q)$ on $\mathbf{1}_B^G$, by right multiplication. This action commutes with the G action.

THEOREM B5.2. (a) *The action of $H_n(q)$ on $\mathbf{1}_B^G$ generates $\mathrm{End}_G(\mathbf{1}_B^G)$.*
(b) *The action of G on $\mathbf{1}_B^G$ generates $\mathrm{End}_{H_n(q)}(\mathbf{1}_B^G)$.*

This theorem gives a "duality" between $\mathrm{GL}(n, \mathbb{F}_q)$ and $H_n(q)$ that is similar to a Schur–Weyl duality, but differs in a crucial way: the representation $\mathbf{1}_B^G$ is not a tensor power representation, and thus this is not yet realizing $H_n(q)$ as a tensor power centralizer.

The following result gives a true analogue of the Schur–Weyl duality for the Iwahori–Hecke algebra of type A: it realizes $H_k(q)$ as a tensor power centralizer. Assume that $q \in \mathbb{C}$ is not 0 and is not a root of unity. Let $U_q\mathfrak{sl}_n$ be the Drinfel'd–Jimbo quantum group of type A_{n-1} and let V be the n-dimensional irreducible representation of $U_q\mathfrak{sl}_n$ with highest weight ω_1. There is an action of $H_k(q^2)$ on $V^{\otimes k}$ which commutes with the $U_q\mathfrak{sl}_n$ action; see [Chari and Pressley 1994].

THEOREM B5.3. (a) *The action of $H_k(q^2)$ on $V^{\otimes k}$ generates $\mathrm{End}_{U_q\mathfrak{sl}_n}(V^{\otimes k})$.*
(b) *The action of $U_q\mathfrak{sl}_n$ on $V^{\otimes k}$ generates $\mathrm{End}_{H_k(q^2)}(V^{\otimes k})$.*

THEOREM B5.4. *The Iwahori–Hecke algebra of type A_{k-1}, $H_k(q)$, is semisimple if and only if $q \neq 0$ and q is not a j-th root of unity for any $2 \leq j \leq n$.*

Partial Results for $H_k(q)$

The following results give answers to the main questions **(Ia)–(Ic)** for the Iwahori–Hecke algebras of type A hold when q is such that $H_k(q)$ is semisimple.

I. What are the irreducible $H_k(q)$-modules?

(a) How do we index/count them?

There is a bijection

$$\text{partitions } \lambda \text{ of } n \overset{1-1}{\longleftrightarrow} \text{irreducible representations } H^\lambda.$$

(b) What are their dimensions?

The **dimension** of the irreducible representation H^λ is given by

$$\dim(H^\lambda) = \# \text{ of standard tableaux of shape } \lambda = \frac{n!}{\prod_{x \in \lambda} h_x},$$

where h_x is the hook length at the box x in λ.

(c) What are their characters?

For each partition $\mu = (\mu_1, \mu_2, \ldots, \mu_l)$ of k, let $\chi^\lambda(\mu)$ be the **character** of the irreducible representation H^λ evaluated at the element T_{γ_μ} where γ_μ is the permutation

Then **the character** $\chi^\lambda(\mu)$ is given by

$$\chi^\lambda(\mu) = \sum_T \mathrm{wt}^\mu(T),$$

where the sum is over all standard tableaux T of shape λ and

$$\mathrm{wt}^\mu(T) = \prod_{i=1}^{n} f(i, T),$$

where

$$f(i, T) = \begin{cases} -1, & \text{if } i \notin B(\mu) \text{ and } i+1 \text{ is SW of } i, \\ 0, & \text{if } i, i+1 \notin B(\mu), \ i+1 \text{ is NE of } i, \text{ and } i+2 \text{ is SW of } i+1, \\ q, & \text{otherwise,} \end{cases}$$

and $B(\mu) = \{\mu_1 + \mu_2 + \cdots + \mu_k | 1 \le k \le l\}$. In the formula for $f(i, T)$, SW means strictly south and weakly west and NE means strictly north and weakly east.

References. The book [Chari and Pressley 1994] contains a treatment of the Schur–Weyl duality type theorem given above. See also the references there. Several basic results on the Iwahori–Hecke algebra are given in [Goodman et al. 1989]. The theorem giving the explicit values of q such that $H_k(q)$ is semisimple is due to Gyoja and Uno [1989]. The character formula given above is due to Roichman [1997]. See [Ram 1998] for an elementary proof.

B6. The Brauer Algebras $B_k(x)$

Fix $x \in \mathbb{C}$. A *Brauer diagram on k dots* is a graph on two rows of k-vertices, one above the other, and k edges such that each vertex is incident to precisely one edge. The product of two k-diagrams d_1 and d_2 is obtained by placing d_1 above d_2 and identifying the vertices in the bottom row of d_1 with the corresponding vertices in the top row of d_2. The resulting graph contains k paths and some number c of closed loops. If d is the k-diagram with the edges that are the paths in this graph but with the closed loops removed, then the product $d_1 d_2$ is given by $d_1 d_2 = x^c d$. For example, if

$$d_1 = \text{[diagram]} \quad \text{and} \quad d_2 = \text{[diagram]},$$

then

$$d_1 d_2 = \text{[diagram]} = x^2 \text{[diagram]}.$$

The *Brauer algebra* $B_k(x)$ is the span of the k-diagrams with multiplication given by the linear extension of the diagram multiplication. The dimension of the Brauer algebra is

$$\dim(B_k(x)) = (2k)!! = (2k-1)(2k-3)\cdots 3 \cdot 1,$$

since the number of k-diagrams is $(2k)!!$.

The diagrams in $B_k(x)$ that have all their edges connecting top vertices to bottom vertices form a symmetric group S_k. The elements

$$s_i = \text{[diagram]} \quad \text{and} \quad e_i = \text{[diagram]},$$

for $1 \leq i \leq k-1$, generate the Brauer algebra $B_k(x)$.

THEOREM B6.1. *The Brauer algebra $B_k(x)$ has a presentation as an algebra by generators $s_1, s_2, \ldots, s_{k-1}, e_1, e_2, \ldots, e_{k-1}$ and relations*

$$s_i^2 = 1, \quad e_i^2 = xe_i, \quad e_i s_i = s_i e_i = e_i, \quad \text{for } 1 \leq i \leq k-1,$$

$$s_i s_j = s_j s_i, \quad s_i e_j = e_j s_i, \quad e_i e_j = e_j e_i, \quad \text{if } |i-j| > 1,$$

$$s_i s_{i+1} s_i = s_{i+1} s_i s_{i+1}, \quad e_i e_{i+1} e_i = e_i, \quad e_{i+1} e_i e_{i+1} = e_{i+1}, \quad \text{for } 1 \leq i \leq k-2,$$

$$s_i e_{i+1} e_i = s_{i+1} e_i, \quad e_{i+1} e_i s_{i+1} = e_{i+1} s_i, \quad \text{for } 1 \leq i \leq k-2.$$

There are two different Brauer algebra analogues of the Schur Weyl duality theorem, Theorem A5.1. In the first one the orthogonal group $O(n, \mathbb{C})$ plays the same role that $GL(n, \mathbb{C})$ played in the S_k-case, and in the second, the symplectic group $Sp(2n, \mathbb{C})$ takes the $GL(n, \mathbb{C})$ role.

Let $O(n, \mathbb{C}) = \{A \in M_n(\mathbb{C}) \mid AA^t = I\}$ be the orthogonal group and let V be the usual n-dimensional representation of the group $O(n, \mathbb{C})$. There is an action of the Brauer algebra $B_k(n)$ on $V^{\otimes k}$ which commutes with the action of $O(n, \mathbb{C})$ on $V^{\otimes k}$.

THEOREM B6.2. (a) *The action of $B_k(n)$ on $V^{\otimes k}$ generates $\mathrm{End}_{O(n)}(V^{\otimes k})$.*
(b) *The action of $O(n, \mathbb{C})$ on $V^{\otimes k}$ generates $\mathrm{End}_{B_k(n)}(V^{\otimes k})$.*

Let $\mathrm{Sp}(2n, \mathbb{C})$ be the symplectic group and let V be the usual $2n$-dimensional representation of the group $\mathrm{Sp}(2n, \mathbb{C})$. There is an action of the Brauer algebra $B_k(-2n)$ on $V^{\otimes k}$ which commutes with the action of $\mathrm{Sp}(2n, \mathbb{C})$ on $V^{\otimes k}$.

THEOREM B6.3. (a) *The action of $B_k(-2n)$ on $V^{\otimes k}$ generates* $\mathrm{End}_{\mathrm{Sp}(2n, \mathbb{C})}(V^{\otimes k})$.
(a) *The action of $\mathrm{Sp}(2n, \mathbb{C})$ on $V^{\otimes k}$ generates* $\mathrm{End}_{B_k(-2n)}(V^{\otimes k})$.

THEOREM B6.4. *The Brauer algebra $B_k(x)$ is semisimple if*

$$x \notin \{-2k + 3, -2k + 2, \ldots, k - 2\}.$$

Partial Results for $B_k(x)$

The following results giving answers to the main questions **(Ia)**–**(Ic)** for the Brauer algebras hold when x is such that $B_k(x)$ is semisimple.

I. What are the irreducible $B_k(x)$-modules?

(a) How do we index/count them?

There is a **bijection**

Partitions of $k - 2h$, for $0 \le h \le \lfloor k/2 \rfloor$
$$\underset{1-1}{\overset{}{\longleftrightarrow}}$$ **Irreducible representations B^λ.**

(b) What are their dimensions?

The **dimension** of the irreducible representation B^λ is

$$\dim(B^\lambda) = \text{\# of up-down tableaux of shape } \lambda \text{ and length } k$$
$$= \binom{k}{2h}(2h-1)!! \frac{(k-2h)!}{\prod_{x \in \lambda} h_x},$$

where h_x is the hook length at the box x in λ. An up-down tableau of shape λ and length k is a sequence $(\varnothing = \lambda^{(0)}, \lambda^{(1)}, \ldots \lambda^{(k)} = \lambda)$ of partitions, such that each partition in the sequence differs from the previous one by either adding or removing a box.

(c) What are their characters?

A Murnaghan–Nakayama type rule for the characters of the Brauer algebras was given in [Ram 1995].

References. The Brauer algebra was defined originally by R. Brauer [1937]; it is also treated in [Weyl 1946]. The Schur–Weyl duality type theorems are also proved in Brauer's original paper [1937]. See also [Ram 1995] for a detailed description of these Brauer algebra actions. The theorem giving values of x for which the Brauer algebra is semisimple is due to Wenzl [1988b].

B7. The Birman–Murakami–Wenzl Algebras $BMW_k(r, q)$

A k-*tangle* is viewed as two rows of k vertices, one above the other, and k strands that connect vertices in such a way that each vertex is incident to precisely one strand. Strands cross over and under one another other in three-space as they pass from one vertex to the next. For example, here are two 7-tangles:

$$t_1 = \quad\raisebox{-1em}{\includegraphics{}} \quad , \quad t_2 = \quad\raisebox{-1em}{\includegraphics{}} \quad .$$

We multiply k-tangles t_1 and t_2 using the concatenation product given by identifying the vertices in the top row of t_2 with the corresponding vertices in the bottom row of t_1 to obtain the product tangle $t_1 t_2$. Then we allow the following "moves":

Reidemeister move II:

Reidemeister move III:

Given a Brauer diagram d we will make a tangle T_d tracing the edges in order from left to right across the top row and then from left to right across the bottom row. Any time an edge that we are tracing crosses and edge that has been already traced we raise the pen briefly so that the edge being traced goes under the edge that is already there. Applying this process to all of the Brauer diagrams on k dots produces a set of $(2k)!!$ tangles.

$$d = \quad\raisebox{-1em}{\includegraphics{}} \qquad T_d = \quad\raisebox{-1em}{\includegraphics{}}$$

Fix numbers $r, q \in \mathbb{C}$. The *Birman–Murakami–Wenzl algebra* $BMW_k(r, q)$ is the span of the $(2k)!!$ tangles produced by tracing the Brauer diagrams with multiplication determined by the tangle multiplication, the Reidemeister moves and the following tangle identities.

$$\raisebox{-1em}{\includegraphics{}} - \raisebox{-1em}{\includegraphics{}} = (q - q^{-1})\left(\raisebox{-1em}{\includegraphics{}} - \raisebox{-1em}{\includegraphics{}} \right),$$

$$\raisebox{-1em}{\includegraphics{}} = r^{-1} \raisebox{-1em}{\includegraphics{}}, \qquad \raisebox{-1em}{\includegraphics{}} = r \raisebox{-1em}{\includegraphics{}},$$

$$\raisebox{-0.5em}{\includegraphics{}} = x, \quad \text{where} \quad x = \frac{r - r^{-1}}{q - q^{-1}} + 1.$$

The Reidemeister moves and the tangle identities can be applied in any appropriate local portion of the tangle.

THEOREM B7.1. *Fix numbers* $r, q \in \mathbb{C}$. *The Birman–Murakami–Wenzl algebra* $\mathrm{BMW}_k(r, q)$ *is the algebra generated over* $\mathbb{C}$ *by* $1, g_1, g_2, \ldots, g_{k-1}$, *which are assumed to be invertible, subject to the relations*

$$g_i g_{i+1} g_i = g_{i+1} g_i g_{i+1},$$

$$g_i g_j = g_j g_i \quad \text{if } |i - j| \geq 2,$$

$$(g_i - r^{-1})(g_i + q^{-1})(g_i - q) = 0,$$

$$E_i g_{i-1}^{\pm 1} E_i = r^{\pm 1} E_i \quad \text{and} \quad E_i g_{i+1}^{\pm 1} E_i = r^{\pm 1} E_i,$$

where E_i *is defined by the equation*

$$(q - q^{-1})(1 - E_i) = g_i - g_i^{-1}.$$

The BMW-algebra is a q-analogue of the Brauer algebra in the same sense that the Iwahori–Hecke algebra of type A is a q-analogue of the group algebra of the symmetric group. If we allow ourselves to be imprecise (about the limit) we can write

$$\lim_{q \to 1} \mathrm{BMW}_k(q^{n+1}, q) = B_k(n).$$

It would be interesting to sharpen the following theorem to make it an if and only if statement.

THEOREM B7.2 [Wenzl 1990]. *The Birman–Murakami–Wenzl algebra is semisimple if* q *is not a root of unity and* $r \neq q^{n+1}$ *for any* $n \in \mathbb{Z}$.

Partial Results for $\mathrm{BMW}_k(r, q)$

The following results hold when r and q are such that $\mathrm{BMW}_k(r, q)$ is semisimple.

I. What are the irreducible $\mathrm{BMW}_k(r, q)$-modules?

(a) How do we index/count them?

There is a **bijection**

partitions of $k - 2h$, **for** $0 \leq h \leq \lfloor k/2 \rfloor$ $\underset{1-1}{\longleftrightarrow}$ **Irreducible representations** W^λ.

(b) What are their dimensions?

The **dimension** of the irreducible representation W^λ is

$$\dim(W^\lambda) = \# \text{ of up-down tableaux of shape } \lambda \text{ and length } k$$

$$= \binom{k}{2h}(2h - 1)!! \frac{(k - 2h)!}{\prod_{x \in \lambda} h_x},$$

where h_x is the hook length at the box x in λ, and up-down tableaux are as in the case **(Ib)** of the Brauer algebra (page 74).

(c) What are their characters?

A Murnaghan–Nakayama rule for the irreducible characters of the BMW-algebras was given in [Halverson and Ram 1995].

References. The Birman–Murakami–Wenzl algebra was defined independently by Birman and Wenzl [1989] and by Murakami [1987]. See [Chari and Pressley 1994] for references to the analogue of Schur–Weyl duality for the BMW-algebras. The articles [Halverson and Ram 1995; Leduc and Ram 1997; Murakami 1990; Reshetikhin 1987; Wenzl 1990] contain further important information about the BMW-algebras.

Kaufmann [1990] first formalized the tangle description of the BMW-algebra.

B8. The Temperley–Lieb Algebras $\mathrm{TL}_k(x)$

A TL_k-*diagram* is a Brauer diagram on k dots that can be drawn with no crossings of edges.

The *Temperley–Lieb algebra* $\mathrm{TL}_k(x)$ is the subalgebra of the Brauer algebra $B_k(x)$ that is the span of the TL_k-diagrams.

THEOREM B8.1. *The Temperley–Lieb algebra* $\mathrm{TL}_k(x)$ *is the algebra over* $\mathbb{C}$ *given by generators* $E_1, E_2, \ldots, E_{k-1}$ *and relations*

$$E_i E_j = E_j E_i \quad \text{if } |i - j| > 1,$$

$$E_i E_{i\pm1} E_i = E_i,$$

$$E_i^2 = x E_i.$$

THEOREM B8.2. *Let* $q \in \mathbb{C}^*$ *be such that* $q + q^{-1} + 2 = 1/x^2$ *and let* $H_k(q)$ *be the Iwahori–Hecke algebra of type* A_{k-1}. *Then the map*

$$
\begin{aligned}
H_k(q) &\longrightarrow \mathrm{TL}_k(x) \\
T_i &\longmapsto \frac{q+1}{x} E_i - 1
\end{aligned}
$$

is a surjective homomorphism and the kernel of this homomorphism is the ideal generated by the elements

$$T_i T_{i+1} T_i + T_i T_{i+1} + T_{i+1} T_i + T_i + T_{i+1} + 1, \quad \text{for } 1 \le i \le n - 2.$$

The Schur–Weyl duality theorem for S_n has the following analogue for the Temperley–Lieb algebras. Let $U_q \mathfrak{sl}_2$ be the Drinfeld–Jimbo quantum group corresponding to the Lie algebra $\mathfrak{sl}_2$ and let V be the 2-dimensional representation of

$U_q\mathfrak{sl}_2$. There is an action, see [CP], of the Temperley–Lieb algebra $\mathrm{TL}_k(q+q^{-1})$ on $V^{\otimes k}$ which commutes with the action of $U_q\mathfrak{sl}_2$ on $V^{\otimes k}$.

THEOREM B8.3. (a) *The action of* $\mathrm{TL}_k(q+q^{-1})$ *on* $V^{\otimes k}$ *generates* $\mathrm{End}_{U_q\mathfrak{sl}_2}(V^{\otimes k})$.
(a) *The action of* $U_q\mathfrak{sl}_2$ *on* $V^{\otimes k}$ *generates* $\mathrm{End}_{\mathrm{TL}_k(q+q^{-1})}(V^{\otimes k})$.

THEOREM B8.4. *The Temperley–Lieb algebra is semisimple if and only if* $1/x^2 \neq 4\cos^2(\pi/l)$, *for any* $2 \leq l \leq k$.

Partial Results for $\mathrm{TL}_k(x)$

The following results giving answers to the main questions **(Ia)–(Ic)** for the Temperley–Lieb algebras hold when x is such that $\mathrm{TL}_k(x)$ is semisimple.

I. What are the irreducible $\mathrm{TL}_k(x)$-modules?

(a) How do we index/count them?

There is a **bijection**

$$\text{partitions of } k \text{ with} \leq 2 \text{ rows} \overset{1-1}{\longleftrightarrow} \text{irreducible representations } T^\lambda.$$

(b) What are their dimensions?

The **dimension** of the irreducible representation $T^{(k-l,l)}$ is

$$\dim(T^{(k-l,l)}) = \# \text{ of standard tableaux of shape } (k-l,l)$$

$$= \binom{k}{l} - \binom{k}{l-1}.$$

(c) What are their characters?

The **character** of the irreducible representation $T^{(k-l,l)}$ evaluated at the element

$$d_{2h} = \underbrace{\left|\,\left|\,\left|\,\left|\,\left|\,\left|\, \cdots \,\right|\right.\right.\right.\right.\right.\right.}_{k-2h} \quad \underbrace{\overset{\smile\ \smile\ \smile}{\underset{\frown\ \frown\ \frown}{}} \cdots \overset{\smile}{\underset{\frown}{}}}_{2h}$$

is

$$\chi^{(k-l,l)}(d_{2h}) = \begin{cases} \dbinom{k-2h}{l-h} - \dbinom{k-2h}{l-h-1}, & \text{if } l \geq h, \\ 0, & \text{if } l < h. \end{cases}$$

There is an algorithm for writing the character $\chi^{(k-l,l)}(a)$ of a general element $a \in \mathrm{TL}_k(x)$ as a linear combination of the characters $\chi^{(k-l,l)}(d_{2h})$.

References. The book [Goodman et al. 1989] contains a comprehensive treatment of the basic results on the Temperley–Lieb algebra. The Schur–Weyl duality theorem is treated in [Chari and Pressley 1994]; see also the references there. The character formula given above is derived in [Halverson and Ram 1995].

B9. Complex Semisimple Lie Groups

We shall not define Lie groups and Lie algebras. We recall simply that a complex Lie group is a differential $\mathbb{C}$-manifold, that a real Lie group is a differential $\mathbb{R}$-manifold and that every Lie group has an associated Lie algebra; see [Carter et al. 1995].

If G is a complex Lie group, the word *representation* usually refers to a *holomorphic representation*, i.e., the homomorphism

$$\rho : G \to \mathrm{GL}(V)$$

determined by the module V should be a morphism of (complex) analytic manifolds. Strictly speaking there are representations which are not holomorphic but there is a good theory only for holomorphic representations, so one usually abuses language and assumes that representation means holomorphic representation. The terms holomorphic representation and *complex analytic representation* are used interchangeably. Similarly, if G is a real Lie group then representation usually means *real analytic representation*. See [Varadarajan 1984, p. 102] for further details. Every holomorphic representation of $\mathrm{GL}(n, \mathbb{C})$ is also a rational representation; see [Fulton and Harris 1991].

A *complex semisimple Lie group* is a connected complex Lie group G such that its Lie algebra $\mathfrak{g}$ is a complex semisimple Lie algebra.

References

[Allen 1997] E. E. Allen, "New bases for the decompositions of the regular representations of wreath products of S_n", preprint, Wake Forest University, 1997.

[Ariki 1995] S. Ariki, "Representation theory of a Hecke algebra of $G(r, p, n)$", *J. Algebra* **177**:1 (1995), 164–185.

[Ariki and Koike 1994] S. Ariki and K. Koike, "A Hecke algebra of $(\mathbb{Z}/r\mathbb{Z}) \wr \mathfrak{S}_n$ and construction of its irreducible representations", *Adv. Math.* **106**:2 (1994), 216–243.

[Barcelo 1990] H. Barcelo, "On the action of the symmetric group on the free Lie algebra and the partition lattice", *J. Combin. Theory Ser. A* **55**:1 (1990), 93–129.

[Barcelo 1993] H. Barcelo, "Young straightening in a quotient S_n-module", *J. Algebraic Combin.* **2**:1 (1993), 5–23.

[Benkart et al. 1990] G. M. Benkart, D. J. Britten, and F. W. Lemire, *Stability in modules for classical Lie algebras — a constructive approach*, Mem. Amer. Math. Soc. **430**, Amer. Math. Soc., 1990.

[Benkart et al. 1994] G. Benkart, M. Chakrabarti, T. Halverson, R. Leduc, C. Lee, and J. Stroomer, "Tensor product representations of general linear groups and their connections with Brauer algebras", *J. Algebra* **166**:3 (1994), 529–567.

[Benkart et al. 1998] G. Benkart, C. L. Shader, and A. Ram, "Tensor product representations for orthosymplectic Lie superalgebras", *J. Pure and Applied Algebra* **130** (1998), 1–48.

[Berele 1986a]   A. Berele, "Construction of Sp-modules by tableaux", *Linear and Multilinear Algebra* **19**:4 (1986), 299–307.

[Berele 1986b]   A. Berele, "A Schensted-type correspondence for the symplectic group", *J. Combin. Theory Ser. A* **43**:2 (1986), 320–328.

[Berele and Regev 1987]   A. Berele and A. Regev, "Hook Young diagrams with applications to combinatorics and to representations of Lie superalgebras", *Adv. in Math.* **64**:2 (1987), 118–175.

[Bernšteĭn et al. 1973]   I. N. Bernšteĭn, I. M. Gel'fand, and S. I. Gel'fand, "Schubert cells and the cohomology of the spaces G/P", *Uspehi Mat. Nauk* **28**:3 (1973), 3–26. In Russian; translation in *Russian Math. Surveys* **28**:3 (1973), 1–26.

[Birman and Wenzl 1989]   J. S. Birman and H. Wenzl, "Braids, link polynomials and a new algebra", *Trans. Amer. Math. Soc.* **313**:1 (1989), 249–273.

[Borel 1953]   A. Borel, "Sur la cohomologie des espaces fibrés principaux et des espaces homogènes de groupes de Lie compacts", *Ann. of Math.* (2) **57** (1953), 115–207.

[Bourbaki 1958]   N. Bourbaki, *Algèbre, Chap. VIII: Modules et anneaux semi-simples*, Actualités Sci. Ind. **1261**, Hermann, Paris, 1958.

[Bourbaki 1968]   N. Bourbaki, *Groupes et algèbres de Lie, Chap. IV: Groupes de Coxeter et systèmes de Tits; Chap. V: Groupes engendrés par des réflexions; Chap. VI: systèmes de racines*, Actualités Sci. Ind. **1337**, Hermann, Paris, 1968.

[Brauer 1937]   R. Brauer, "On algebras which are connected with the semisimple continous groups", *Ann. Math.* **38** (1937), 854–872.

[Brenti 1994]   F. Brenti, "A combinatorial formula for Kazhdan-Lusztig polynomials", *Invent. Math.* **118**:2 (1994), 371–394.

[Bröcker and tom Dieck 1985]   T. Bröcker and T. tom Dieck, *Representations of compact Lie groups*, Graduate Texts in Mathematics **98**, Springer, New York, 1985.

[Broué and Malle 1993]   M. Broué and G. Malle, "Zyklotomische Heckealgebren", pp. 119–189 in *Représentations unipotentes génériques et blocs des groupes réductifs finis*, Astérisque **212**, Soc. math. France, Paris, 1993.

[Carter 1985]   R. W. Carter, *Finite groups of Lie type: Conjugacy classes and complex characters*, John Wiley, New York, 1985.

[Carter et al. 1995]   R. Carter, G. Segal, and I. Macdonald, *Lectures on Lie groups and Lie algebras*, London Math. Soc. Student Texts **32**, Cambridge University Press, Cambridge, 1995. With a foreword by Martin Taylor.

[Chari and Pressley 1994]   V. Chari and A. Pressley, *A guide to quantum groups*, Cambridge University Press, Cambridge, 1994.

[Chriss and Ginzburg 1997]   N. Chriss and V. Ginzburg, *Representation theory and complex geometry*, Birkhäuser, Boston, 1997.

[Clifford 1937]   A. H. Clifford, "Representations induced in an invariant subgroup", *Ann. of Math.* **38** (1937), 533–550.

[Curtis 1988]   C. W. Curtis, "Representations of Hecke algebras", pp. 13–60 in *Orbites unipotentes et représentations, I*, Astérisque **168**, Soc. math. France, Paris, 1988.

[Curtis and Reiner 1962]   C. W. Curtis and I. Reiner, *Representation theory of finite groups and associative algebras*, Pure and Applied Mathematics **11**, Wiley/Interscience, New York, 1962. Reprinted by Wiley, 1998.

[Curtis and Reiner 1981] C. W. Curtis and I. Reiner, *Methods of representation theory*, vol. I, Pure and Applied Mathematics, John Wiley, New York, 1981.

[Curtis and Reiner 1987] C. W. Curtis and I. Reiner, *Methods of representation theory*, vol. II, Pure and Applied Mathematics, John Wiley, New York, 1987.

[De Concini and Procesi 1981] C. De Concini and C. Procesi, "Symmetric functions, conjugacy classes and the flag variety", *Invent. Math.* **64**:2 (1981), 203–219.

[Deligne and Lusztig 1976] P. Deligne and G. Lusztig, "Representations of reductive groups over finite fields", *Ann. of Math.* (2) **103**:1 (1976), 103–161.

[Digne and Michel 1991] F. Digne and J. Michel, *Representations of finite groups of Lie type*, London Math. Soc. Student Texts **21**, Cambridge University Press, Cambridge, 1991.

[Dipper and James 1986] R. Dipper and G. James, "Representations of Hecke algebras of general linear groups", *Proc. London Math. Soc.* (3) **52**:1 (1986), 20–52.

[Dipper and James 1992] R. Dipper and G. James, "Representations of Hecke algebras of type B_n", *J. Algebra* **146**:2 (1992), 454–481.

[Dipper et al. 1995] R. Dipper, G. James, and E. Murphy, "Hecke algebras of type B_n at roots of unity", *Proc. London Math. Soc.* (3) **70**:3 (1995), 505–528.

[Duchamp et al. 1995] G. Duchamp, D. Krob, A. Lascoux, B. Leclerc, T. Scharf, and J.-Y. Thibon, "Euler–Poincaré characteristic and polynomial representations of Iwahori–Hecke algebras", *Publ. Res. Inst. Math. Sci.* **31**:2 (1995), 179–201.

[Feĭgin and Fuchs 1990] B. L. Feĭgin and D. B. Fuchs, "Representations of the Virasoro algebra", pp. 465–554 in *Representation of Lie groups and related topics*, edited by A. M. Vershik and D. P. Zhelobenko, Adv. Stud. Contemp. Math. **7**, Gordon and Breach, New York, 1990.

[Fomin and Greene 1998] S. Fomin and C. Greene, "Noncommutative Schur functions and their applications", *Discrete Math.* **193**:1-3 (1998), 179–200.

[Frame et al. 1954] J. S. Frame, G. d. B. Robinson, and R. M. Thrall, "The hook graphs of the symmetric groups", *Canadian J. Math.* **6** (1954), 316–324.

[Frobenius 1900] F. G. Frobenius, "Über die Charaktere der symmetrischen Gruppe", *Sitzungsber. König. Preuss. Akad. Wiss. Berlin* (1900), 516–534. Reprinted as pp. 148–166 in his *Gesammelte Abhandlungen*, vol. 3, edited by J.-P. Serre, Berlin, Springer, 1968.

[Fulton and Harris 1991] W. Fulton and J. Harris, *Representation theory*, Graduate Texts in Mathematics **129**, Springer, New York, 1991.

[Garsia 1990] A. M. Garsia, "Combinatorics of the free Lie algebra and the symmetric group", pp. 309–382 in *Analysis, et cetera*, edited by P. H. Rabinowitz and E. Zehnder, Academic Press, Boston, MA, 1990.

[Garsia and McLarnan 1988] A. M. Garsia and T. J. McLarnan, "Relations between Young's natural and the Kazhdan–Lusztig representations of S_n", *Adv. in Math.* **69**:1 (1988), 32–92.

[Garsia and Procesi 1992] A. M. Garsia and C. Procesi, "On certain graded S_n-modules and the q-Kostka polynomials", *Adv. Math.* **94**:1 (1992), 82–138.

[Garsia and Remmel 1986] A. M. Garsia and J. B. Remmel, "q-counting rook configurations and a formula of Frobenius", *J. Combin. Theory Ser. A* **41**:2 (1986), 246–275.

[Garsia and Wachs 1989] A. M. Garsia and M. L. Wachs, "Combinatorial aspects of skew representations of the symmetric group", *J. Combin. Theory Ser. A* **50**:1 (1989), 47–81.

[Gel'fand and Tsetlin 1950a] I. M. Gel'fand and M. L. Tsetlin, "Finite-dimensional representations of the group of unimodular matrices", *Doklady Akad. Nauk SSSR (N.S.)* **71** (1950), 825–828. In Russian.

[Gel'fand and Tsetlin 1950b] I. M. Gel'fand and M. L. Tsetlin, "Finite-dimensional representations of groups of orthogonal matrices", *Doklady Akad. Nauk SSSR (N.S.)* **71** (1950), 1017–1020. In Russian.

[Ginsburg 1987] V. Ginsburg, " "Lagrangian" construction for representations of Hecke algebras", *Adv. in Math.* **63**:1 (1987), 100–112.

[Goodman et al. 1989] F. M. Goodman, P. de la Harpe, and V. F. R. Jones, *Coxeter graphs and towers of algebras*, Mathematical Sciences Research Institute Publications **14**, Springer, New York, 1989.

[Graham and Lehrer 1996] J. J. Graham and G. I. Lehrer, "Cellular algebras", *Invent. Math.* **123**:1 (1996), 1–34.

[Green 1955] J. A. Green, "The characters of the finite general linear groups", *Trans. Amer. Math. Soc.* **80** (1955), 402–447.

[Gyoja 1986] A. Gyoja, "A q-analogue of Young symmetrizer", *Osaka J. Math.* **23**:4 (1986), 841–852.

[Gyoja and Uno 1989] A. Gyoja and K. Uno, "On the semisimplicity of Hecke algebras", *J. Math. Soc. Japan* **41**:1 (1989), 75–79.

[Halverson 1995] T. Halverson, "A q-rational Murnaghan-Nakayama rule", *J. Combin. Theory Ser. A* **71**:1 (1995), 1–18.

[Halverson 1996] T. Halverson, "Characters of the centralizer algebras of mixed tensor representations of $\mathrm{GL}(r, \mathbb{C})$ and the quantum group $\mathcal{U}_q(\mathfrak{gl}(r, \mathbb{C}))$", *Pacific J. Math.* **174**:2 (1996), 359–410.

[Halverson and Ram 1995] T. Halverson and A. Ram, "Characters of algebras containing a Jones basic construction: the Temperley–Lieb, Okasa, Brauer, and Birman–Wenzl algebras", *Adv. Math.* **116**:2 (1995), 263–321.

[Halverson and Ram 1998] T. Halverson and A. Ram, "Murnaghan–Nakayama rules for characters of Iwahori–Hecke algebras of the complex reflection groups $G(r, p, n)$", *Canad. J. Math.* **50**:1 (1998), 167–192.

[Hanlon 1981] P. Hanlon, "The fixed point partition lattices", *Pacific J. Math.* **96**:2 (1981), 319–341.

[Hanlon 1984] P. Hanlon, "An introduction to the complex representations of the symmetric group and general linear Lie algebra", pp. 1–18 in *Combinatorics and algebra* (Boulder, CO, 1983), edited by C. Greene, Contemp. Math. **34**, Amer. Math. Soc., Providence, 1984.

[Hanlon and Wales 1989] P. Hanlon and D. Wales, "Eigenvalues connected with Brauer's centralizer algebras", *J. Algebra* **121**:2 (1989), 446–476.

[Hoefsmit 1974] P. N. Hoefsmit, *Representations of Hecke algebras of finite groups with BN-pairs of classical type*, Ph.D. thesis, University of British Columbia, 1974.

[Hotta and Springer 1977] R. Hotta and T. A. Springer, "A specialization theorem for certain Weyl group representations and an application to the Green polynomials of unitary groups", *Invent. Math.* **41**:2 (1977), 113–127.

[Howe 1994] R. Howe, "Hecke algebras and p-adic GL_n", pp. 65–100 in *Representation theory and analysis on homogeneous spaces* (New Brunswick, NJ, 1993), edited by S. Gindikin et al., Contemp. Math. **177**, Amer. Math. Soc., Providence, RI, 1994.

[Howe 1995] R. Howe, "Perspectives on invariant theory: Schur duality, multiplicity-free actions and beyond", pp. 1–182 in *The Schur lectures* (Tel Aviv, 1992), edited by I. Piatetski-Shapiro and S. Gelbart, Israel Math. Conf. Proc. **8**, Bar-Ilan Univ., Ramat Gan, 1995.

[Humphreys 1972] J. E. Humphreys, *Introduction to Lie algebras and representation theory*, Graduate Texts in Mathematics **9**, Springer, New York, 1972.

[Humphreys 1990] J. E. Humphreys, *Reflection groups and Coxeter groups*, Cambridge Studies in Advanced Mathematics **29**, Cambridge University Press, Cambridge, 1990.

[Iwahori 1964] N. Iwahori, "On the structure of a Hecke ring of a Chevalley group over a finite field", *J. Fac. Sci. Univ. Tokyo Sect. I* **10** (1964), 215–236.

[Jantzen 1996] J. C. Jantzen, *Lectures on quantum groups*, vol. 6, Graduate Studies in Math., Amer. Math. Soc., Providence, RI, 1996.

[Jimbo 1986] M. Jimbo, "A q-analogue of $U(\mathfrak{gl}(N+1))$, Hecke algebra, and the Yang-Baxter equation", *Lett. Math. Phys.* **11**:3 (1986), 247–252.

[Jones 1983] V. F. R. Jones, "Index for subfactors", *Invent. Math.* **72**:1 (1983), 1–25.

[Jones 1994] V. F. R. Jones, "The Potts model and the symmetric group", pp. 259–267 in *Subfactors: proceedings of the Taniguchi Symposium on Operator Algebras* (Kyuzeso, 1993), edited by H. Araki et al., World Sci. Publishing, River Edge, NJ, 1994.

[Joyal 1986] A. Joyal, "Foncteurs analytiques et espèces de structures", pp. 126–159 in *Combinatoire énumérative* (Montreal, 1985), edited by G. L. et P. Leroux, Lecture Notes in Math. **1234**, Springer, Berlin, 1986.

[Kac 1990] V. G. Kac, *Infinite-dimensional Lie algebras*, Third ed., Cambridge University Press, Cambridge, 1990.

[Kashiwara 1990] M. Kashiwara, "Crystalizing the q-analogue of universal enveloping algebras", *Comm. Math. Phys.* **133**:2 (1990), 249–260.

[Kato 1983] S.-i. Kato, "A realization of irreducible representations of affine Weyl groups", *Nederl. Akad. Wetensch. Indag. Math.* **45**:2 (1983), 193–201.

[Kauffman 1987] L. H. Kauffman, "State models and the Jones polynomial", *Topology* **26**:3 (1987), 395–407.

[Kauffman 1990] L. H. Kauffman, "An invariant of regular isotopy", *Trans. Amer. Math. Soc.* **318**:2 (1990), 417–471.

[Kazhdan and Lusztig 1979] D. Kazhdan and G. Lusztig, "Representations of Coxeter groups and Hecke algebras", *Invent. Math.* **53**:2 (1979), 165–184.

[Kazhdan and Lusztig 1987] D. Kazhdan and G. Lusztig, "Proof of the Deligne-Langlands conjecture for Hecke algebras", *Invent. Math.* **87**:1 (1987), 153–215.

[Kerov 1992] S. V. Kerov, "Characters of Hecke and Birman–Wenzl algebras", pp. 335–340 in *Quantum groups* (Leningrad, 1990), edited by P. P. Kulish, Lecture Notes in Math. **1510**, Springer, Berlin, 1992.

[King and Welsh 1993] R. C. King and T. A. Welsh, "Construction of orthogonal group modules using tableaux", *Linear and Multilinear Algebra* **33**:3-4 (1993), 251–283.

[King and Wybourne 1992] R. C. King and B. G. Wybourne, "Representations and traces of the Hecke algebras $H_n(q)$ of type A_{n-1}", *J. Math. Phys.* **33**:1 (1992), 4–14.

[Klyachko 1974] A. A. Klyachko, "Елементы Ли в тенсорои алгебре (Lie elements in the tensor algebra)", *Sib. Mat. Zh.* **15** (1974), 1296–1304. Translated in *Siberian Math. J.*, **15** (1974), 1296–1304.

[Knapp and Vogan 1995] A. W. Knapp and J. Vogan, David A., *Cohomological induction and unitary representations*, Princeton Mathematical Series **45**, Princeton University Press, Princeton, NJ, 1995.

[Koike 1989] K. Koike, "On the decomposition of tensor products of the representations of the classical groups: by means of the universal characters", *Adv. Math.* **74**:1 (1989), 57–86.

[Kostant and Kumar 1986] B. Kostant and S. Kumar, "The nil Hecke ring and cohomology of G/P for a Kac–Moody group G", *Adv. in Math.* **62**:3 (1986), 187–237.

[Kostka 1882] C. Kostka, "Über den Zusammenhang zwischen einigen Formen von symmetrischen Funktionen", *J. reine ungew. math.* **93** (1882), 89–123.

[Kosuda and Murakami 1992] M. Kosuda and J. Murakami, "The centralizer algebras of mixed tensor representations of $U_q(\mathfrak{gl}_n)$ and the HOMFLY polynomial of links", *Proc. Japan Acad. Ser. A Math. Sci.* **68**:6 (1992), 148–151.

[Kosuda and Murakami 1993] M. Kosuda and J. Murakami, "Centralizer algebras of the mixed tensor representations of quantum group $U_q(\mathfrak{gl}(n, \mathbb{C}))$", *Osaka J. Math.* **30**:3 (1993), 475–507.

[Kraft 1981] H. Kraft, "Conjugacy classes and Weyl group representations", pp. 191–205 in *Young tableaux and Schur functions in algebra and geometry* (Toruń, Poland, 1980), Astérisque **87–88**, Soc. Math. France, Paris, 1981.

[Kuperberg 1994a] G. Kuperberg, "The quantum G_2 link invariant", *Internat. J. Math.* **5**:1 (1994), 61–85.

[Kuperberg 1994b] G. Kuperberg, "Self-complementary plane partitions by Proctor's minuscule method", *European J. Combin.* **15**:6 (1994), 545–553.

[Kuperberg 1994c] G. Kuperberg, "Symmetries of plane partitions and the permanent-determinant method", *J. Combin. Theory Ser. A* **68**:1 (1994), 115–151.

[Kuperberg 1996a] G. Kuperberg, "Four symmetry classes of plane partitions under one roof", *J. Combin. Theory Ser. A* **75**:2 (1996), 295–315.

[Kuperberg 1996b] G. Kuperberg, "Spiders for rank 2 Lie algebras", *Comm. Math. Phys.* **180**:1 (1996), 109–151.

[Lakshmibai and Seshadri 1991] V. Lakshmibai and C. S. Seshadri, "Standard monomial theory", pp. 279–322 in *Proceedings of the Hyderabad Conference on Algebraic Groups* (Hyderabad, 1989), edited by S. Ramanan, Manoj Prakashan, Madras, 1991.

[Lascoux 1991] A. Lascoux, "Cyclic permutations on words, tableaux and harmonic polynomials", pp. 323–347 in *Proceedings of the Hyderabad Conference on Algebraic Groups* (Hyderabad, 1989), edited by S. Ramanan, Manoj Prakashan, Madras, 1991.

[Leduc 1994] R. Leduc, *A two-parameter version of the centralizer algebra of the mixed tensor representations of the general linear group and quantum general linear group*, Ph.D. thesis, University of Wisconsin–Madison, 1994.

[Leduc and Ram 1997] R. Leduc and A. Ram, "A ribbon Hopf algebra approach to the irreducible representations of centralizer algebras: the Brauer, Birman–Wenzl, and type A Iwahori–Hecke algebras", *Adv. Math.* **125**:1 (1997), 1–94.

[Lehrer 1995] G. I. Lehrer, "Poincaré polynomials for unitary reflection groups", *Invent. Math.* **120**:3 (1995), 411–425.

[Littelmann 1994] P. Littelmann, "A Littlewood-Richardson rule for symmetrizable Kac–Moody algebras", *Invent. Math.* **116**:1-3 (1994), 329–346.

[Littelmann 1995] P. Littelmann, "Paths and root operators in representation theory", *Ann. of Math.* (2) **142**:3 (1995), 499–525.

[Littlewood 1940] D. E. Littlewood, *The theory of group characters and matrix representations of groups*, Oxford University Press, New York, 1940.

[Lusztig 1974] G. Lusztig, *The discrete series of* GL_n *over a finite field*, Annals of Mathematics Studies **81**, Princeton University Press, Princeton, NJ, 1974.

[Lusztig 1977] G. Lusztig, "Classification des représentations irréductibles des groupes classiques finis", *C. R. Acad. Sci. Paris Sér. A-B* **284**:9 (1977), A473–A475.

[Lusztig 1978] G. Lusztig, *Representations of finite Chevalley groups*, CBMS Regional Conference Series in Mathematics **39**, Amer. Math. Soc., Providence, 1978.

[Lusztig 1984] G. Lusztig, *Characters of reductive groups over a finite field*, Annals of Mathematics Studies **107**, Princeton University Press, Princeton, NJ, 1984.

[Lusztig 1987a] G. Lusztig, "Cells in affine Weyl groups, II", *J. Algebra* **109**:2 (1987), 536–548.

[Lusztig 1987b] G. Lusztig, "Cells in affine Weyl groups, III", *J. Fac. Sci. Univ. Tokyo Sect. IA Math.* **34**:2 (1987), 223–243.

[Lusztig 1988] G. Lusztig, "Cuspidal local systems and graded Hecke algebra", *Inst. Hautes Études Sci. Publ. Math.* **67** (1988), 145–202.

[Lusztig 1989a] G. Lusztig, "Affine Hecke algebras and their graded version", *J. Amer. Math. Soc.* **2**:3 (1989), 599–635.

[Lusztig 1989b] G. Lusztig, "Representations of affine Hecke algebras", pp. 73–84 in *Orbites unipotentes et représentations, II*, Astérisque **171-172**, Soc. math. France, Paris, 1989.

[Lusztig 1995a] G. Lusztig, "Classification of unipotent representations of simple p-adic groups", *Internat. Math. Res. Notices* **11** (1995), 517–589.

[Lusztig 1995b] G. Lusztig, "Cuspidal local systems and graded Hecke algebras, II", pp. 217–275 in *Representations of groups* (Banff, AB, 1994), edited by B. N. Allison and G. H. Cliff, CMS Conf. Proc. **16**, Amer. Math. Soc., Providence, RI, 1995.

[Lusztig and Spaltenstein 1985] G. Lusztig and N. Spaltenstein, "On the generalized Springer correspondence for classical groups", pp. 289–316 in *Algebraic groups and related topics* (Kyoto/Nagoya, 1983), edited by R. Hotta, Adv. Stud. Pure Math. **6**, North-Holland, Amsterdam, 1985.

[Macdonald 1995] I. G. Macdonald, *Symmetric functions and Hall polynomials*, 2nd ed., Oxford Univ. Press, 1995.

[Malle 1995] G. Malle, "Unipotente Grade imprimitiver komplexer Spiegelungsgruppen", *J. Algebra* **177**:3 (1995), 768–826.

[Martin 1996] P. Martin, "The structure of the partition algebras", *J. Algebra* **183**:2 (1996), 319–358.

[Mathas 1997] A. Mathas, "Canonical bases and the decomposition matrices of Ariki–Koike algebras", preprint, 1997.

[Murakami 1987] J. Murakami, "The Kauffman polynomial of links and representation theory", *Osaka J. Math.* **24**:4 (1987), 745–758.

[Murakami 1990] J. Murakami, "The representations of the q-analogue of Brauer's centralizer algebras and the Kauffman polynomial of links", *Publ. Res. Inst. Math. Sci.* **26**:6 (1990), 935–945.

[Murnaghan 1937] F. D. Murnaghan, "The characters of the symmetric group", *Amer. J. Math.* **59** (1937), 739–753.

[Murphy 1992] G. E. Murphy, "On the representation theory of the symmetric groups and associated Hecke algebras", *J. Algebra* **152**:2 (1992), 492–513.

[Murphy 1995] G. E. Murphy, "The representations of Hecke algebras of type A_n", *J. Algebra* **173**:1 (1995), 97–121.

[Murphy $\geq$ 1999] G. E. Murphy, "Representations of Hecke algebras of $C_r \wr S_n$ at roots of unity". In preparation.

[Nakayama 1941] T. Nakayama, "On some modular properties of irreducible representations of symmetric groups, I and II", *Jap. J. Math.* **17** (1941), 165–184 and 411–423.

[Nazarov 1996] M. Nazarov, "Young's orthogonal form for Brauer's centralizer algebra", *J. Algebra* **182**:3 (1996), 664–693.

[Orlik and Solomon 1980] P. Orlik and L. Solomon, "Unitary reflection groups and cohomology", *Invent. Math.* **59**:1 (1980), 77–94.

[Osima 1954] M. Osima, "On the representations of the generalized symmetric group", *Math. J. Okayama Univ.* **4** (1954), 39–56.

[Pallikaros 1994] C. Pallikaros, "Representations of Hecke algebras of type D_n", *J. Algebra* **169**:1 (1994), 20–48.

[Penrose 1969] R. Penrose, "Angular momentum: an approach to combinatorial space-time", in *Quantum theory and beyond*, edited by T. A. Bastin, Cambridge Univ. Press, 1969.

[Penrose 1971] R. Penrose, "Applications of negative dimensional tensors", pp. 221–244 in *Combinatorial mathematics and its applications* (Oxford, 1969), edited by D. J. A. Welsh, Academic Press, London, 1971.

[Procesi 1976] C. Procesi, "The invariant theory of $n \times n$ matrices", *Advances in Math.* **19**:3 (1976), 306–381.

[Proctor 1984] R. A. Proctor, "Bruhat lattices, plane partition generating functions, and minuscule representations", *European J. Combin.* **5**:4 (1984), 331–350.

[Ram 1991] A. Ram, "A Frobenius formula for the characters of the Hecke algebras", *Invent. Math.* **106**:3 (1991), 461–488.

[Ram 1995] A. Ram, "Characters of Brauer's centralizer algebras", *Pacific J. Math.* **169**:1 (1995), 173–200.

[Ram 1998] A. Ram, "An elementary proof of Roichman's rule for irreducible characters of Iwahori–Hecke algebras of type A", pp. 335–342 in *Mathematical essays in honor* of Gian-Carlo Rota (Cambridge, MA, 1996), edited by B. E. Sagan and R. P. Stanley, Progress in Mathematics **161**, Birkhäuser, Boston, 1998.

[Remmel 1992] J. B. Remmel, "Formulas for the expansion of the Kronecker products $S_{(m,n)} \otimes S_{(1^{p-r},r)}$ and $S_{(1^k 2^l)} \otimes S_{(1^{p-r},r)}$", *Discrete Math.* **99**:1-3 (1992), 265-287.

[Reshetikhin 1987] N. Reshetikhin, "Quantized universal enveloping algebras, the Yang–Baxter equation and invariants of links I", Preprint E-4-87, LOMI, Leningrad, 1987.

[Roichman 1997] Y. Roichman, "A recursive rule for Kazhdan–Lusztig characters", *Adv. Math.* **129**:1 (1997), 25–45.

[Rumer et al. 1932] G. Rumer, E. Teller, and H. Weyl, "Eine für die Valenztheorie geeignete Basis der binären Vektor-invarianten", *Nachr. Ges. Wiss. Göttingen Math.-Phys. Kl.* (1932), 499–504. Reprinted as pp. 380–385 in Weyl's *Gesammelte Abhandlungen*, v. 3, edited by K. Chandrasekharan, Springer, Berlin, 1968.

[Rutherford 1948] D. E. Rutherford, *Substitutional Analysis*, Edinburgh U. Press, 1948.

[Scheunert 1979] M. Scheunert, *The theory of Lie superalgebras: an introduction*, Lecture Notes in Math. **716**, Springer, Berlin, 1979.

[Schur 1901] I. Schur, *Über eine Klasse von Matrizen, die sich einer gegeben Matrix zuordnen lassen*, Dissertation, 1901. Reprinted as pp. 1–72 in his *Gesammelte Abhandlungen*, vol. 1, edited by A. Brauer and H. Rohrbach, Springer, Berlin, 1973.

[Schur 1927] I. Schur, "Über die rationalen Darstellungen der allgemeinen linearen Gruppe", *Sitzungsber. Preuss. Akad. Wiss. Berlin Math.-Phys. Kl.* (1927), 58–75. Reprinted as pp. 68–85 in his *Gesammelte Abhandlungen*, vol. 3, edited by A. Brauer and H. Rohrbach, Springer, Berlin, 1973.

[Serganova 1996] V. Serganova, "Kazhdan–Lusztig polynomials and character formula for the Lie superalgebra $\mathfrak{gl}(m|n)$", *Selecta Math. (N.S.)* **2**:4 (1996), 607–651.

[Sergeev 1984] A. N. Sergeev, "Representations of the Lie superalgebras $\mathfrak{gl}(n, m)$ and $Q(n)$ in a space of tensors", *Funktsional. Anal. i Prilozhen.* **18**:1 (1984), 80–81. In Russian; translation in *Funct. Anal. App.* **18**:1 (1984), 70–72.

[Serre 1971] J.-P. Serre, *Représentations linéaires des groupes finis*, 2nd ed., Hermann, Paris, 1971. Translated as *Linear representations of finite groups*, Graduate Texts in Math. **42**, Springer, New York, 1977.

[Serre 1987] J.-P. Serre, *Complex semisimple Lie algebras*, Springer, New York, 1987.

[Shafarevich 1994] I. R. Shafarevich, *Basic algebraic geometry, 2 : Schemes and complex manifolds*, Second ed., Springer, Berlin, 1994.

[Shephard and Todd 1954] G. C. Shephard and J. A. Todd, "Finite unitary reflection groups", *Canadian J. Math.* **6** (1954), 274–304.

[Shoji 1979] T. Shoji, "On the Springer representations of the Weyl groups of classical algebraic groups", *Comm. Algebra* **7**:16 (1979), 1713–1745. Correction in **7**:18 (1979), 2027–2033.

[Solomon 1990] L. Solomon, "The Bruhat decomposition, Tits system and Iwahori ring for the monoid of matrices over a finite field", *Geom. Dedicata* **36**:1 (1990), 15–49.

[Solomon 1995] L. Solomon, "Abstract No. 900-16-169", *Abstracts American Math. Soc.* **16**:2 (Spring 1995).

[Spaltenstein 1976] N. Spaltenstein, "On the fixed point set of a unipotent transformation on the flag manifold", *Proc. Kon. Nederl. Akad. Wetensch.* **79** (1976), 452–456.

[Specht 1932] W. Specht, "Eine Verallgemeinerung der symmetrischen Gruppe", *Schriften Math. Seminar (Berlin)* **1** (1932), 1–32.

[Springer 1978] T. A. Springer, "A construction of representations of Weyl groups", *Invent. Math.* **44**:3 (1978), 279–293.

[Stanley 1971] R. Stanley, "Theory and application of plane partitions", *Stud. Appl. Math.* **50** (1971), 167–188 and 259–279.

[Stanley 1982] R. P. Stanley, "Some aspects of groups acting on finite posets", *J. Combin. Theory Ser. A* **32**:2 (1982), 132–161.

[Stanley 1983] R. P. Stanley, "GL$(n, \mathbb{C})$ for combinatorialists", pp. 187–199 in *Surveys in combinatorics* (Southampton, 1983), edited by E. K. Lloyd, London Math. Soc. Lecture Note Ser. **82**, Cambridge Univ. Press, Cambridge, 1983.

[Stembridge 1987] J. R. Stembridge, "Rational tableaux and the tensor algebra of $\mathfrak{gl}_n$", *J. Combin. Theory Ser. A* **46**:1 (1987), 79–120.

[Stembridge 1989a] J. R. Stembridge, "A combinatorial theory for rational actions of GL$_n$", pp. 163–176 in *Invariant theory* (Denton, TX, 1986), edited by R. Fossum et al., Contemp. Math. **88**, Amer. Math. Soc., Providence, RI, 1989.

[Stembridge 1989b] J. R. Stembridge, "On the eigenvalues of representations of reflection groups and wreath products", *Pacific J. Math.* **140**:2 (1989), 353–396.

[Stembridge 1994a] J. R. Stembridge, "On minuscule representations, plane partitions and involutions in complex Lie groups", *Duke Math. J.* **73**:2 (1994), 469–490.

[Stembridge 1994b] J. R. Stembridge, "Some hidden relations involving the ten symmetry classes of plane partitions", *J. Combin. Theory Ser. A* **68**:2 (1994), 372–409.

[Stembridge 1995] J. R. Stembridge, "The enumeration of totally symmetric plane partitions", *Adv. Math.* **111**:2 (1995), 227–243.

[Sundaram 1990a] S. Sundaram, "The Cauchy identity for Sp$(2n)$", *J. Combin. Theory Ser. A* **53**:2 (1990), 209–238.

[Sundaram 1990b] S. Sundaram, "Orthogonal tableaux and an insertion algorithm for SO$(2n + 1)$", *J. Combin. Theory Ser. A* **53**:2 (1990), 239–256.

[Sundaram 1990c] S. Sundaram, "Tableaux in the representation theory of the classical Lie groups", pp. 191–225 in *Invariant theory and tableaux* (Minneapolis, 1988), IMA Vol. Math. Appl. **19**, Springer, New York, 1990.

[Sundaram 1994] S. Sundaram, "The homology representations of the symmetric group on Cohen-Macaulay subposets of the partition lattice", *Adv. Math.* **104**:2 (1994), 225–296.

[Tanisaki 1982] T. Tanisaki, "Defining ideals of the closures of the conjugacy classes and representations of the Weyl groups", *Tôhoku Math. J.* (2) **34**:4 (1982), 575–585.

[Temperley and Lieb 1971] H. N. V. Temperley and E. H. Lieb, "Relations between the 'percolation' and 'colouring' problem and other graph-theoretical problems associated with regular planar lattices: some exact results for the 'percolation' problem", *Proc. Roy. Soc. London Ser. A* **322** (1971), 251–280.

[Ueno and Shibukawa 1992] K. Ueno and Y. Shibukawa, "Character table of Hecke algebra of type A_{N-1} and representations of the quantum group $U_q(\mathfrak{gl}_{n+1})$", pp. 977–984 in *Infinite analysis* (Kyoto, 1991), vol. 2, edited by A. Tsuchiya et al., Adv. Ser. Math. Phys. **16**, World Sci. Publishing, River Edge, NJ, 1992.

[Van der Jeugt 1991] J. Van der Jeugt, "An algorithm for characters of Hecke algebras $H_n(q)$ of type A_{n-1}", *J. Phys. A* **24**:15 (1991), 3719–3725.

[Varadarajan 1984] V. S. Varadarajan, *Lie groups, Lie algebras, and their representations*, Graduate Texts in Math. **102**, Springer, New York, 1984. Reprint of the 1974 edition.

[Vershik and Kerov 1988] A. M. Vershik and S. V. Kerov, "Characters and realizations of representations of the infinite-dimensional Hecke algebra, and knot invariants", *Dokl. Akad. Nauk SSSR* **301**:4 (1988), 777–780. In Russian; translated in *Soviet Math. Dokl.* **38** (1989), 134–137.

[Vogan 1981] J. Vogan, David A., *Representations of real reductive Lie groups*, Progress in Mathematics **15**, Birkhäuser, Boston, 1981.

[Vogan 1987] J. Vogan, David A., *Unitary representations of reductive Lie groups*, Annals of Mathematics Studies **118**, Princeton University Press, Princeton, NJ, 1987.

[Wallach 1988] N. R. Wallach, *Real reductive groups*, vol. 1, Academic Press, Boston, MA, 1988.

[Wallach 1992] N. R. Wallach, *Real reductive groups*, vol. 2, Academic Press, Boston, MA, 1992.

[Wenzl 1988a] H. Wenzl, "Hecke algebras of type A_n and subfactors", *Invent. Math.* **92**:2 (1988), 349–383.

[Wenzl 1988b] H. Wenzl, "On the structure of Brauer's centralizer algebras", *Ann. of Math.* (2) **128**:1 (1988), 173–193.

[Wenzl 1990] H. Wenzl, "Quantum groups and subfactors of type B, C, and D", *Comm. Math. Phys.* **133**:2 (1990), 383–432.

[Weyl 1925] H. Weyl, "Theorie der Darstellung kontinuerlicher halb-einfacher Gruppen durch lineare Transformationen I", *Mathematische Zeitschrift* **23** (1925), 271–309.

[Weyl 1926] H. Weyl, "Theorie der Darstellung kontinuerlicher halb-einfacher Gruppen durch lineare Transformationen II, III", *Mathematische Zeitschrift* **24** (1926), 328–376, 377–395.

[Weyl 1946] H. Weyl, *The classical groups: their invariants and representations*, 2nd ed., Princeton University Press, Princeton, NJ, 1946.

[Weyman 1989] J. Weyman, "The equations of conjugacy classes of nilpotent matrices", *Invent. Math.* **98**:2 (1989), 229–245.

[Young 1901] A. Young, "On quantitative substitutional analysis, I", *Proc. London Math. Soc.* **33** (1901), 97–146. Reprinted as pp. 42–91 in [Young 1977].

[Young 1902] A. Young, "On quantitative substitutional analysis, II", *Proc. London Math. Soc.* **34** (1902), 361–397. Reprinted as pp. 92–128 in [Young 1977].

[Young 1928] A. Young, "On quantitative substitutional analysis, III", *Proc. London Math. Soc.* (2) **28** (1928), 255–292. Reprinted as pp. 352–389 in [Young 1977].

[Young 1930a] A. Young, "On quantitative substitutional analysis, IV", *Proc. London Math. Soc.* (2) **31** (1930), 353–372. Reprinted as pp. 396–415 in [Young 1977].

[Young 1930b] A. Young, "On quantitative substitutional analysis, V", *Proc. London Math. Soc.* (2) **31** (1930), 273–288. Reprinted as pp. 416–431 in [Young 1977].

[Young 1931] A. Young, "On quantitative substitutional analysis, VI", *Proc. London Math. Soc.* (2) **34** (1931), 196–230. Reprinted as pp. 432–466 in [Young 1977].

[Young 1934a] A. Young, "On quantitative substitutional analysis, VII", *Proc. London Math. Soc.* (2) **36** (1934), 304–368. Reprinted as pp. 494–558 in [Young 1977].

[Young 1934b] A. Young, "On quantitative substitutional analysis, VIII", *Proc. London Math. Soc.* (2) **37** (1934), 441–495. Reprinted as pp. 559–613 in [Young 1977].

[Young 1977] A. Young, *The Collected Papers of Alfred Young,* 1873–1940, Mathematical Exposition **21**, U. of Toronto Press, 1977.

[Zhelobenko 1970] D. P. Zhelobenko, Компактные группы Ли и их представления, Nauka, Moscow, 1970. Translated as *Compact Lie groups and their representations,* Translations of Mathematical Monographs **40**, Amer. Math. Soc., Providence, 1973.

[Zhelobenko 1987] D. P. Zhelobenko, "An analogue of the Gel'fand–Tsetlin basis for symplectic Lie algebras", *Uspekhi Mat. Nauk* **42**:6 (= 258) (1987), 193–194. In Russian; translation in *Russ. Math. Surveys* **42**:6 (1987), 247–248.

HÉLÈNE BARCELO
DEPARTMENT OF MATHEMATICS
ARIZONA STATE UNIVERSITY
TEMPE, AZ 85287-1804
UNITED STATES
helene@nelligan.la.asu.edu

ARUN RAM
DEPARTMENT OF MATHEMATICS
PRINCETON UNIVERSITY
PRINCETON, NJ 08544
UNITED STATES
rama@math.Princeton.edu

After August 1999:
DEPARTMENT OF MATHEMATICS
UNIVERSITY OF WISCONSIN–MADISON
MADISON, WI 53706
UNITED STATES
rama@math.wisc.edu

New Perspectives in Geometric Combinatorics
MSRI Publications
Volume **38**, 1999

An Algorithmic Theory of Lattice Points in Polyhedra

ALEXANDER BARVINOK AND JAMES E. POMMERSHEIM

ABSTRACT. We discuss topics related to lattice points in rational polyhedra, including efficient enumeration of lattice points, "short" generating functions for lattice points in rational polyhedra, relations to classical and higher-dimensional Dedekind sums, complexity of the Presburger arithmetic, efficient computations with rational functions, and others. Although the main slant is algorithmic, structural results are discussed, such as relations to the general theory of valuations on polyhedra and connections with the theory of toric varieties. The paper surveys known results and presents some new results and connections.

Contents

Key words and phrases. integer points, lattice, algorithms, polyhedra, toric varieties, generating functions, valuations.

1. Introduction:
"A Formula for the Number of Lattice Points..."

Let $\mathbb{R}^d$ be Euclidean d-space of all d-tuples $x = (\xi_1, \ldots, \xi_d)$ of real numbers with the standard scalar product $\langle x, y \rangle = \xi_1 \eta_1 + \cdots + \xi_d \eta_d$, where $x = (\xi_1, \ldots, \xi_d)$ and $y = (\eta_1, \ldots, \eta_d)$. The first main object of this paper is the integer lattice $\mathbb{Z}^d \subset \mathbb{R}^d$ consisting of the points with integer coordinates. We define the second main object.

DEFINITION 1.1. A *rational polyhedron* $P \subset \mathbb{R}^d$ is the set of solutions of a finite system of linear inequalities with integer coefficients:

$$P = \left\{ x \in \mathbb{R}^d : \langle c_i, x \rangle \le \beta_i \text{ for } i = 1, \ldots, m \right\}, \quad \text{where } c_i \in \mathbb{Z}^d \text{ and } \beta_i \in \mathbb{Z}.$$

A bounded rational polyhedron is called a *polytope*. A polytope $P \subset \mathbb{R}^d$ is called a *lattice polytope* or an *integer polytope* if its vertices are points from $\mathbb{Z}^d$.

We are interested in the set $P \cap \mathbb{Z}^d$ of lattice points belonging to a given rational polyhedron P. For example, we may be interested in finding a "formula" for the number of lattice points in a given rational or integer polytope P. But what does it mean to "find a formula"? We consider a few examples.

EXAMPLE 1.2. Suppose that $d = 2$ and $P \subset \mathbb{R}^2$ is an integer polygon. The famous formula of G. Pick [1899] states that

$$|P \cap \mathbb{Z}^2| = \text{area}(P) + \frac{|\partial P \cap \mathbb{Z}^d|}{2} + 1,$$

or in words: the number of integer points in an integer polygon is equal to the area of the polygon plus half the number of integer points on the boundary of the polygon plus 1. See [Lagarias 1995] as a general reference. Nearly every mathematician would agree that Pick's formula is a beautiful and useful formula.

EXAMPLE 1.3. Let $P \subset \mathbb{R}^d$ be a polytope. We can write the number of lattice points in P as

$$|P \cap \mathbb{Z}^d| = \sum_{x \in \mathbb{Z}^d} \delta(x, P), \quad \text{where } \delta(x, P) = \begin{cases} 1 & \text{if } x \in P, \\ 0 & \text{if } x \notin P, \end{cases}$$

but this formula is not very interesting or useful.

In most cases, the formulae one can get are neither so nice and simple as Pick's formula (Example 1.2), nor so tautological as the formula from Example 1.3. We consider a few more examples.

EXAMPLE 1.4. Let $\Delta \subset \mathbb{R}^3$ be the tetrahedron with the vertices $(0,0,0)$, $(a,0,0)$, $(0,b,0)$, and $(0,0,c)$, where a, b and c are pairwise coprime positive integers. Then the number of lattice points in Δ can be expressed as

$$|\Delta \cap \mathbb{Z}^3| = \frac{abc}{6} + \frac{ab + ac + bc + a + b + c}{4}$$

$$+ \frac{1}{12}\left(\frac{ac}{b} + \frac{bc}{a} + \frac{ab}{c} + \frac{1}{abc}\right) - s(bc,a) - s(ac,b) - s(ab,c) + 2,$$

where $s(p,q)$ is the *Dedekind sum* defined for coprime positive integers p and q by

$$s(p,q) = \sum_{i=1}^{q} \left(\left(\frac{i}{q}\right)\right)\left(\left(\frac{pi}{q}\right)\right) \quad \text{and} \quad ((x)) = \begin{cases} x - \lfloor x \rfloor - \frac{1}{2} & \text{if } x \notin \mathbb{Z}, \\ 0 & \text{if } x \in \mathbb{Z}; \end{cases}$$

here as usual $\lfloor \cdot \rfloor$ is the floor function. See [Mordell 1951; Pommersheim 1993; Dyer 1991].

EXAMPLE 1.5. Let $P \subset \mathbb{R}^d$ be a nonempty integer polytope. For a positive integer n, let $nP = \{nx : x \in P\}$ denote the dilated polytope P. As E. Ehrhart discovered [1977], there is a polynomial $p(n)$, now called the *Ehrhart polynomial* of P, such that

$$|nP \cap \mathbb{Z}^d| = p(n), \quad \text{where } p(n) = a_d n^d + a_{d-1} n^{d-1} + \cdots + a_0.$$

Furthermore, $a_0 = 1$ and $a_d = \text{vol}_d(P)$, the volume of P. The following *reciprocity law* holds:

$$p(-n) = (-1)^{\dim(P)} |\text{relint}(nP) \cap \mathbb{Z}^d|, \quad \text{for positive integers } n.$$

That is, the value of p at a negative integer $-n$ equals, up to a sign, the number of integer points in the relative interior of nP. See [Stanley 1997, Section 4.6], for example.

We will argue that both Example 1.4 and Example 1.5 are useful and beautiful.

To navigate the sea of "lattice points formulae" which can be found in the literature and which are to be discovered in the future, we have to set up some criteria for beauty and usefulness. Of course, like all such criteria, ours is purely subjective. We look at the *computational complexity* of the formula.

Fix the space $\mathbb{R}^d$. Suppose that $P \subset \mathbb{R}^d$ is a rational polytope. There is an obvious way to count integer points in P: we consider a sufficiently large box $B = \{x = (\xi_1, \ldots, \xi_d) : \alpha_i \leq \xi_i \leq \beta_i \text{ for } i = 1, \ldots, d\}$ which contains P, and check integer points from B one by one to see if they are contained in P. In other words, this is an "effective" version of the formula of Example 1.3. We will measure the "usefulness" and "niceness" of the formula for the number of lattice points by how much time it allows us to save compared with this straightforward procedure of enumeration. In particular, we will be interested in a special class of formulae whose complexity is bounded by a polynomial in the *input size* of the polytope P. A polytope P may be given by its description as a set of linear inequalities (as a rational polyhedron: see Definition 1.1). The input size of this *facet* description of P is the total size in binary encoding of the coefficients of the inequalities needed to describe P. For example, the input size of the description $I = \{x : 0 \le x \le a\}$ of an interval, where a is a positive integer, is $O(\log a)$. A polytope $P \subset \mathbb{R}^d$ may be given as the convex hull of its vertices; the input size of such a *vertex description* is defined in a similar way: the total size of the coordinates of the vertices of P in binary encoding. It is well understood that if the dimension d is fixed and not a part of the input, then the facet description and the vertex description are polynomially equivalent; that is, the length of one is bounded by a polynomial in the length of the other. See any of [Grötschel et al. 1993; Lovász 1986; Schrijver 1986], for example. Sometimes we talk about formulae that can be applied to polytopes from some particular class. In this case, we are looking at the computational complexity of the formula relative to the input size of the description of P within the class.

From this perspective, the formula of Example 1.2 is very nice: it is much more efficient than the direct enumeration of integer points in a polygon. Indeed, it is easy to compute the area of P by triangulating the polygon. Furthermore, the boundary ∂P is a union of finitely many straight line intervals, and counting lattice points in intervals is easy. Formula of Example 1.3 is bad since it has exponential complexity even in dimension 1. Indeed, the input size of the interval $[0, a]$ is $O(\log a)$, whereas the straightforward counting of Example 1.3 would give us $O(a)$ complexity, which is exponentially large in the input size. The formula of Example 1.4 is nice, because it reduces counting to the computation of the Dedekind sums, which can be done efficiently. Indeed, by recursively applying the reciprocity relation

$$s(p, q) + s(q, p) = -\frac{1}{4} + \frac{1}{12}\left(\frac{p}{q} + \frac{q}{p} + \frac{1}{pq}\right)$$

and the obvious identity

$$s(p, q) = s(r, q), \quad \text{where} \quad r \equiv p \pmod{q} \text{ and } 0 \le r < q,$$

one can compute $s(p, q)$ in time polynomial in the input size of p and q, by a procedure resembling the Euclidean algorithm. See [Rademacher and Grosswald 1972], for example. Finally, the formula of Example 1.5 is also nice since its allows us to save time counting integer points in nP, where n is a large positive integer. Indeed, we can apply the "brute force" counting of Example 1.3 to find the number of integer points in polytopes $P, 2P, \ldots, \lfloor d/2 \rfloor P$ and their relative interiors, and then interpolate the polynomial p. Once p is found, it is easy to find $|nP \cap \mathbb{Z}^d|$ for any positive integer n.

To summarize, we approach every formula in this paper primarily from the point of view of computational complexity. Of course, there are different philosophies that are equally legitimate. The topic of this paper is "lattices and polyhedra," as opposed to a close, but somewhat different in spirit, topic "lattices and convex bodies." This is why we omit many interesting results on integer programming, lattice reduction algorithms, and counting lattice points in general convex bodies; see [Cook et al. 1992; Lovász 1986; Grötschel et al. 1993; Schrijver 1986]. Similarly, we do not discuss rather interesting results concerning lattice points in nonrational polyhedra [Skriganov 1998]. Our approach is algebraic and we don't cover recent advances in probabilistic methods of counting, such as those in [Dyer et al. 1993; 1997] (see [Bollobás 1997] for a survey). In short, the paper presents "an algorithmic theory," one of many possible.

This area of the research has been quite active. Along with such activity, one expects independent discoveries of certain results, and with unequal publication delays, there is often confusion about who did what first. We have tried to be accurate in the chronology, but, unfortunately, inaccuracies are possible.

This paper is meant to be a survey. However, it does contain some new results: Theorem 4.4 (especially the second part), Theorem 5.3, Theorem 9.6, results in Section 10, and possibly some results in Section 7. In addition, some of the links in Section 8 are new. Whenever possible, we have tried to provide the reader with sketches of proofs.

2. Preliminaries. Algebra of Polyhedra

The number of integer points in a polytope is a *valuation*; that is, it satisfies the inclusion-exclusion property. The theory of valuations, with the theory of the polytope algebra as its basis, was developed by many authors; see [McMullen and Schneider 1983; McMullen 1993] for a survey. Several inequivalent definitions and approaches have been used, each having its own advantages. For example, one can either choose to consider arbitrary polytopes, or to consider lattice polytopes only. In addition, one can decide either to identify or not to identify two polytopes which differ by a (lattice) translation. Also, valuations can be defined via the inclusion-exclusion principle for the union of two or several polytopes. See [Kantor and Khovanskii 1992; Pukhlikov and Khovanskii 1992a; Lawrence 1988; McMullen 1989; Morelli 1993c; 1993d]. Here we employ an approach which is convenient for us.

Let $A \subset \mathbb{R}^d$ be a set. The *indicator function* $[A] : \mathbb{R}^d \to \mathbb{R}$ of A is defined by

$$[A](x) = \begin{cases} 1 & \text{if } x \in A, \\ 0 & \text{if } x \notin A. \end{cases}$$

The *algebra of polyhedra* $\mathcal{P}(\mathbb{R}^d)$ is the vector space (over $\mathbb{Q}$) spanned by the indicator functions $[P]$ of all polyhedra $P \subset \mathbb{R}^d$. The space $\mathcal{P}(\mathbb{R}^d)$ is closed under pointwise multiplication of functions: for any two functions $f, g \in \mathcal{P}(\mathbb{R}^d)$, we have

$fg \in \mathcal{P}(\mathbb{R}^d)$, since $[P][Q] = [P \cap Q]$. Hence $\mathcal{P}(\mathbb{R}^d)$ is a commutative algebra under pointwise multiplication. We will be interested in some particular subspaces of the algebra of polyhedra. The *polytope algebra* $\mathcal{P}_c(\mathbb{R}^d)$ is the subspace spanned by the indicator functions of all polytopes in $\mathbb{R}^d$, and the *algebra of cones* $\mathcal{P}_K(\mathbb{R}^d)$ is the subspace spanned by the indicator functions of the polyhedral cones in $\mathbb{R}^d$. (A nonempty polyhedron P is called a *cone* if $\lambda x \in P$ whenever $x \in P$ and $\lambda \geq 0$.)

Clearly, $\mathcal{P}_c(\mathbb{R}^d)$ and $\mathcal{P}_K(\mathbb{R}^d)$ are subalgebras.

DEFINITION 2.1. A linear transformation

$$\Phi : \mathcal{P}(\mathbb{R}^d) \to V,$$

where V is a vector space over $\mathbb{Q}$, is called a *valuation*. Similarly, linear transformations defined on $\mathcal{P}_c(\mathbb{R}^d)$ and $\mathcal{P}_K(\mathbb{R}^d)$ are also called valuations.

One particular valuation is very important.

THEOREM 2.2. *There exists a unique valuation* $\mu : \mathcal{P}(\mathbb{R}^d) \to \mathbb{Q}$, *called the Euler characteristic, such that* $\mu([P]) = 1$ *for each nonempty polyhedron* $P \subset \mathbb{R}^d$.

Note that we cannot simply define μ by letting $\mu([P]) = 1$, because the indicator functions of polyhedra are *not* linearly independent (for $d > 0$). Since the indicator functions $[P]$ span $\mathcal{P}(\mathbb{R}^d)$, the uniqueness is immediate. The following proof belongs to H. Hadwiger. See also [McMullen and Schneider 1983].

SKETCH OF PROOF. We use induction on d to establish the existence of $\mu = \mu_d$. We have $\mathcal{P}(\mathbb{R}^0) = \mathbb{Q}[0]$, and we can define μ_0 by letting $\mu(a[0]) = a$. Suppose that $d \geq 1$. First, we define μ_d on the polytope subalgebra $\mathcal{P}_c(\mathbb{R}^d)$. Choose a nonzero linear function $l : \mathbb{R}^d \to \mathbb{R}$ and slice $\mathbb{R}^d$ into level hyperplanes $H^\alpha = \{x \in \mathbb{R}^d : l(x) = \alpha\}$, with $\alpha \in \mathbb{R}$. For a function $f \in \mathcal{P}_c(\mathbb{R}^d)$, let $f^\alpha : H^\alpha \to \mathbb{R}$ be the restriction of f to H^α. We note that H^α can be identified with a $(d-1)$-dimensional Euclidean space, so we can consider the algebra of polytopes $\mathcal{P}_c(H^\alpha)$ and the Euler characteristic $\mu^\alpha : \mathcal{P}_c(H^\alpha) \to \mathbb{Q}$. We note that $f^\alpha \in \mathcal{P}_c(H^\alpha)$, and we can define $\mu^\alpha(f^\alpha)$, which we denote by $\mu^\alpha(f)$. For $\alpha \in \mathbb{R}$ and $f \in \mathcal{P}_c(\mathbb{R}^d)$, define

$$\mu^{\alpha-}(f) = \lim_{\varepsilon \to +0} \mu^{\alpha-\varepsilon}(f).$$

Suppose that

$$f = \sum_{i \in I} a_i[P_i],$$

where $P_i \subset \mathbb{R}^d$ are polytopes and $a_i \in \mathbb{Q}$ are numbers. It is easy to see that $\mu^{\alpha-}(f)$ is always well-defined, and that $\mu^{\alpha-}(f) = \mu^\alpha(f)$ unless α is the minimum value of the linear function l on some P_i. In particular, for any f, there are only finitely many α's for which $\mu^{\alpha-}(f) \neq \mu^\alpha(f)$. Now we can define $\mu = \mu_d$ on $\mathcal{P}_c(\mathbb{R}^d)$ by

$$\mu_d(f) = \sum_{\alpha \in \mathbb{R}} \left(\mu^\alpha(f) - \mu^{\alpha-}(f) \right).$$

The sum is well-defined since there are only finitely many nonzero terms. Now we are ready to extend μ_d to $\mathcal{P}(\mathbb{R}^d)$. Choose a polytope Q containing the origin as an interior point, and let $Q(t) = \{tx : x \in Q\}$ be the dilatation of Q by a factor of t. For any $t > 0$ and any $f \in \mathcal{P}(\mathbb{R}^d)$, we have $f_t = [Q(t)]f \in \mathcal{P}_c(\mathbb{R}^d)$ and we let

$$\mu_d(f) = \lim_{t \to +\infty} \mu_d(f_t).$$

It is very easy to see that μ_d is well-defined and that it satisfies the condition $\mu_d([P]) = 1$ for any nonempty polyhedron $P \subset \mathbb{R}^d$. □

The Euler characteristic μ allows us to interpret various important valuations as integral transforms with respect to μ *as a measure.* See [Khovanskii and Pukhlikov 1993].

THEOREM 2.3. *Suppose that $A : \mathbb{R}^n \to \mathbb{R}^m$ is an affine transformation. Then there is a unique valuation $\mathcal{A} : \mathcal{P}(\mathbb{R}^n) \to \mathcal{P}(\mathbb{R}^m)$ such that $\mathcal{A}([P]) = [A(P)]$ for each polyhedron $P \subset \mathbb{R}^d$.*

PROOF. Define a *kernel* $K : \mathbb{R}^n \times \mathbb{R}^m \to \mathbb{R}$ by

$$K(x, y) = \begin{cases} 1 & \text{if } y = Ax, \\ 0 & \text{if } y \neq Ax. \end{cases}$$

Then for each fixed y and each $f \in \mathcal{P}(\mathbb{R}^n)$, we have $K(x, y)f \in \mathcal{P}(\mathbb{R}^n)$, so we can apply the Euler characteristic on $\mathcal{P}(\mathbb{R}^n)$, which we denote by μ_x (to stress the variable x). Now we let

$$g = \mathcal{A}(f), \quad \text{where } g(y) = \mu_x\big(K(x,y)f(x)\big). \qquad □$$

In addition to pointwise multiplication, there is a commutative and associative bilinear operation $\star$ on $\mathcal{P}(\mathbb{R}^d)$, which we call *convolution*, because it can be considered as the convolution with respect to the Euler characteristic as a measure. Many authors [Lawrence 1988; McMullen 1989; 1993] consider $\star$ as the true multiplication in the algebra of polyhedra, and perhaps rightly so, because it has many interesting properties.

DEFINITION 2.4. Let P and Q be polyhedra in $\mathbb{R}^d$. The *Minkowski sum* $P + Q$ is defined as

$$P + Q = \{x + y : x \in P, \, y \in Q\}.$$

THEOREM 2.5. *There is a unique bilinear operation $\star : \mathcal{P}(\mathbb{R}^d) \times \mathcal{P}(\mathbb{R}^d) \to \mathcal{P}(\mathbb{R}^d)$ such that $[P] \star [Q] = [P + Q]$ for any two polyhedra P and Q.*

PROOF. Fix a decomposition $\mathbb{R}^{2d} = \mathbb{R}^d \times \mathbb{R}^d$. Let $A : \mathbb{R}^d \times \mathbb{R}^d \to \mathbb{R}^d$ be the linear transformation $A(x, y) = x + y$ and let $\mathcal{A} : \mathcal{P}(\mathbb{R}^{2d}) \to \mathcal{P}(\mathbb{R}^d)$ be the corresponding valuation whose existence is asserted by Theorem 2.3. For functions $f, g \in \mathcal{P}(\mathbb{R}^d)$, define their outer product $f \times g \in \mathcal{P}(\mathbb{R}^{2d})$ by $(f \times g)(x, y) = f(x)g(y)$. Then $f \star g = \mathcal{A}(f \times g)$. □

COROLLARY 2.6. *Suppose that $P_1, \ldots, P_k \subset \mathbb{R}^d$ are polyhedra such that*

$$\alpha_1[P_1] + \cdots + \alpha_k[P_k] = 0$$

for certain rational numbers $\alpha_1, \ldots, \alpha_k$. Then for any polyhedron $Q \subset \mathbb{R}^d$, one has

$$\alpha_1[P_1 + Q] + \cdots + \alpha_k[P_k + Q] = 0.$$

PROOF. We have

$$0 = 0 \star [Q] = (\alpha_1[P_1] + \cdots + \alpha_k[P_k]) \star [Q] = \alpha_1[P_1 + Q] + \cdots + \alpha_k[P_k + Q]. \quad \square$$

Convolution $\star$ has many interesting properties; see [McMullen and Schneider 1983]. For example, it is easy to see that $[0]$ plays the role of the identity. It turns out that $[P]$ is invertible for any polytope P, and that the inverse element is $(-1)^{\dim(P)}[-\text{relint } P]$, the indicator function (up to sign) of the relative interior of the centrally symmetric image $-P$ of P [McMullen and Schneider 1983].

Next we discuss *duality* in the algebra of cones $\mathcal{P}_K(\mathbb{R}^d)$ (see [Lawrence 1988]). If $K \subset \mathbb{R}^d$ is a cone, then

$$K^* = \{x \in \mathbb{R}^d : \langle x, y \rangle \geq 0 \text{ for each } y \in K\}$$

is called the *dual cone* to K.

THEOREM 2.7. *There exists a valuation $\mathcal{D} : \mathcal{P}_K(\mathbb{R}^d) \to \mathcal{P}_K(\mathbb{R}^d)$ such that*

$$\mathcal{D}([K]) = [K^*]$$

for each cone $K \subset \mathbb{R}^d$.

SKETCH OF PROOF. Define the kernel $K(x, y) : \mathbb{R}^d \times \mathbb{R}^d \to \mathbb{R}$ by

$$K(x, y) = \begin{cases} 1 & \text{if } \langle x, y \rangle = -1, \\ 0 & \text{otherwise.} \end{cases}$$

Then, for each $f \in \mathcal{P}_K(\mathbb{R}^d)$ and for each y, we have $K(x, y)f \in \mathcal{P}_K(\mathbb{R}^d)$ and we can apply the Euler characteristic $\mu = \mu_x$. Now we let

$$g = \mathcal{D}(f), \quad \text{where } g(y) = \mu(f) - \mu_x(K(x, y)f(x)).$$

It is straightforward to show that $\mathcal{D}$ satisfies the required properties. $\quad \square$

Theorem 2.7 has an interesting corollary, which says that if a linear identity holds for cones, the same identity holds for their dual cones.

COROLLARY 2.8. *Suppose that $K_1, \ldots, K_m \subset \mathbb{R}^d$ are cones such that*

$$\sum_{i=1}^{m} \alpha_i [K_i] = 0$$

for certain rational numbers α_i. Then

$$\sum_{i=1}^{m} \alpha_i [K_i^*] = 0.$$

PROOF. We apply the valuation $\mathcal{D}$ of Theorem 2.7 to both sides of the identity. □

The valuation $\mathcal{D}$ plays the role of the Fourier Transform with respect to the Euler characteristic μ as a measure. The valuation $\mathcal{D}$ transforms pointwise products into the convolutions:

$$\mathcal{D}(fg) = \mathcal{D}(f) \star \mathcal{D}(g) \quad \text{and} \quad \mathcal{D}(f \star g) = \mathcal{D}(f) \cdot \mathcal{D}(g) \quad \text{for } f, g \in \mathcal{P}_K(\mathbb{R}^d).$$

It suffices to check the identity for $f = [K]$ and $g = [C]$, where $K, C \subset \mathbb{R}^d$ are cones. We have $fg = [K \cap C]$, $\mathcal{D}(f) = [K^*]$, $\mathcal{D}(g) = [C^*]$, and

$$\mathcal{D}(f) \star \mathcal{D}(g) = [K^*] \star [C^*] = [K^* + C^*] = [(K \cap C)^*] = \mathcal{D}(fg).$$

Similarly,

$$\mathcal{D}(f) \cdot \mathcal{D}(g) = [K^*] \cdot [C^*] = [K^* \cap C^*] = [(K + C)^*] = \mathcal{D}(f \star g).$$

Finally, we describe an important valuation on the polytope algebra $\mathcal{P}_c(\mathbb{R}^d)$ associated with a vector $u \in \mathbb{R}^d$. Let $P \subset \mathbb{R}^d$ be a polytope and let $u \in \mathbb{R}^d$ be a vector. Let

$$\max(u, P) = \max\{\langle u, x \rangle : x \in P\}$$

be the maximal value of the linear function $\langle u, x \rangle$ on the polytope P and let

$$P_u = \{x \in P : \langle u, x \rangle = \max(u, P)\}$$

be the face of P where this maximum is attained.

THEOREM 2.9. *For any $u \in \mathbb{R}^d$ there is a valuation*

$$T_u : \mathcal{P}_c(\mathbb{R}^d) \to \mathcal{P}_c(\mathbb{R}^d)$$

such that

$$T_u([P]) = [P_u]$$

for any polytope $P \subset \mathbb{R}^d$.

SKETCH OF PROOF. For $\varepsilon > 0$ and $\delta > 0$, define the kernel

$$K_{\varepsilon\delta}(x, y) = \begin{cases} 1 & \text{if } \langle u, x - y \rangle \geq \delta \text{ and } \|x - y\|_\infty \leq \varepsilon, \\ 0 & \text{otherwise.} \end{cases}$$

Then for $f \in \mathcal{P}_c(\mathbb{R}^d)$, we let $T_u(f) = g$, where

$$g(y) = f(y) - \lim_{\varepsilon \to +0} \lim_{\delta \to +0} \mu_x\big(K_{\varepsilon\delta}(x, y) f(x)\big).$$

□

The valuation T_u commutes with convolution: $T_u(f \star g) = T_u(f) \star T_u(g)$. This identity is easy to check on indicator functions of polytopes, and it can then be extended by linearity.

3. Generating Functions for Integer Points in Rational Polyhedra

In this section, we consider the subalgebra $\mathcal{P}(\mathbb{Q}^d) \subset \mathcal{P}(\mathbb{R}^d)$ spanned by the indicator functions $[P]$, where $P \subset \mathbb{Q}^d$ is a *rational* polyhedron. Let $\mathbb{Q}(x)$ be the algebra of rational functions in d complex variables $x = (x_1, \ldots, x_d)$ with rational coefficients. We discuss a very interesting valuation

$$\mathfrak{F} : \mathcal{P}(\mathbb{Q}^d) \to \mathbb{Q}(x).$$

This valuation first was described by J. Lawrence [1991]. At about the same time, A. Khovanskii and A. Pukhlikov gave an independent description [Pukhlikov and Khovanskii 1992b]. See also [Brion and Vergne 1997c].

Let $P \subset \mathbb{R}^d$ be a rational polyhedron. To the set $P \cap \mathbb{Z}^d$ of integral points in P, we associate the generating function

$$f(P \cap \mathbb{Z}^d; x) = \sum_{m \in P \cap \mathbb{Z}^d} x^m, \quad \text{where } x^m = x_1^{\mu_1} \cdots x_d^{\mu_d} \text{ for } m = (\mu_1, \ldots, \mu_d)$$

in d complex variables $x = (x_1, \ldots, x_d)$. We often write $f(P; x)$ instead of $f(P \cap \mathbb{Z}^d; x)$.

THEOREM 3.1. *There is a map $\mathfrak{F}$ which, to each rational polyhedron $P \subset \mathbb{R}^d$ associates a rational function $f(P; x)$ in d complex variables $x \in \mathbb{C}^d$, $x = (x_1, \ldots, x_d)$, such that the following properties are satisfied:*

(i) *The map $\mathfrak{F}$ is a valuation: if $P_1, \ldots, P_k \subset \mathbb{R}^d$ are rational polyhedra whose indicator functions satisfy a linear identity*

$$\alpha_1[P_1] + \cdots + \alpha_k[P_k] = 0,$$

then the functions $f(P_i; x)$ satisfy the same identity:

$$\alpha_1 f(P_1; x) + \cdots + \alpha_k f(P_k; x) = 0.$$

(ii) *If $m + P$ is a translation of P by an integer vector $m \in \mathbb{Z}^d$, then*

$$f(P + m; x) = x^m f(P; x).$$

(iii) *We have*

$$f(P; x) = \sum_{m \in P \cap \mathbb{Z}^d} x^m$$

for any $x \in \mathbb{C}^d$ such that the series converges absolutely.
(iv) *If P contains a straight line then $f(P; x) \equiv 0$.*

We consider some examples.

EXAMPLE 3.2. Define four rational polyhedra in $\mathbb{R}^1$: $P = \mathbb{R}^1$, $P_+ = \{x : x \geq 0\}$, $P_- = \{x : x \leq 0\}$, and $P_0 = \{0\}$. Part 3 of Theorem 3.1 implies that $f(P_0, x) = x^0 = 1$. Now

$$f(P_+; x) = \sum_{k \geq 0} x^k = \frac{1}{1-x} \quad \text{for } |x| < 1,$$

so by Part 3, we must have $f(P_+; x) = 1/(1-x)$. Similarly,

$$f(P_-; x) = \sum_{k \leq 0} x^k = \frac{1}{1-x^{-1}} = -\frac{x}{1-x} \quad \text{for } |x| > 1,$$

so by Part 3 we must have $f(P_-; x) = -x/(1-x)$. By Part 4, $f(P; x) \equiv 0$. Finally, since $[P] = [P_+] + [P_-] - [P_0]$, Part 1 implies that $f(P_+; x) + f(P_-; x) - f(P_0; x) = 0$, which is indeed the case.

EXAMPLE 3.3. Choose $k \leq d$ linearly independent integer vectors $u_1, \ldots, u_k$ and let $K = \mathrm{co}\{u_1, \ldots, u_k\}$ be the cone generated by $u_1, \ldots, u_k$. In other words,

$$K = \{\lambda_1 u_1 + \cdots + \lambda_k u_k : \lambda_i \geq 0 \text{ for } i = 1, \ldots, k\}.$$

Let

$$\Pi = \{\lambda_1 u_1 + \cdots + \lambda_k u_k : 1 > \lambda_i \geq 0 \text{ for } i = 1, \ldots, k\}$$

be the "fundamental parallelepiped" generated by $u_1, \ldots, u_k$. As is well-known (see, for example, [Stanley 1997, Lemma 4.6.7]), for each integer point $m \in K \cap \mathbb{Z}^d$, there is a unique representation

$$m = n + a_1 u_1 + \cdots + a_k u_k,$$

where $n \in \Pi \cap \mathbb{Z}^d$ and $a_1, \ldots, a_k$ are nonnegative integers. Let

$$U_K = \{x \in \mathbb{C}^d : |x^{u_i}| < 1 \quad \text{for } i = 1, \ldots, k\}.$$

Then $U_K \subset \mathbb{C}^d$ is a nonempty open set, and for each $x \in U_k$ we have

$$\sum_{m \in K \cap \mathbb{Z}^d} x^m = \left(\sum_{n \in \Pi \cap \mathbb{Z}^d} x^n\right) \prod_{i=1}^k \frac{1}{1-x^{u_i}},$$

where the series converges absolutely for each $x \in U_K$. Part 3 of Theorem 3.1 implies that we must have

$$f(K; x) = \left(\sum_{n \in \Pi \cap \mathbb{Z}^d} x^n\right) \prod_{i=1}^k \frac{1}{1-x^{u_i}}.$$

An important particular case arises when the fundamental parallelepiped Π contains only one integer point, the origin. This happens if and only if $u_1, \ldots, u_k$ form a basis of the k-dimensional lattice $\mathrm{Span}\{u_1, \ldots, u_k\} \cap \mathbb{Z}^d$. In this case, the cone K is called *unimodular*, and the function $f(K; x)$ has the especially simple form:

$$f(K; x) = \prod_{i=1}^{k} \frac{1}{1 - x^{u_i}}.$$

Theorem 3.1 is very general and powerful, and therefore it has a simple proof. We follow [Pukhlikov and Khovanskii 1992b], with some changes.

SKETCH OF THE PROOF OF THEOREM 3.1. We show that if $P \subset \mathbb{R}^d$ is a rational polyhedron without straight lines (or, equivalently, with a vertex), there exists a nonempty open subset $U_P \subset \mathbb{C}^d$ such that the series

$$\sum_{m \in P \cap \mathbb{Z}^d} x^m$$

converges absolutely for all $x \in U_P$ to a rational function $f(P; x)$. First, suppose that P is a pointed rational cone, that is

$$P = \big\{x : \langle c_i, x \rangle \leq 0 \text{ for } i = 1, \dots, m\big\}, \quad \text{where} \quad c_i \in \mathbb{Z}^d$$

and 0 is the vertex of P. Then P can be represented as the conic hull $P = \operatorname{co}\{u_1, \dots, u_n\}$ of finitely many points $u_i \in \mathbb{Z}^d$, which belong to some open halfspace in $\mathbb{R}^d$. Then

$$U_P = \big\{x \in \mathbb{C}^d : |x^{u_i}| < 1 \text{ for } i = 1, \dots, n\big\}$$

is a nonempty open set, and for each $x \in U_P$, the series converges absolutely to some function $f(P, x)$. We triangulate P into finitely many simple cones K_i and use the inclusion-exclusion principle to express $f(P; x)$ as a linear combination of $f(K_i; x)$. From Example 3.3, we conclude that $f(P; x)$ is a rational function. Suppose now that $P \subset \mathbb{R}^d$ is an arbitrary rational polyhedron without straight lines. Embed $\mathbb{R}^d \subset \mathbb{R}^{d+1}$ by $x \mapsto (x, 1)$ as the flat $\xi_{d+1} = 1$. Let $K = \operatorname{co}\{P\}$ be the conic hull of P in $\mathbb{R}^{d+1}$. Then K is a pointed rational cone in $\mathbb{R}^{d+1}$, so the function $f\big(K; (x, t)\big)$: $x \in \mathbb{C}^d, t \in \mathbb{C}$ is well-defined. Now we observe that

$$f(P; x) = \frac{\partial f\big(K; (x, t)\big)}{\partial t}\bigg|_{t=0}.$$

So far, we have defined the map $\mathfrak{F}$ on rational polyhedra $P \subset \mathbb{R}^d$ without straight lines so that Part 3 of the theorem is satisfied, and so that for every such polyhedron P, the series converges absolutely for all x in some nonempty open set $U_P \subset \mathbb{C}^d$. Part 2 is then satisfied as well, because it is clearly satisfied for $x \in U_P$. Finally, if $P_1, \dots, P_k$ are rational polyhedra without straight lines, Part 1 is satisfied as long as $U_{P_1} \cap U_{P_2} \cap \cdots \cap U_{P_k} \neq \varnothing$. That is, the functions $f(P_i; x)$ converge for all x in some nonempty open set in $\mathbb{C}^d$. Now we want to extend $\mathfrak{F}$ to all rational polyhedra P. Let P be an arbitrary rational polyhedron. We represent P as a union $P = P_1 \cup P_2 \cup \cdots \cup P_k$, where the P_i are polyhedra without straight lines. For $I \subset \{1, \dots, k\}$, let $P_I = \bigcap_{i \in I} P_i$. Then the P_I are rational polyhedra without straight lines. *Define $f(P; x)$ as a linear combination of $f(P_I; x)$ via the inclusion-exclusion principle.* Proving that $f(P; x)$ is well-defined boils down to proving that we get a consistent definition of $f(P; x)$ if P itself does not contain straight lines. This is true since there is a nonempty open set $U_P \subset \mathbb{C}^d$, where all the series defining $f(P; x)$ and $f(P_I; x)$ converge absolutely. It then follows that the properties (1)–(3) are satisfied, and it remains to check (4). If P contains a straight line, then for some nonzero $m \in \mathbb{Z}^d$ we have $P + m = P$. Therefore, $f(P; x) = x^m f(P; x)$ and $f(P; x)$ must be identically zero. $\qquad\square$

The map $\mathfrak{F}$ can be extended to a valuation $\mathfrak{F} : \mathcal{P}(\mathbb{Q}^d) \to \mathbb{Q}(x)$, sending every function in $\mathcal{P}_d$ to a rational function in d complex variables such that

$$\mathfrak{F}(f) = \sum_{m \in \mathbb{Z}^d} f(m) x^m,$$

provided the series converges absolutely. Furthermore, if $g(x) = f(x - m)$ for some $m \in \mathbb{Z}^d$, then

$$\mathfrak{F}(g) = x^m \mathfrak{F}(f).$$

Finally, the kernel of this valuation contains the subspace spanned by the indicator functions of rational polyhedra with straight lines.

We are going to present in Theorem 3.5 a very interesting and important corollary to Theorem 3.1. It was proved by M. Brion *before* Theorem 3.1 and its first proof used algebraic geometry [Brion 1988]. Elementary proofs were published in [Lawrence 1991; Pukhlikov and Khovanskii 1992b; Barvinok 1993] and elsewhere. First, we need a definition.

DEFINITION 3.4. Let $P \subset \mathbb{R}^d$ be a polyhedron and let $v \in P$ be a vertex of P. The *supporting* or *tangent* cone $\mathrm{cone}(P, v)$ of P at v is defined as follows: suppose that

$$P = \left\{ x \in \mathbb{R}^d : \langle c_i, x \rangle \le \beta_i \text{ for } i = 1, \ldots, m \right\}$$

is a representation of P as the set of solutions of a system of linear inequalities, where $c_i \in \mathbb{R}^d$ and $\beta_i \in \mathbb{R}$. Let $I_v = \{i : \langle c_i, v \rangle = \beta_i\}$ be the set of constraints that are active on v. Then

$$\mathrm{cone}(P, v) = \left\{ x \in \mathbb{R}^d : \langle c_i, x \rangle \le \beta_i \text{ for } i \in I_v \right\}.$$

Of course, the cone $\mathrm{cone}(P, v)$ does not depend on a particular system of inequalities chosen to represent P. If P is a rational polyhedron then $\mathrm{cone}(P, v)$ is a rational pointed cone with vertex v. More generally, if $F \subset P$ is a face, we define

$$\mathrm{cone}(P, F) = \left\{ x \in \mathbb{R}^d : \langle c_i, x \rangle \le \beta_i \text{ for } i \in I_F \right\},$$

where I_F is the set of inequalities that are active on F. If $\dim P = d$ and $\dim F = k$, the apex of $\mathrm{cone}(P, F)$ is a k-dimensional affine subspace in $\mathbb{R}^d$.

THEOREM 3.5. *Let P be a rational polyhedron. Then*

$$f(P; x) = \sum_{v \in \text{Vert}(P)} f(\text{cone}(P, v); x),$$

where the sum is taken over all vertices v of P.

EXAMPLE 3.6. Suppose that $P = \{x : 0 \le x \le 1\} \subset \mathbb{R}^1$ is an interval. Then $f(P, x) = x^0 + x^1 = 1 + x$. The polyhedron P has two vertices, 0 and 1, with supporting cones $\text{cone}(P, 0) = [0, +\infty)$ and $\text{cone}(P, 1) = (-\infty, 1]$, respectively. Furthermore,

$$f(\text{cone}(P, 0); x) = \sum_{m=0}^{+\infty} x^m = \frac{1}{1-x}, \quad \text{and}$$

$$f(\text{cone}(P, 1); x) = \sum_{m=-\infty}^{1} x^m = \frac{x}{1-x^{-1}} = -\frac{x^2}{1-x}.$$

We observe that indeed

$$f(P; x) = 1 + x = \frac{1}{1-x} - \frac{x^2}{1-x} = f(\text{cone}(P, 0); x) + f(\text{cone}(P, 1); x).$$

SKETCH OF PROOF OF THEOREM 3.5. Let $\mathcal{L} \subset \mathcal{P}(\mathbb{R}^d)$ be the subspace in the polyhedral algebra spanned by the indicator functions $[P]$ of polyhedra that contain straight lines. The theorem follows from Theorem 3.1 and an identity in the polyhedral algebra $\mathcal{P}(\mathbb{R}^d)$: for any polyhedron $P \subset \mathbb{R}^d$

$$[P] \equiv \sum_{v \in \text{Vert}(P)} [\text{cone}(P, v)] \pmod{\mathcal{L}}.$$

First, we demonstrate this identity in the case where $P = \Delta$ is a d-dimensional simplex. We represent Δ as the intersection of $d + 1$ closed halfspaces $H_1^+, \ldots, H_{d+1}^+$ bounded by flats $H_1, \ldots, H_{d+1}$:

$$\Delta = H_1^+ \cap \cdots \cap H_{d+1}^+.$$

Each tangent cone $\text{cone}(\Delta, v)$ is the intersection of some d halfspaces $H_i^+ : i \in I_v$. Let $H_i^- = \mathbb{R}^d \setminus H_i^+$ be the complementary open halfspaces and let $K_v^- = \bigcap_{i \in I_v} H_i^-$ be the open cone "vertical" to the supporting cone $\text{cone}(\Delta, v)$. Then

$$\mathbb{R}^d \setminus \Delta = \bigcup_{i=1}^{d+1} H_i^-.$$

We rewrite $\bigcup_{i=1}^{d+1} H_i^-$ using the inclusion-exclusion formula. Since the intersection of fewer than d halfspaces contains a straight line, we can write

$$[\mathbb{R}^d \setminus \Delta] \equiv (-1)^{d+1} \sum_{v \in \text{Vert}(\Delta)} [K_v^-] \pmod{\mathcal{L}}.$$

Therefore,

$$[\Delta] \equiv (-1)^d \sum_{v \in \mathrm{Vert}(\Delta)} [K_v^-] \quad (\mathrm{mod}\ \mathcal{L}).$$

It remains to show that

$$[\mathrm{cone}(\Delta, v)] \equiv (-1)^d [K_v^-] \quad (\mathrm{mod}\ \mathcal{L}).$$

The last assertion follows by comparing $\mathrm{cone}(\Delta, v)$ and K_v^- via a chain of intermediate cones $K_v^\varepsilon = \bigcap_{i \in I_v} H_i^{\varepsilon_i}$, where $\varepsilon_i \in \{-, +\}$ and $\varepsilon = (\varepsilon_1, \ldots, \varepsilon_d)$. Note that $\mathrm{cone}(\Delta, v) = K_v^{+, \cdots, +}$ and $K_v^- = K_v^{-, \cdots, -}$. Once we get

$$[\Delta] \equiv \sum_{v \in \mathrm{Vert}(\Delta)} [\mathrm{cone}(\Delta, v)] \quad (\mathrm{mod}\ \mathcal{L})$$

for any simplex Δ, using triangulations we can show that

$$[P] \equiv \sum_{v \in \mathrm{Vert}(P)} [\mathrm{cone}(P, v)] \quad (\mathrm{mod}\ \mathcal{L})$$

for any polytope P.

Suppose that $P \subset \mathbb{R}^d$ is an arbitrary polyhedron. As is well known, P can be represented as the Minkowski sum $P = Q + K + L$, where Q is a polytope, K is a cone without straight lines, and L is a subspace in $\mathbb{R}^d$. If $L \neq \{0\}$, then P has no vertices, and $f(P; x) = 0$, so the statement of the theorem is true. If $L = \{0\}$, we may write

$$[Q] \equiv \sum_{v \in \mathrm{Vert}(Q)} [\mathrm{cone}(Q, v)] \quad (\mathrm{mod}\ \mathcal{L}),$$

and by Corollary 2.6,

$$[P] \equiv [Q + K] \equiv \sum_{v \in \mathrm{Vert}(Q)} [\mathrm{cone}(Q, v) + K] \quad (\mathrm{mod}\ \mathcal{L}).$$

Each vertex v of P is a vertex of Q, and $\mathrm{cone}(P, v) = \mathrm{cone}(Q, v) + K$. Furthermore, a vertex v of Q is a vertex of P if and only if the cone $\mathrm{cone}(Q, v) + K$ does not contain straight lines. The proof now follows. $\qquad \square$

Following [Brion and Vergne 1997c], instead of $\mathfrak{F} : \mathcal{P}(\mathbb{Q}^d) \to \mathbb{Q}(x)$, one can consider a valuation $\mathfrak{F}' : \mathcal{P}(\mathbb{Q}^d) \to \mathbb{Q}((x))$ with the values in the space of formal Laurent power series in $x = (x_1, \ldots, x_d)$ with rational coefficients:

$$\mathfrak{F}'(f) = \sum_{m \in \mathbb{Z}^d} f(m) x^m.$$

This leads to essentially the same theory: the space $\mathbb{Q}((x))$ has the natural structure of a module over the ring of polynomials $\mathbb{Q}[x_1, \ldots, x_d]$. A formal power series f is identified with a rational function $g(x)/h(x)$, where $h(x) = (1 - x^{a_1}) \cdots (1 - x^{a_m})$ provided $hf = g$. For example, the series $f = \sum_{k=-\infty}^{+\infty} x^k$ in $\mathbb{Q}((x))$ is identified with zero, since $f(x)(1 - x) = 0$. In this approach, the valuation $\mathfrak{F}'$ can be viewed as an *algebra homomorphism*, provided the ring structure on $\mathbb{Q}((\boldsymbol{x}))$ is given by the Hadamard product $\star$:

$$f \star g = \sum_{m \in \mathbb{Z}^d} (a_m b_m) \boldsymbol{x}^m, \quad \text{where } f = \sum_{m \in \mathbb{Z}^d} a_m \boldsymbol{x}^m \text{ and } g = \sum_{m \in \mathbb{Z}^d} b_m \boldsymbol{x}^m.$$

4. Complexity of Generating Functions

Let $P \subset \mathbb{R}^d$ be a rational polyhedron. In this section, we show that if the dimension d is fixed, the generating function $f(P; \boldsymbol{x})$ has a representation whose complexity is bounded by a polynomial in the input size of P. If P does not contain straight lines, then $f(P; \boldsymbol{x})$, "in principle", encodes all the information about the set $P \cap \mathbb{Z}^d$ of integer points in P. We claim that this information can be encoded in a compact way. In particular, our results imply that in any fixed dimension there is a polynomial time algorithm for counting integer points in a given rational polytope. First, we find a formula for $f(K; \boldsymbol{x})$, where K is a rational unimodular cone (with vertex not necessarily at the origin).

LEMMA 4.1. *Let*

$$K = \{x \in \mathbb{R}^d : \langle c_i, x \rangle \le \beta_i \text{ for } i = 1, \ldots, d\},$$

where $c_1, \ldots, c_d$ *is a basis of the integer lattice* $\mathbb{Z}^d$ *and* $\beta_i \in \mathbb{Q}$ *are rational numbers. Let* $u_1, \ldots, u_d$ *be the (negative) dual basis of* $\mathbb{Z}^d$:

$$\langle u_i, c_j \rangle = \begin{cases} -1 & \text{if } i = j, \\ 0 & \text{if } i \ne j. \end{cases}$$

Then

$$f(K; \boldsymbol{x}) = \boldsymbol{x}^v \prod_{i=1}^{d} \frac{1}{1 - \boldsymbol{x}^{u_i}}, \quad \text{where } v = -\sum_{i=1}^{d} \lfloor \beta_i \rfloor u_i.$$

PROOF. Let $K_0 = \text{co}\{u_1, \ldots, u_d\}$. This is a unimodular cone with vertex at the origin, and by Example 3.3, we have

$$f(K_0; \boldsymbol{x}) = \prod_{i=1}^{d} \frac{1}{1 - \boldsymbol{x}^{u_i}}.$$

A point $x \in \mathbb{R}^d$ is an integer point in K if and only if for $i = 1, \ldots, d$, the value $\langle c_i, x \rangle$ is an integer which does not exceed β_i, and hence does not exceed $\lfloor \beta_i \rfloor$. In other words, the set of integer points in K and in the translation $K_0 + v$ coincide. Therefore, $f(K, \boldsymbol{x}) = f(K_0 + v, \boldsymbol{x}) = \boldsymbol{x}^v f(K_0, x)$ (by Part 2 of Theorem 3.1), and the proof is complete.   □

REMARK. If we fix $c_1, \ldots, c_d$ and allow $\beta_1, \ldots, \beta_d$ to vary, the denominator of the rational function $f(K, \boldsymbol{x})$ does not change.

The following result is proved in [Barvinok 1994b].

THEOREM 4.2. *Fix d. There exists a polynomial time algorithm, which, given a rational polyhedral cone $K \subset \mathbb{R}^d$, computes unimodular cones $K_i : i \in I$ and numbers $\varepsilon_i \in \{-1, 1\}$ such that*

$$[K] = \sum_{i \in I} \varepsilon_i [K_i].$$

In particular, the number $|I|$ of cones in the decomposition is bounded by a polynomial in the input size of K.

SKETCH OF PROOF. Using triangulation, we reduce the problem to the case of a simple cone $K = \mathrm{co}\{u_1, \dots, u_d\}$, where $u_1, \dots, u_d \subset \mathbb{Z}^d$ are linearly independent integer points (we leave aside lower-dimensional cones, which can be treated in a similar way). We introduce the *index* $\mathrm{ind}(K)$ which measures how far is K from being unimodular: $\mathrm{ind}(K) = |u_1 \wedge \dots \wedge u_d|$ is the volume of the parallelepiped spanned by the generators $u_1, \dots, u_d$. One can show that $\mathrm{ind}(K)$ is the number of integer points in the fundamental parallelepiped Π of K (cf. Example 3.3). Thus the index of a cone is a positive integer which equals 1 if and only if K is unimodular. One can show that $\log \mathrm{ind}(K)$ is bounded by a polynomial in the input size of K. We are going to iterate a procedure which replaces $[K]$ by a linear combination of $[K_j]$, where K_j are rational cones with smaller indices. Consider a parallelepiped

$$B = \{\alpha_1 u_1 + \dots + \alpha_d u_d : |\alpha_j| \leq \big(\mathrm{ind}(K)\big)^{-1/d} \text{ for } j = 1, \dots, d\}.$$

We observe that B is centrally symmetric and has volume 2^d. Therefore, by the Minkowski convex body theorem (see, for example, [Lagarias 1995]), there is a nonzero integer point $w \in B$ (such a vector can be constructed efficiently using, for example, integer programming in dimension d: see [Schrijver 1986]). For $j = 1, \dots, d$, let

$$K_j = \mathrm{co}\{u_1, \dots, u_{j-1}, w, u_{j+1}, \dots, u_d\}.$$

If $w = \alpha_1 u_1 + \dots + \alpha_d u_d$, then

$$\begin{aligned}
\mathrm{ind}(K_j) &= |u_1 \wedge \dots \wedge u_{j-1} \wedge w \wedge u_{j+1} \wedge \dots \wedge u_d| \\
&= |\alpha_j| |u_1 \wedge \dots \wedge u_{j-1} \wedge u_j \wedge u_{j+1} \wedge \dots \wedge u_d| \\
&= |\alpha_j| \mathrm{ind}(K) \leq \big(\mathrm{ind}(K)\big)^{(d-1)/d}.
\end{aligned}$$

Furthermore, there is a decomposition

$$[K] = \sum_{j \in J} \varepsilon_j [K_j] + \sum_F \varepsilon_F [F],$$

where F ranges over lower-dimensional faces of K_j, and $\varepsilon_j, \varepsilon_F \in \{-1, 1\}$. If we iterate this procedure, we observe that the indices of the cones involved decrease doubly exponentially, whereas the number of cones increases only exponentially. Therefore, iterating the procedure $O(\log \log \mathrm{ind}(K))$ times, we end up with a decomposition of K into a linear combination of unimodular cones, with the number of cones bounded by a polynomial in $\log(\operatorname{ind}(K))$. □

REMARK (TRIANGULATIONS ARE NOT ENOUGH). As is well-known (see [Fulton 1993, Section 2.6], for example), every rational cone can be triangulated into unimodular cones. However, to ensure polynomial time complexity, it is important to use signed decompositions; triangulations alone are not enough, as the following simple example shows.

Let $K = \operatorname{co}\{u_1, u_2\} \subset \mathbb{R}^2$ be the planar cone spanned by vectors $u_1 = (1,0)$ and $u_2 = (1,n)$, where n is a positive integer. To triangulate K into unimodular cones, we have to draw a line through each point $(1,i)$, $1 \leq i < n$. Thus the number of cones will grow linearly in n, that is, exponentially in the size $O(\log n)$ of the input. However, to write a signed decomposition we need only three unimodular cones: let $w = (0,1)$, $K_1 = \operatorname{co}\{w, u_1\}$, $K_2 = \operatorname{co}\{w, u_2\}$, and $K_3 = \operatorname{co}\{w\}$. Then K_1, K_2 and K_3 are unimodular cones, and $[K] = [K_1] - [K_2] + [K_3]$.

REMARK 4.3 (THE DUALITY TRICK). Once we have a decomposition

$$[K] = \sum_{i \in I} \varepsilon_i [K_i],$$

where the K_i are unimodular cones, we can write

$$f(K; \boldsymbol{x}) = \sum_{i \in I} \varepsilon_i f(K_i; \boldsymbol{x}) \qquad (4.3.1)$$

(see Theorem 3.1). Since there are explicit formulas for $f(K_i; \boldsymbol{x})$ (see Example 3.3), we get an explicit formula for $f(K; \boldsymbol{x})$. The complexity of such a formula, as asserted by Theorem 4.2, is bounded by a polynomial in the input size of K. There is a trick which allows us to obtain a decomposition of the type (4.3.1), where all K_i are unimodular *and* d-dimensional. The idea can already be found in the seminal paper [Brion 1988]. Sometimes this trick significantly reduces the computational complexity of the formula. Namely, let K^* be the dual cone to K. We apply the iterative procedure of Theorem 4.2 to K^*, discarding lower-dimensional cones on every step. Thus we get a decomposition

$$[K^*] = \sum_{i \in I} \varepsilon_i [K_i] \quad \text{modulo lower-dimensional cones,}$$

where the K_i are d-dimensional unimodular cones. The dual of a lower-dimensional cone contains a straight line. Therefore, by Corollary 2.8 we get that

$$[K] = \sum_{i \in I} \varepsilon_i [K_i^*] \quad \text{modulo cones with straight lines.}$$

From Theorem 3.1, we get that

$$f(K; \boldsymbol{x}) = \sum_{i \in I} \varepsilon_i f(K_i^*; \boldsymbol{x}).$$

Note that the K_i^* are unimodular, provided that the K_i are unimodular. If the cone K is defined by linear inequalities in $\mathbb{R}^d$, then the complexity of the algorithm and the resulting formula is $\mathcal{L}^{O(d)}$, where $\mathcal{L}$ is the input size of K. The complexity of the algorithm of Theorem 4.2, as stated, can be as bad as $\mathcal{L}^{\Omega(d^2)}$ if K is given as a conic hull of integer vectors in $\mathbb{Z}^d$, and $\mathcal{L}^{\Omega(d^3)}$ if K is given by a set of linear inequalities. The savings in computational complexity comes from the fact that if we iterate the procedure of Theorem 4.2 as stated, the number of cones in every step grows by a factor of 2^d. If we discard lower-dimensional cones, the number of cones in every step grows by a factor of d only.

We are ready to state the main result of this section. We not only compute the expression for $f(P; x)$, but we also describe how it changes when the facets of P are moved parallel to themselves so that the combinatorial structure of P does not change.

THEOREM 4.4. *Fix d. There exists a polynomial time algorithm, which, for a given rational polyhedron $P \subset \mathbb{R}^d$,*

$$P = \{x \in \mathbb{R}^d : \langle c_i, x \rangle \leq \beta_i \text{ for } i = 1, \dots, m\}, \quad \text{where } c_i \in \mathbb{Z}^d \text{ and } \beta_i \in \mathbb{Q},$$

computes the generating function $f(P; x) = \sum_{m \in P \cap \mathbb{Z}^d} x^m$ in the form

$$f(P; x) = \sum_{i \in I} \varepsilon_i \frac{x^{a_i}}{(1 - x^{b_{i1}}) \cdots (1 - x^{b_{id}})}, \tag{4.4.1}$$

where $\varepsilon_i \in \{-1, 1\}$, $a_i \in \mathbb{Z}^d$, and $b_{i1}, \dots, b_{id}$ is a basis of $\mathbb{Z}^d$ for each i.

Suppose that the vectors $c_i : i = 1, \dots, m$ are fixed and the β_i vary in such a way that the combinatorial structure of the polyhedron $P = P(\beta) : \beta = (\beta_1, \dots, \beta_m)$ stays the same. Then the exponents b_{ij} in the denominator of each fraction remain the same, whereas the exponents $a_i = a_i(\beta)$ in the numerator change with $\beta \in \mathbb{Q}^m$ as

$$a_i = \sum_{j=1}^{d} \lfloor l_{ij}(\beta) \rfloor b_{ij}, \tag{4.4.2}$$

where the $l_{ij} : \mathbb{Q}^m \to \mathbb{Q}$ are linear functions. If β is such that $P(\beta)$ is an integer polytope, then $l_{ij}(\beta) \in \mathbb{Z}$ for each pair i, j.

The computational complexity of the algorithm for finding (4.4.1) and (4.4.2) is $\mathcal{L}^{O(d)}$, where $\mathcal{L}$ is the input size of P. In particular, the number $|I|$ of terms in (4.4.1) is $\mathcal{L}^{O(d)}$.

PROOF. Let $\mathrm{Vert}(P)$ be the set of vertices of P. For $v \in \mathrm{Vert}(P)$, let $I_v = \{i \in I : \langle c_i, v \rangle = \beta_i\}$ be the set of those inequalities that are active at v, and let $N(P, v) = \mathrm{co}\{c_i : i \in I_v\}$ be the conic hull of the normals of the facets containing v. Then for the tangent cone $\mathrm{cone}(P, v)$, we have

$$\mathrm{cone}(P, v) = -N^*(P, v) + v.$$

Using Theorem 4.2 and Remark 4.3, we construct d-dimensional unimodular cones $K(v,j) : j \in J_v$ and numbers $\varepsilon(v,j) \in \{-1,1\}$ such that

$$[N(P,v)] = \sum_{j \in J_v} \varepsilon(v,j)[K(v,j)] \quad \text{modulo lower-dimensional cones.}$$

Therefore, for the supporting cone of P at v, we have

$$[\text{cone}(P,v)] = [v - N^*(P,v)]$$
$$= \sum_{j \in J_v} \varepsilon(v,j)[v - K^*(v,j)] \quad \text{modulo cones with straight lines.}$$

Thus,

$$f\big(\text{cone}(P,v); x\big) = \sum_{j \in J_v} \varepsilon(v,j) f\big(v - K^*(v,j); x\big)$$

by Part 4 of Theorem 3.1. Now, each cone $v - K^*(v,j)$ is a unimodular cone with vertex v. If the c_i are fixed, as long as the combinatorial structure of P does not change, each vertex $v = v(\beta)$ of P changes linearly with $\beta \in \mathbb{R}^m$. Therefore, we can apply Lemma 4.1 to find $f(v - K^*(v,j); x)$. Applying Theorem 3.5, we get

$$f(P;x) = \sum_{v \in \text{Vert}(P)} f\big(\text{cone}(P,v); x\big),$$

and the proof is complete. $\qquad\qquad\qquad\qquad\qquad\qquad\qquad\qquad\square$

If $P \subset \mathbb{R}^d$ is a polytope, then

$$f(P;x) = \sum_{m \in P \cap \mathbb{Z}^d} x^m$$

is a polynomial, whose expression as the sum of monomials can be very long. Theorem 4.4 asserts that if P is a rational polytope, then this polynomial can be written as a short rational function. A typical example is provided by a polynomial $\sum_{k=1}^{n} x^k$, containing n monomials, which can be written as the rational function $(1 - x^{n+1})/(1 - x)$, which one needs only $O(\log n)$ bits to write.

Theorem 4.4 implies that if the dimension d is fixed, the valuation

$$\mathfrak{F} : \mathcal{P}(\mathbb{Q}^d) \to \mathbb{Q}(x)$$

is computable in polynomial time.

5. Efficient Counting of Lattice Points

If $P \subset \mathbb{R}^d$ is a rational polytope, the generating function

$$f(P;x) = \sum_{m \in P \cap \mathbb{Z}^d} x^m$$

is a polynomial, and its value at $x = (1, \ldots, 1)$ is the number of integer points $|P \cap \mathbb{Z}^d|$ in P. Note that if we compute $f(P;x)$ as a short rational function, as provided by Theorem 4.4, then the point $x = (1, \ldots, 1)$ is a pole of each fraction in the representation (4.4.1). This can be can be handled by taking an appropriate residue or by computing the value of $f(P; x)$ at a point x close to $1 = (1, \ldots, 1)$ and rounding the answer to the nearest integer as in [Dyer and Kannan 1997]. We use the "residue" approach, suggested in [Brion 1988] and also used in [Barvinok 1994b], where the first polynomial time algorithm for counting integer points in a rational polytope was constructed.

DEFINITION 5.1. Consider the function

$$F(\tau; \xi_1, \ldots, \xi_d) = \prod_{i=1}^{d} \frac{\tau \xi_i}{1 - \exp(-\tau \xi_i)}$$

in $d + 1$ complex variables τ and $\xi_1, \ldots, \xi_d$. It is easy to see that F is analytic in a neighborhood of the origin $\tau = \xi_1 = \ldots = \xi_d = 0$ and therefore there exists an expansion

$$F(\tau; \xi_1, \ldots, \xi_d) = \sum_{k=0}^{+\infty} \tau^k \operatorname{td}_k(\xi_1, \ldots, \xi_d),$$

where $\operatorname{td}_k(\xi_1, \ldots, \xi_d)$ is a homogeneous polynomial of degree k, called the k-th *Todd polynomial* in $\xi_1, \ldots, \xi_d$. It is easy to check that td_k is a symmetric polynomial with rational coefficients; see [Hirzebruch 1966] or [Fulton 1993, Section 5.3]).

ALGORITHM 5.2 (A POLYNOMIAL TIME ALGORITHM FOR COUNTING INTEGER POINTS IN RATIONAL POLYTOPES WHEN THE DIMENSION IS FIXED). Suppose the dimension d is fixed and $P \subset \mathbb{R}^d$ is a rational polytope. We use Theorem 4.4 to compute

$$f(P; x) = \sum_{i \in I} \varepsilon_i \frac{x^{a_i}}{(1 - x^{b_{i1}}) \cdots (1 - x^{b_{id}})}.$$

We construct a vector $l \in \mathbb{Z}^d$ such that $\langle l, b_{ij} \rangle \neq 0$ for each i and j. To do this efficiently, we consider the "moment curve" $g(\tau) = (1, \tau, \tau^2, \ldots, \tau^{d-1}) \in \mathbb{R}^d$. For each b_{ij}, the function $\langle g(\tau), b_{ij} \rangle : \tau \in \mathbb{R}$ is a nonzero polynomial of degree at most $d - 1$ in τ, and thus this function has at most $d - 1$ zeros. Therefore, we can select l from the set of integer vectors $\{g(0), g(1), \ldots, g(m)\}$, where $m = d(d-1)|I| + 1$.

Let $l = (\lambda_1, \ldots, \lambda_d)$. For $\tau > 0$, let $x_\tau = (\exp(\tau \lambda_1), \ldots, \exp(\tau \lambda_d))$ and let $\xi_{ij} = \langle l, b_{ij} \rangle$ and $\eta_i = \langle l, a_i \rangle$. Then

$$|P \cap \mathbb{Z}^d| = \lim_{\tau \to 0} f(P; x_\tau) = \lim_{\tau \to 0} \sum_{i \in I} \varepsilon_i \frac{\exp(\tau \langle l, a_i \rangle)}{(1 - \exp(\tau \langle l, b_{i1} \rangle)) \cdots (1 - \exp(\tau \langle l, b_{id} \rangle))}$$

$$= \lim_{\tau \to 0} \frac{1}{\tau^d} \sum_{i \in I} \varepsilon_i \frac{\tau^d \exp(\tau \eta_i)}{(1 - \exp(\tau \xi_{i1})) \cdots (1 - \exp(\tau \xi_{id}))}.$$

Now, each fraction

$$h_i(\tau) = \frac{\tau^d \exp(\tau \eta_i)}{(1 - \exp(\tau \xi_{i1})) \cdots (1 - \exp(\tau \xi_{id}))}$$

is a holomorphic function in a neighborhood of $\tau = 0$ and the d-th coefficient of its Taylor series is

$$\frac{1}{\xi_{i1} \cdots \xi_{id}} \sum_{k=0}^{d} \frac{\eta_i^k}{k!} \, \mathrm{td}_{d-k}(\xi_{i1}, \dots, \xi_{id}).$$

Finally, we get an efficient formula for the number of integer points in P:

$$|P \cap \mathbb{Z}^d| = \sum_{i \in I} \frac{\varepsilon_i}{\xi_{i1} \cdots \xi_{id}} \sum_{k=0}^{d} \frac{\eta_i^k}{k!} \, \mathrm{td}_{d-k}(\xi_{i1}, \dots, \xi_{id}). \qquad (5.2.1)$$

The construction of Algorithm 5.2 allows us to find out how the number of integer points in a rational polytope changes when the facets of the polytope are moved parallel to themselves, as long as the combinatorial type of the polytope does not change.

THEOREM 5.3.   *Fix vectors* $c_1, \dots, c_m \in \mathbb{Z}^d$ *such that for any* $\beta \in \mathbb{Q}^m$: $\beta = (\beta_1, \dots, \beta_m)$, *the set*

$$P(\beta) = \left\{ x \in \mathbb{R}^d : \langle c_i, x \rangle \leq \beta_i \text{ for } i = 1, \dots, m \right\}$$

is a rational polytope in $\mathbb{R}^d$, *if nonempty. Let* $\mathcal{B} \subset \mathbb{Q}^m$ *be a set such that for any* $\beta_1, \beta_2 \in \mathcal{B}$, *the polytopes* $P(\beta_1)$ *and* $P(\beta_2)$ *have the same combinatorial type. Then there exist linear functions* $l_{ij} : \mathbb{Q}^m \to \mathbb{Q}$ *and rational numbers* α_{ik} *and* γ_{ij} *such that for any* $\beta \in \mathcal{B}$,

$$|P(\beta) \cap \mathbb{Z}^d| = \sum_{i \in I} \sum_{k=1}^{d} \alpha_{ik} \left(\sum_{j=1}^{d} \gamma_{ij} \lfloor l_{ij}(\beta) \rfloor \right)^k. \qquad (5.3.1)$$

If for some $\beta \in \mathcal{B}$, $P(\beta)$ *is an integer polytope, then* $l_{ij}(\beta) \in \mathbb{Z}$. *Furthermore, for any fixed dimension* d, *there is a polynomial time algorithm for computing the formula* (5.3.1).

PROOF.   The theorem follows from Theorem 4.4 and Algorithm 5.2. Note that in (5.2.1), the numbers ξ_{ij} and ε_i do not change as long as $\beta \in \mathcal{B}$. Hence we let

$$\alpha_{ik} = \frac{\varepsilon_i}{k! \, \xi_{i1} \cdots \xi_{id}} \, \mathrm{td}_{d-k}(\xi_{i1}, \dots, \xi_{id}).$$

For $\beta \in \mathcal{B}$, we have $\eta_i = \eta_i(\beta) = \langle l, a_i(\beta) \rangle$ for some fixed $l \in \mathbb{Z}^d$, where by Theorem 4.4,

$$a_i(\beta) = \sum_{j=1}^{d} \lfloor l_{ij}(\beta) \rfloor b_{ij}$$

for some fixed $b_{ij} \in \mathbb{Z}^d$. Hence we let $\gamma_{ij} = \langle l, b_{ij} \rangle$. $\square$

REMARK (RELATION TO INTEGER PROGRAMMING ALGORITHMS). Integer programming is concerned with optimizing a given linear function on the set of integer points in a given rational polyhedron in $\mathbb{R}^d$. By using a standard trick of dichotomy, one can reduce an integer programming problem to a sequence of feasibility problems: given a rational polyhedron $P \subset \mathbb{R}^d$, decide whether P contains an integer point, and if so, find such a point. Integer Programming is difficult (NP-hard) in general, but it admits a polynomial time algorithm if the dimension d is fixed and not a part of the input. The first integer programming algorithm having polynomial time complexity in fixed dimension was constructed by H. W. Lenstra [1983]; see [Grötschel et al. 1993; Lovász 1986; Schrijver 1986] for a survey and subsequent improvements. Of course, if we can count integer points in P, we can decide whether $P \cap \mathbb{Z}^d = \varnothing$. The catch is that in one of the crucial ingredients of Algorithm 5.2, namely in the proof of Theorem 4.2, we refer to integer programming in fixed dimension. Therefore, employing Algorithm 5.2 to solve an integer programming problem may seem to result in a vicious circle. However, as M. Dyer and R. Kannan [1997] show, one can avoid the dependence on Lenstra's algorithm in Theorem 4.2 by using an appropriate lattice reduction algorithm. Hence Algorithm 5.2 gives rise to a new linear programming algorithm, whose complexity is polynomial time when the dimension d is fixed and which uses lattice reduction (cf. [Lenstra et al. 1982]), but does not use "rounding" of a given convex body in $\mathbb{R}^d$, which is the second main ingredient in Lenstra's algorithm and its subsequent improvements (see [Grötschel et al. 1993; Lovász 1986]). This rounding seems to be quite time-consuming and it would be interesting to find out if Algorithm 5.2 can compete with the known integer programming algorithms. On the other hand, Lenstra's algorithm can be naturally extended to "convex integer" problems, whereas Algorithm 5.2 heavily uses the polyhedral structure.

We conclude this section with a description of a very general "trick" which sometimes allows one to count lattice points efficiently even if the dimension is large.

REMARK (CHANGING THE LATTICE). Suppose that $\Lambda \subset \mathbb{Z}^d$ is a sublattice of a finite index $|\mathbb{Z}^d : \Lambda|$. Let Λ^* be the dual lattice. Therefore, $\mathbb{Z}^d \subset \Lambda^* \subset \mathbb{Q}^d$ and $|\Lambda^* : \mathbb{Z}^d| = |\mathbb{Z}^d : \Lambda|$.

For $x = (x_1, \ldots, x_d) \in \mathbb{C}^d$ and any vector $l = (\lambda_1, \ldots, \lambda_d) \in \mathbb{Q}^d$, define

$$e^{2\pi i l} x = \big(\exp(2\pi i \lambda_1) x_1, \ldots, \exp(2\pi i \lambda_d) x_d\big).$$

Then the value of

$$\big(e^{2\pi i l} x\big)^m = \exp\big(2\pi i \langle l, m \rangle\big) x^m$$

depends only on the coset $\Lambda^* : \mathbb{Z}^d$ represented by l.

Furthermore, we have

$$\frac{1}{|\Lambda^* : \mathbb{Z}^d|} \sum_{l \in \Lambda^* : \mathbb{Z}^d} \left(e^{2\pi i l} \boldsymbol{x}\right)^m = \begin{cases} \boldsymbol{x}^m & \text{if } m \in \Lambda, \\ 0 & \text{otherwise.} \end{cases}$$

Therefore, if $P \subset \mathbb{R}^d$ is a rational polytope with a known function $f(P; \boldsymbol{x})$, one can compute the modified function

$$f(P, \Lambda; \boldsymbol{x}) = \sum_{m \in P \cap \Lambda} \boldsymbol{x}^m$$

by the formula

$$f(P, \Lambda; \boldsymbol{x}) = \frac{1}{|\Lambda^* : \mathbb{Z}^d|} \sum_{l \in \Lambda^* : \mathbb{Z}^d} f(P; e^{2\pi i l} \boldsymbol{x}).$$

Therefore, if the function $f(P; \boldsymbol{x})$ is known and the index $|\mathbb{Z}^d : \Lambda|$ is not very large, one can compute $f(P, \Lambda; \boldsymbol{x})$ efficiently.

The computational complexity of the algorithm above is exponential in the input size of Λ even when the dimension d is fixed, but for lattices Λ of small index $|\mathbb{Z}^d : \Lambda|$, it may appear computationally useful. Suppose, for example, that $P = [0, n_1] \times [0, n_2] \times \ldots \times [0, n_d]$ is the d-dimensional integer "box". Then there is an explicit formula for the generating function $f(P; \boldsymbol{x})$:

$$f(P; \boldsymbol{x}) = \prod_{i=1}^{d} \frac{1 - x_i^{n_i+1}}{1 - x_i}.$$

Thus there is an algorithm for counting points in $P \cap \Lambda$ whose complexity is polynomial in the index $|\mathbb{Z}^d : \Lambda|$ (d need not be fixed).

This approach was used in [Brion and Vergne 1997c] and in [Barvinok 1993] in a particular situation.

This construction can be dualized. Suppose that $\Lambda \subset \mathbb{Q}^d$ is a lattice such that $\mathbb{Z}^d \subset \Lambda$. Now $m \in \Lambda$ no longer needs to be integer vector, so in order to resolve the ambiguity of $\boldsymbol{x}^m$, it is convenient to make the substitution $x_j = e^{\lambda_j}$, $j = 1, \ldots, d$ and hence interpret the monomial $\boldsymbol{x}^m$ as the function $\mathbb{C}^d \to \mathbb{C}$, $l \mapsto \exp(\langle l, m \rangle)$, where $l = (\lambda_1, \ldots, \lambda_d)$.

Then

$$f(P, \Lambda; \boldsymbol{x}) = \sum_{m \in P \cap \Lambda} \boldsymbol{x}^m = \sum_{l \in \Lambda : \mathbb{Z}^d} \boldsymbol{x}^l f(P; \boldsymbol{x}).$$

(Clearly, the sum does not depend on a particular choice of coset representatives l). Therefore, if $f(P; \boldsymbol{x})$ is known and the index $|\Lambda : \mathbb{Z}^d|$ is small, then one can compute $f(P, \Lambda; \boldsymbol{x})$ efficiently. We can iterate the two constructions: first we take a sublattice $\Lambda_1 \subset \mathbb{Z}^d$ of a small index, then a superlattice $\Lambda_2 \supset \Lambda_1$ of a small index, then a sublattice $\Lambda_3 \subset \Lambda_2$, and so forth.

6. Existence of "Local Formulae"

The results of Sections 4 and 5 provide a satisfactory solution of the counting problem when the dimension of the ambient space is fixed. If the dimension d is allowed to grow, the algorithms can become less efficient than straightforward enumeration. If the dimension is allowed to grow, the problem of "efficient counting" seems to be ill-posed, since much depends on the particulars of the polytope. For example, it becomes relevant whether the polytope is given by the list of its vertices or by the list of its facets. In this section, we explain our approach to what the "right" counting problem is when the dimension is allowed to grow.

P. McMullen [1983] proved that the number of integer points in a rational polytope P can be expressed as a linear combination of the volumes of the faces of the polytope, where the coefficient of $\mathrm{vol}(F)$, where F is a face of P, depends only on the translation class mod $\mathbb{Z}^d$ of the supporting cone $\mathrm{cone}(P, F)$ of P at F (see Definition 3.4).

THEOREM 6.1 ("LOCAL FORMULA"). *For every rational cone $K \subset \mathbb{R}^d$ one can define a rational number $\phi(K)$, such that*

(i) *The function ϕ is invariant under lattice translations:*

$$\phi(K) = \phi(K + u) \quad \text{for any } u \in \mathbb{Z}^d;$$

(ii) *For any rational polytope $P \subset \mathbb{R}^d$,*

$$|P \cap \mathbb{Z}^d| = \sum_F \mathrm{vol}(F)\phi\big(\mathrm{cone}(P, F)\big),$$

where the sum is taken over all faces F of P, and $\mathrm{vol}(F)$ is the volume of F, measured intrinsically in its affine span.

Theorem 6.1 immediately implies that the number of integer points $|kP \cap \mathbb{Z}^d|$ in the dilated polytope kP, where k is positive integer, is a quasipolynomial $\sum_{i=1}^d f_i(k)k^d$, where the $f_i(k)$ are periodic functions. If P is integral, then we get a genuine polynomial, namely the Ehrhart polynomial of P, see Example 1.5.

The function ϕ fails badly to be unique. Essentially, one can get the existence of ϕ using a Hahn–Banach type reasoning: see [McMullen 1993]. In the next section, we sketch a more constructive approach, also due to McMullen, via what we call the "Combinatorial Stokes Formula" (the formula belongs to McMullen, whereas its name is our invention). In many important cases, ϕ can be chosen to be a valuation on rational cones.

Note that, if $\dim P = d$ and $\dim F = k$, the apex of $\mathrm{cone}(P, F)$ is a k-dimensional affine subspace. Thus the cone is just the product of a $(d - k)$-dimensional pointed cone with the apex at the origin and a rational subspace.

Hence $\mathrm{cone}(P, F)$ looks "simple" when $d - k$ is small. In our opinion, the right problem to consider is the following:

PROBLEM 6.2. *If k is fixed and d is allowed to grow, find a computationally efficient choice of ϕ on rational cones with $(d - k)$-dimensional apex.*

In other words, we are interested in computing the highest terms of the expression for the Ehrhart (quasi)polynomial of P. Problem 6.2 is still not completely solved. In Section 8, we will see this problem solved for integer polytopes: computation of any fixed number of the highest coefficients of the Ehrhart polynomial of an integer polytope reduces in polynomial time to computation of volumes of faces; see [Barvinok 1994a]. However, the problem of finding computable functions ϕ for rational polytope is not yet solved.

The supporting cone $\mathrm{cone}(P, F)$ is "unnecessarily large", as it contains an affine subspace. Sometimes it is desirable to get a more "condensed" representation for the function ϕ.

DEFINITION 6.3. Let $P \subset \mathbb{R}^d$ be a full-dimensional polyhedron and let $F \subset P$ be a face. The *normal cone* $N(P, F)$ is the cone spanned by the outer normals of facets of P that contain F. In other words, if

$$P = \{x \in P : \langle c_i, x \rangle \le \beta_i \text{ for } i = 1, \ldots, m\},$$

F is a face of P, and $I_F = \{i : \langle c_i, x \rangle = \beta_i \text{ for every } x \in F\}$ is the set of inequalities that are active on F, then

$$N(P, F) = \mathrm{co}\{c_i : i \in I_F\}.$$

If $\dim F = k$, then $N(P, F)$ is a $(d - k)$-dimensional pointed cone with vertex at the origin.

Thus, for a full-dimensional integer polytope $P \subset \mathbb{R}^d$, Theorem 6.1 asserts that

$$|P \cap \mathbb{Z}^d| = \sum_F \phi(N(P, F)) \, \mathrm{vol}(F),$$

where ϕ is a function on pointed rational cones. The values of ϕ on lower-dimensional cones determine higher-dimensional coefficients of the Ehrhart polynomial. If P is a rational polytope, which is not an integer polytope, the value of ϕ depends not only on the normal cone $N(P, F)$, but also on the translation class mod $\mathbb{Z}^d$ of the affine hull $\mathrm{aff}(F)$ of the face F.

7. The Combinatorial Stokes Formula and Applications

Roughly speaking, the main idea of this section is to express the number $|P \cap \mathbb{Z}^d|$ of integer points in a polytope P as a sum of a main term, which is the volume of P, and a correction term, associated with the boundary ∂P of P. Naturally, we will have to consider lower-dimensional sublattices of $\mathbb{Z}^d$.

This explains why it is convenient to consider right from the beginning a general lattice $\Lambda \subset \mathbb{R}^d$ (discrete additive subgroup of $\mathbb{R}^d$), rather than only $\mathbb{Z}^d$.

DEFINITIONS 7.1. Fix a lattice $\Lambda \subset \mathbb{R}^d$. A polytope $P \subset \mathbb{R}^d$ is called a *lattice* or Λ-polytope provided that its vertices belong to Λ. A polytope $P \subset \mathbb{R}^d$ is called a *rational* or $\mathbb{Q}\Lambda$-polytope provided that for some positive integer n, $nP = \{nx : x \in P\}$ is a lattice polytope. Let $\mathcal{P}_c(\mathbb{Q}\Lambda)$ be the subspace (subalgebra) of $\mathcal{P}_c(\mathbb{R}^d)$ spanned by the indicator functions $[P]$ of rational polytopes. A valuation $\Phi : \mathcal{P}_c(\mathbb{Q}\Lambda) \to V$ is called *simple* provided $\Phi([P]) = 0$ when $\dim P < d$. A valuation Φ is called Λ-*invariant* provided $\Phi([P + \lambda]) = \Phi([P])$ for any $\lambda \in \Lambda$. A valuation Φ is called *centrally symmetric* provided $\Phi([P]) = \Phi([-P])$ for every polytope P. Often we write $\Phi(P)$ instead of $\Phi([P])$.

7.2 (CONES AND ANGLES). Let $K \subset \mathbb{R}^d$ be a cone. Let S^{d-1} be the unit sphere centered at the apex of K. We define $\alpha_d(K)$ to be the *spherical measure* of the intersection $K \cap S^{d-1}$ normalized in such a way that the spherical measure of the whole sphere S^{d-1} is 1. We also agree that if $d = 0$, then $\alpha_0(0) = 1$. Clearly, $\alpha_d(K) = 0$ if K is not a full-dimensional cone. The *intrinsic measure* $\alpha(K)$ is defined as the spherical measure $\alpha_k(K)$ in the affine hull of K, where $k = \dim K$.

Finally, let $P \subset \mathbb{R}^d$ be a d-dimensional polyhedron, and let $F \subset K$ be a k-dimensional face of P. The *exterior angle* $\gamma(P, F)$ of P at F is the intrinsic measure of the normal cone of P at F (see Definition 6.3); that is, $\gamma(P, F) = \alpha(N(P, F))$.

If $K \subset \mathbb{R}^d$ is a d-dimensional polyhedral cone, then

$$\sum_F \alpha(F)\gamma(K, F) = 1, \tag{7.2.1}$$

where the sum is taken over all nonempty faces F of K; see [McMullen 1975]. The proof, also due to McMullen, is immediate: for every point $x \in \mathbb{R}^d$, let $n(x) \in K$ be the (unique) point closest to x in the Euclidean metric. Then the summands of (7.2.1) correspond to a dissection of $\mathbb{R}^d$ into pieces, each of which consists of those points x such that $n(x)$ is in the relative interior of a given face F.

An important example of a simple lattice-invariant valuation related to lattice point counting arises when we count lattice points with their "solid angles."

EXAMPLE 7.3 (THE SOLID ANGLE VALUATION ρ). For a polyhedron $P \subset \mathbb{R}^d$ and a point $x \in \mathbb{R}^d$, define the *solid angle* $\beta(P, x)$ of P at x in the following way: let $B_r(x)$ denote the ball centered at x of radius r. We let

$$\beta(P, x) = \lim_{r \to 0} \frac{\mathrm{vol}(P \cap B_r(x))}{\mathrm{vol}(B_r(x))},$$

where "vol" is the usual volume in $\mathbb{R}^d$. For example, if $x \notin P$ then $\beta(P, x) = 0$. Similarly, if $\dim P < d$ then $\beta(P, x) = 0$ for any x. Furthermore, $\beta(P, x) = 1$ if and only if x is in the interior of P. If P is d-dimensional and x lies in the relative interior of a facet of P, then $\beta(P, x) = \frac{1}{2}$. Generally, if x lies in the relative interior of a face F of P, then $\beta(P, x)$ is the spherical measure of the supporting cone of P at F.

Let

$$\rho(P) = \sum_{x \in \Lambda} \beta(P, x).$$

It is easy to see that ρ extends to a simple centrally-symmetric Λ-invariant valuation

$$\rho : \mathcal{P}_c(\mathbb{Q}\Lambda) \to \mathbb{R}.$$

From (7.2.1), one can deduce that

$$|P \cap \Lambda| = \sum_F \rho(F)\gamma(P, F), \qquad (7.3.1)$$

where the sum is taken over all faces F of P and $\rho(F)$ is defined intrinsically in the affine span of F.

Fix a lattice $\Lambda \subset \mathbb{R}^d$, with rank $\Lambda = d$. A subspace $L \subset \mathbb{R}^d$ is called a *lattice subspace*, if it is spanned by lattice points. Similarly, an affine subspace $A \subset \mathbb{R}^d$ is called a *lattice subspace* provided it is a lattice translation of a linear lattice subspace. A *rational* subspace is a translation of a lattice linear subspace by a vector in $\mathbb{Q}\Lambda$.

Suppose we are given a simple Λ-invariant valuation Φ. The idea of the Combinatorial Stokes Formula is to construct a family of valuations $\{\phi_A\}$ concentrated on affine rational hyperplanes $A \subset \mathbb{R}^d$. It turns out that each valuation ϕ_A is simple, considered as a valuation in the $(d-1)$-dimensional space A. If $F \subset A$ is a polytope, instead of writing $\phi_A(F)$, we write simply $\phi(F)$. Our theorem follows [McMullen 1978/79].

THEOREM 7.4. *Fix a lattice $\Lambda \subset \mathbb{R}^d$ and a simple lattice invariant valuation Φ on rational polytopes in $\mathbb{R}^d$. Then there exist a number $\alpha \in \mathbb{R}$, a function $\kappa : \mathbb{R}^d \to \mathbb{R}$, and a family of simple valuations $\{\phi_A\}$, associated with rational hyperplanes $A \subset \mathbb{R}^d$, such that the following properties are satisfied:*

(i) *Valuations ϕ_A are Λ-invariant: if P and Q are rational $(d-1)$-dimensional polytopes and P is a lattice translation of Q, then $\phi(P) = \phi(Q)$.*

(ii) *If $P \subset \mathbb{R}^d$ is a rational polytope, then*

$$\Phi(P) = \alpha \operatorname{vol}(P) + \sum_F \kappa(n_F)\phi(F),$$

where the sum is taken over all facets of P and n_F is the outer unit normal to F;

(iii) *κ is an odd function: $\kappa(-u) = -\kappa(u)$ for each $\in \mathbb{R}^d$.*

SKETCH OF PROOF. Without loss of generality, we assume that $\Lambda = \mathbb{Z}^d$. We proceed by induction on d. For $d = 0$, the result is clear.

Suppose that $d \geq 1$, and consider $\mathbb{R}^{d-1}$ as a hyperplane in $\mathbb{R}^d$ (the last coordinate is 0). We define a valuation Ψ on $\mathbb{R}^{d-1}$ by $\Psi(Q) = \Phi(Q \times [0,1])$. Clearly, Ψ is a simple $\mathbb{Z}^{d-1}$-invariant valuation in $\mathbb{R}^{d-1}$, so we can apply the induction hypothesis to Ψ. Let $\alpha = \alpha_\Psi$ be the corresponding number, $\kappa = \kappa_\Psi : \mathbb{R}^{d-1} \to \mathbb{R}$ be the corresponding function, and let $\{\psi_H\}$ be the corresponding family of valuations for rational hyperplanes $H \subset \mathbb{R}^{d-1}$. If $H \subset \mathbb{R}^{d-1}$ is a rational hyperplane, then $A = H \oplus \mathbb{R}^1$ is a rational hyperplane in $\mathbb{R}^d$ whose normal vector is in $\mathbb{R}^{d-1}$. Using the induction hypothesis, one can show that the valuations ψ_H can be extended to valuations ϕ_A in such a way that $\phi_A(Q \times [0,1]) = \psi_H(Q)$ for any rational polytope $Q \in H$. Hence we have constructed valuations ϕ_A on the hyperplanes whose normal vectors are in $\mathbb{R}^{d-1}$. For a rational polytope $P \subset \mathbb{R}^d$, let

$$\bar{\Phi}(P) = \Phi(P) - \alpha \, \mathrm{vol}_d(P) - \sum_F \kappa(n_F)\phi(F),$$

where the sum is taken over all facets of P whose normal vectors are in $\mathbb{R}^{d-1}$. Using Theorem 2.9, one can show that $\bar{\Phi}$ is in fact a simple Λ-invariant valuation.

It is clear that $\bar{\Phi}(Q \times [0,1]) = 0$, where Q is a lattice polytope in $\mathbb{R}^{d-1}$. Since $\bar{\Phi}$ is a simple valuation, we conclude that $\bar{\Phi}(P) = 0$ if P is a "lattice prism" $Q \times [m,n]$, where $m, n \in \mathbb{Z}$.

Now we are ready to define ϕ_A for any rational hyperplane $A \subset \mathbb{R}^d$ and κ for all $u \in \mathbb{R}^d$. Let $A \subset \mathbb{R}^d$ be a rational affine hyperplane and let $Q \subset A$ be a rational polytope. Translating by a lattice vector, if necessary, we can always assume that Q is in the upper halfspace of $\mathbb{R}^d$ (the halfspace with positive last coordinate). Let Q' be the projection of Q down onto $\mathbb{R}^{d-1}$, and let $\Pi(Q) = \mathrm{conv}(Q \cup Q')$ be the "skewed prism" with bottom facet Q' and top facet Q. Let

$$\phi_A(Q) = \bar{\Phi}(\Pi(Q)).$$

One can show that $\phi_A(Q)$ is well-defined and that the family $\{\phi_A\}$ is Λ-invariant, since if we choose a lattice translation P of Q, the difference between the values of $\bar{\Phi}$ on the skewed prisms $\Pi(Q)$ and $\Pi(P)$ will be the value of $\bar{\Phi}$ on a right prism $Q \times [m,n]$, which is zero. Let u_d denote the last coordinate of a vector $u \in \mathbb{R}^d$. If $u_d = 0$ we can think of u as a vector in $\mathbb{R}^{d-1}$. Define κ by

$$\kappa(u) = \begin{cases} 1 & \text{if} \quad u_d > 0, \\ -1 & \text{if} \quad u_d < 0, \\ \kappa_\Psi(u) & \text{if} \quad u_d = 0. \end{cases}$$

Now the theorem follows since

$$[P] = \sum_F \kappa(n_F)[\Pi(F)] \quad \text{modulo lower-dimensional polytopes,}$$

where the sum is taken over all facets F of P.   □

The following example justifies, in our opinion, the name "Combinatorial Stokes Formula" for Theorem 7.4.

EXAMPLE 7.5 (EXPONENTIAL VALUATIONS). Let Λ^* be the lattice dual to Λ:

$$\Lambda^* = \{x \in \mathbb{R}^d : \langle x, u \rangle \in \mathbb{Z} \text{ for each } u \in \Lambda\}.$$

To each $l \in \Lambda^*$ one can associate a simple Λ-invariant valuation Φ_l defined by

$$\Phi_l(P) = \int_P \exp\big(2\pi i \langle l, x \rangle\big)\, dx.$$

It is clear that if $l = 0$ then $\Phi_l(P) = \mathrm{vol}(P)$. If $l \neq 0$ then (the ordinary) Stokes formula implies that

$$\int_P \exp\big(2\pi i \langle l, x \rangle\big)\, dx = \frac{1}{\langle l, c \rangle} \sum_F \langle c, n_F \rangle \int_F \exp\big(2\pi i \langle l, x \rangle\big)\, dx_F,$$

where c is any vector such that $\langle l, c \rangle \neq 0$. Here the sum is taken over all facets F of P, and dx_F is Lebesgue measure on the supporting hyperplane of F; see [Barvinok 1993]. Therefore, Theorem 7.4 holds with $\kappa(u) = \langle c, u \rangle / \langle l, c \rangle$ and $\{\phi_A\}$ being a family of exponential valuations on rational hyperplanes in $\mathbb{R}^d$.

Theorem 7.4 has an interesting corollary. We say that a set $X \subset \mathbb{R}^d$ is centrally symmetric provided there is a point y such that $2y - x \in X$ for any $x \in X$.

COROLLARY 7.6. *Let Φ be a Λ-invariant simple centrally symmetric valuation. There exists a constant α such that for each lattice polytope P whose facets are centrally symmetric,*

$$\Phi(P) = \alpha \,\mathrm{vol}\, P.$$

PROOF. We have

$$\Phi(P) = \frac{\Phi(P) + \Phi(-P)}{2}.$$

Expressing $\Phi(P)$ and $\Phi(-P)$ by Theorem 7.4, we notice that all the terms except the main one cancel each other out. □

Applications to the solid angle valuation ρ. We consider the valuation ρ of Example 7.3, which counts every lattice point in P with weight equal to the solid angle at that point.

COROLLARY 7.7. *Let $\Lambda \subset \mathbb{R}^d$ be a lattice of rank d and let $P \subset \mathbb{R}^d$ be a lattice polytope whose facets are centrally symmetric. Then*

$$\rho(P) = \frac{\mathrm{vol}(P)}{\det \Lambda}.$$

PROOF. By Corollary 7.6, it follows that $\rho(P) = \alpha \,\mathrm{vol}(P)$ for some α. Let $nP = \{nx : x \in P\}$ be a dilatation of P. Then

$$\alpha = \lim_{n \to +\infty} \frac{\rho(nP)}{\mathrm{vol}(nP)} = \frac{1}{\det \Lambda}. \qquad \Box$$

Corollary 7.7 can be considered as a "101st" generalization of Pick's formula (Example 1.2). Indeed, all facets of a polygon are centrally symmetric. Pick's formula is equivalent to saying that if we count every integer point in a polygon P with weight equal to the angle at this point, we get the area of the polygon.

An interesting example of a polytope with centrally symmetric faces is provided by a *zonotope*, that is, the Minkowski sum of finitely many lattice intervals. To remove various normalizing factors, it is convenient to measure volumes of polytopes intrinsically, with respect to a given lattice. Namely, fix a lattice $\Lambda \subset \mathbb{R}^d$. Suppose that $P \subset \mathbb{R}^d$ is a lattice polytope and suppose that $k = \dim P$. Without loss of generality, we may assume that the affine hull A of P contains the origin. Then $\Lambda_A = \Lambda \cap A$ is a lattice of rank k, and we normalize the volume form in A in such a way that $\det \Lambda_A = 1$.

COROLLARY 7.8. *Let $P \subset \mathbb{R}^d$ be an integer zonotope, that is, the Minkowski sum of finitely many integer intervals. Then the number of integer points in P is expressed by the formula:*

$$|P \cap \mathbb{Z}^d| = \sum_F \mathrm{vol}(F)\gamma(P,F),$$

where the sum is taken over all faces F of P, $\gamma(P,F)$ is the exterior angle of P at F, and the volume of a face is measured intrinsically with respect to the lattice.

PROOF. Since every face of a zonotope is centrally symmetric, the result follows from Corollary 7.7 and formula (7.3.1). Of course, it is quite easy to find a simple alternative proof which does not use Theorem 7.4. For a full-dimensional lattice parallelepiped Π (that is, the Minkowski sum of d linearly independent intervals) we have $\rho(\Pi) = \mathrm{vol}(\Pi)$, since lattice translates of Π tile the space $\mathbb{R}^d$ and both the volume and the valuation ρ are simple and lattice-invariant. Since a lattice zonotope can be dissected into lattice parallelepipeds (see, for example, [Ziegler 1995, Lecture 7]), we get $\rho(P) = \mathrm{vol}(P)$ for any lattice zonotope P. Since every face of a lattice zonotope is a lattice zonotope itself, we use (7.3.1) to complete the proof.   □

The boundary of a convex polytope can be represented as a union of lower-dimensional polytopes. Therefore, we can apply the Stokes formula of Theorem 7.4 recursively, first to the polytope P, then to its facets, then to its ridges, and so forth. We will end up with a decomposition involving volumes of faces and some "local" functions, depending only on the supporting cones at the faces (see [McMullen 1978/79]). Thus using Theorem 7.4, one can prove Theorem 6.1 first for the simple valuation ρ, and then, applying (7.3.1), for the number of integer points.

EXAMPLE 7.9 (COEFFICIENTS OF THE EHRHART POLYNOMIAL). For any rational polytope $P \subset \mathbb{R}^d$, there is a positive integer m such that the dilatation

$Q_m = mP$ is an integer polytope. Consider the Ehrhart polynomial of Q_m (see Example 1.5)

$$|nQ_m \cap \mathbb{Z}^d| = a_d(Q_m)n^d + \cdots + a_0(Q_m),$$

and define $a_k(P) = a_k(Q_m)/m^k$ for $k = 0, \ldots, d$. It is easy to see that the numbers $a_k(P)$ are well-defined, that is independent on the choice of the scaling factor m. Furthermore, one can see that $a_k : \mathcal{P}_c(\mathbb{Q}^d) \to \mathbb{Q}$ is a valuation. As proved by U. Betke and M. Kneser [1985], the coefficients a_k constitute a basis of the vector space of all valuations $\mathcal{P}_c(\mathbb{Q}^d) \to \mathbb{Q}$ that are invariant under the affine transformations of $\mathbb{R}^d$ that map the lattice $\mathbb{Z}^d$ onto itself. Obviously, $a_k(P)$ are homogeneous, that is $a_k(rP) = r^k a_k(P)$, where $r > 0$ is a rational number. Corollary 7.8 implies that if P is a rational zonotope, then

$$a_k(P) = \sum_{F:\dim F=k} \operatorname{vol}(F)\gamma(P,F).$$

In particular, the coefficients of the Ehrhart polynomial of a zonotope are always nonnegative. Curiously, if $k = 0$, $d-1$, or d, the formula holds true for *any* rational polytope P. One can show that in fixed dimension, one can compute the spherical measure of a given polyhedral cone within any given error $\varepsilon > 0$ in polynomial time. Furthermore, even if the dimension is allowed to be a part of the input, using the technique of [Dyer et al. 1991] (see [Kannan et al. 1997] for recent improvements), one can come up with a randomized algorithm which, given an $\varepsilon > 0$, approximates the spherical measure of a given polyhedral cone with relative error ε in time which is polynomial in the input size and ε^{-1}. The cone may be given as the convex hull of rays or as the intersection of subspaces. Thus in the class of integer zonotopes, Problem 6.2 has a satisfactory solution.

For general rational polytopes $P \subset \mathbb{R}^d$, Theorem 6.1 asserts that

$$a_k(P) = \sum_{F:\dim F=k} \operatorname{vol}(F)\phi\big(N(P,F)\big),$$

where ϕ is "some function" on the normal cones $N(P,F)$ of P at the k-dimensional faces F. There are 3-dimensional integer polytopes P, such that $a_1(P) < 0$. (Example: the tetrahedron $\Delta \subset \mathbb{R}^3$ with the vertices $(0,0,0), (1,0,0), (0,1,0)$ and $(1,1,13)$; see [Fulton 1993, Section 5.3].) Hence, in general, ϕ can hardly be any geometric measure, it must reflect arithmetic structure. Computationally efficient choices of ϕ are discussed further in Sections 8 and 9.

Let $P \subset \mathbb{R}^d$ be an integer polytope. Instead of considering the coefficients $a_k(P)$, it can be useful to consider their linear combinations, $h_0^*(P), \ldots, h_d^*(P)$, defined by the following expression of a power series in one variable x as a rational function:

$$\sum_{n=0}^{\infty}\Big(a_0(P) + a_1(P)n + \cdots + a_d(P)n^d\Big)x^n = \frac{h_0^*(P) + h_1^*(P)x + \cdots + h_d^*(P)x^d}{(1-x)^{d+1}}.$$

Unlike the coefficients $a_k(P)$, the numbers $h_k^*(P)$ are monotone, that is, $h_k^*(P) \geq h_k^*(Q)$ if Q and P are integer polytopes with $Q \subset P$ [Stanley 1993]. In particular, the $h_k^*(P)$ are always nonnegative. However, they are not homogeneous.

From the proof of Theorem 7.4, it is not at all clear what the functions ϕ in Theorem 6.1 might look like. The problem appears somewhat easier for exponential valuations Φ_l (Example 7.5). Since the main term is 0 unless $l = 0$ we conclude that if the highest term of the Ehrhart (quasi)polynomial of the valuation Φ_l is k, then l is orthogonal to some k-dimensional face of P.

A possible approach would be to relate the exponential valuations Φ_l of Example 7.5 and the solid angle valuation ρ (see Section 7.3). Let

$$\theta_\tau(x) = \tau^{d/2} \sum_{u \in \Lambda} \exp\left(-\tau\pi\|x - u\|^2\right) = \frac{1}{\det \Lambda} \sum_{l \in \Lambda^*} \exp\left(-\pi\|l\|^2/\tau\right) \exp\left(2\pi i \langle l, x \rangle\right),$$

where $\tau > 0$ is a parameter. It is then easy to show (see [Barvinok 1992]) that

$$\lim_{\tau \to +\infty} \int_P \theta_\tau(x) \, dx = \rho(P).$$

Thus we get a decomposition of ρ into the Fourier series of the valuations Φ_l:

$$\rho(P) = \lim_{\tau \to +\infty} \sum_{l \in \Lambda^*} \exp\left(-\pi\|l\|^2/\tau\right) \Phi_l(P).$$

This approach was discovered independently and is being developed by R. Diaz and S. Robins [$\geq$ 1999].

8. Using Algebraic Geometry to Count Lattice Points

In recent years, many authors [Brion and Vergne 1997a; Cappell and Shaneson 1994; Ginzburg et al. $\geq$ 1999; Kantor and Khovanskii 1993; Infirri 1992; Morelli 1993a; 1993b; Pommersheim 1993; 1996; 1997; Pukhlikov and Khovanskii 1992b] have studied the problem of lattice point enumeration using the subject of toric varieties. In this section, we describe some of these results. In particular, we show how the valuation $\mathfrak{F}$ introduced in Section 3 plays a key role in various formulas for counting lattice points and for the closely-related problem of finding the Todd class of a toric variety.

Toric varieties, though very special and somewhat simple from the point of view of algebraic geometry, provide a powerful link between the theory of lattice polytopes and algebraic geometry. Early researchers in the field of toric varieties realized that finding a formula for a toric variety's *Todd class*, a characteristic class living in homology, would yield a formula for the number of lattice points in an integral polytope. Details of this connection, a direct result of the Riemann–Roch Theorem, are sketched following the statement of Algorithm 8.7. For a more complete discussion, see [Fulton 1993, Section 5.3]. This connection between polytopes and toric varieties has been very fruitful, especially over the last ten years. Indeed, much of the recent progress in counting lattice points is the result of a better understanding of the Todd class of a toric variety.

Interestingly, many of the lattice point formulas arising from the theory of toric varieties were later found to have elementary proofs, completely independent of algebraic geometry. A good example of this is Theorem 3.5 above, which Brion first proved by applying a localization theorem for equivariant K-theory to toric varieties. However, we have seen in Section 3 (as have others before us) that it is not difficult to give an elementary proof of this simple and beautiful formula. Another example is Brion–Vergne's generalization of Khovanskii–Pukhlikov's Euler–Maclaurin summation formula for polytopes. Again, toric varieties provided motivation and an initial line of attack for these authors, though the final proofs in [Brion and Vergne 1997b] are entirely elementary. One is left wondering what stopped these beautiful formulas from appearing long ago.

Let K be a d-dimensional rational, simple cone in $\mathbb{R}^d$. Here *simple* means that K is the convex hull of d rational rays from the origin. We denote the rays of K by $v_1, \ldots, v_d$, and identify each ray v_i with the first nonzero lattice point on that ray. By abuse of notation, we will use K to denote the d-by-d matrix whose ith row is v_i. Let K^* denote the dual cone, introduced in Section 2. Finally, if $x = (x_1, \ldots, x_d)$, the following abbreviation will be convenient:

$$e^x = (e^{x_1}, \ldots, e^{x_n}).$$

We are now ready to introduce a certain Laurent series $\mathfrak{s}_K$ in d variables $y = (y_1, \ldots, y_d)$ corresponding to the rays of K. This Laurent series is a reparametrization of the function f of Theorem 3.1:

DEFINITION 8.1. To any d-dimensional rational, simple cone K in $\mathbb{R}^d$, we define

$$\mathfrak{s}_K(y) = f(K^*; e^{-yK}),$$

which represents a rational function in the variables $e^{y_1}, \ldots, e^{y_n}$. It will also be convenient to have a modified version $\mathfrak{t}_K$ of $\mathfrak{s}_K$. We define

$$\mathfrak{t}_K(y) = \operatorname{ind}(K) y_1 y_2 \cdots y_d \cdot \mathfrak{s}_K(y).$$

It is precisely this variant $\mathfrak{t}_K$ of the function f that will express for us the Todd class of a simplicial toric variety: we shall shortly see that $\mathfrak{t}_K$ defines a *power series* in the variables $y_1, \ldots, y_n$, and we can express the Todd class by taking the variables y_i in $\mathfrak{t}_K$, which correspond to rays of K, and substituting the corresponding divisor class. Before stating this precisely, we make some easy observations about $\mathfrak{s}_K$.

First notice that the expression for $\mathfrak{s}_K$ alters the function f of Theorem 3.1 in two ways: exponentials are substituted for the variables, and a linear change of coordinates is made. The linear change of coordinates has the pleasant effect of making $\mathfrak{s}_K$ invariant under lattice automorphisms. This is expressed in the following proposition, along with two other important properties of $\mathfrak{s}_K$.

PROPOSITION 8.2.   *The assignment of $\mathfrak{s}_K$ to cones K satisfies the following properties:*

(i) (INVARIANCE UNDER LATTICE AUTOMORPHISMS) *If K and L are equivalent under an automorphism of $\mathbb{Z}^d$ which preserves the prescribed ordering of the rays of K and L, then*

$$\mathfrak{s}_K(y) = \mathfrak{s}_L(y).$$

(ii) (ADDITIVITY UNDER SUBDIVISION) *If $K, K_1, \ldots, K_l$ are d-dimensional rational simple cones whose indicator functions satisfy*

$$[K] = \alpha_1[K_1] + \cdots + \alpha_l[K_l]$$

modulo cones of smaller dimension, then

$$\mathfrak{s}_K(y) = \alpha_1 \mathfrak{s}_{K_1}(yKK_1^{-1}) + \cdots + \alpha_l \mathfrak{s}_{K_l}(yKK_l^{-1}).$$

(iii) (SUMMATION FORMULA) *Let Π be the fundamental parallelepiped for K^* and let $u_1, \ldots, u_d$ be the primitive generators for the cone K^*. Then*

$$\mathfrak{s}_K(y) = \left(\sum_{u \in \Pi \cap \mathbb{Z}^d} e^{-\langle u, yK \rangle} \right) \prod_{i=1}^{d} \frac{1}{1 - e^{-\langle u_i, v_i \rangle y_i}}.$$

Thus the function $\mathfrak{t}_K$ is regular at the origin and hence defines a power series in the variables $y_1, \ldots, y_n$.

SKETCH OF PROOF. Invariance under lattice automorphisms can be easily checked from the definition of $\mathfrak{s}$; it is also a consequence of the summation formula (Part 3 above), which is clearly invariant under lattice automorphisms.

To check additivity under subdivisions, let $K, K_1, \ldots, K_l$ be as above, and apply Corollary 2.8, which implies that the indicator functions $[K_i^*]$ of the dual cones sum to the indicator function of K^* modulo straight lines. But by Part 2 of Theorem 3.1, f vanishes on cones containing straight lines. The desired identity is then a consequence of the additivity formula for f given in Part 1 of Theorem 3.1.

The summation formula follows easily from the formula of Example 3.3 and the definitions, bearing in mind that $\langle u_i, yK \rangle = \langle u_i, v_i \rangle y_i$. Finally, the assertion about the power series $\mathfrak{t}_K(y)$ follows from the summation formula for $\mathfrak{s}_K(y)$ and the fact the function

$$g(z) = \frac{z}{1 - e^{-z}}$$

is regular at $z = 0$.   □

One further useful property of $\mathfrak{t}$ is its compatibility along common faces:

LEMMA 8.3. *If the cones K and L meet at a common face F, the power series $\mathfrak{t}_K$ and $\mathfrak{t}_L$ agree when restricted to F; that is, if in $\mathfrak{t}_K$ and $\mathfrak{t}_L$ we set to zero all variables corresponding to rays outside of F, we obtain identical power series in the remaining variables.*

SKETCH OF PROOF. When the variables in t_K corresponding to rays not in F are set to zero, one checks that we obtain t_F, computed with respect to the linear subspace $F + (-F)$. The result is therefore dependent only on F, not on all of K.   □

We are now ready to state the Todd class formula. Let Σ be a complete simplicial fan with rays $v_1, \ldots, v_l$. We introduce the Stanley–Reisner ring of the fan

$$A_\Sigma = \frac{\mathbb{Q}[y_1, \ldots, y_l]}{I_\Sigma},$$

where I_Σ denotes the ideal generated by all monomials $y_{i_1} \cdots y_{i_k}$ such that $\langle v_{i_1}, \ldots, v_{i_n} \rangle$ is not a cone in the fan Σ.

The power series t_K corresponding to the d-dimensional cones of Σ fit together to form a single power series:

DEFINITION 8.4. Let σ be a complete simplicial fan as above. Since the power series t_K where K is a maximal cone of Σ are compatible in the sense of Lemma 8.3, they patch together to form a single power series

$$t_\Sigma(y_1, \ldots, y_l)$$

in variables $y_1, \ldots y_l$ corresponding to the rays $v_1, \ldots, v_l$ of Σ. Since monomials corresponding to cones not in the fan do not appear in $t_\Sigma(y_1, \ldots, y_l)$, this power series lives naturally in the completion of the Stanley–Reisner ring A_Σ.

The main theorem of this section states that t_Σ computes the Todd class of the toric variety X_Σ. This theorem was proved in the context of equivariant cohomology by Brion and Vergne [1997a]. It is also equivalent to the formulas of [Pommersheim 1997]. In that context, the Todd class formula was a consequence of a local push-forward formula for products of cycles on a toric variety.

THEOREM 8.5. *For any complete simplicial fan Σ, the Todd class of X_Σ is obtained by evaluating $t_\Sigma(y_1, \ldots, y_l)$ at $y_i = D_i$, where $D_i \in A^1 X_\Sigma$ is the divisor class corresponding to the ray v_i. That is:*

$$\mathrm{Td}\, X_\Sigma = t_\Sigma(D_1, \ldots D_l).$$

We remark here that if all cones of Σ are unimodular, then the equation above expresses the well-known fact that the Todd classes of a nonsingular toric variety are found by taking the Todd polynomials, introduced in Definition 5.1, in the classes D_i of the torus-invariant divisors. This results from the general fact that the Todd classes of any nonsingular variety can be computed from the Chern classes of the tangent bundle using the Todd polynomials.

In order to determine the Todd class of any simplicial X_Σ of dimension d, only those terms of t_Σ of degree less than or equal to d need to be computed. This is because any product of degree larger than d represents 0 in the Chow group. With this in mind, it is not hard to see that the expression above for the Todd class may be computed in polynomial time in fixed dimension:

THEOREM 8.6. *For a fixed dimension d, there exists a polynomial time algorithm which, given a complete fan Σ, computes t_Σ up to degree d. Thus the Todd class Σ can be expressed as a polynomial in the torus-invariant divisors in polynomial time.*

SKETCH OF PROOF. The idea is very similar to the proof of Theorem 4.4. Again, the key is the result of [Barvinok 1994b], stated as Theorem 4.2 above, that a polynomial time algorithm exists to write an arbitrary rational cone as the difference of unimodular cones. To compute the Todd class, one applies this theorem to all d-dimensional cones of Σ, and uses the additivity property of Proposition 8.2 to express the power series t_K up to degree d. At this point we are done, since these expressions for t_K determine t_Σ.   □

The well-known dictionary between polytopes and toric varieties allows us to use this result to obtain a polynomial-time algorithm for computing the number of lattice points in an integral convex polytope. The following algorithm appeared in [Pommersheim 1997]:

ALGORITHM 8.7. Given a simple integral convex polytope P, the following algorithm computes the number of lattice points in P:

(1) Suppose that P is represented as the solution to the finitely many inequalities as follows:

$$P = \{x \in \mathbb{R}^d : \langle v_i, x \rangle \geq \beta_i \text{ for } i = 1, \ldots l\},$$

where the v_i are the rays of the inner normal fan Σ of P.

(2) Using (8.6), compute the Todd power series $t_\Sigma(y_1, \ldots, y_d)$ in degree up to d. Denote this degree d polynomial by T.

(3) Let

$$C = \exp\left(\sum -\beta_i y_i\right),$$

and let Q be the degree d part of the product TC.

(4) In the Stanley–Reisner ring A_Σ, let J be the ideal generated by the set

$$\left\{\sum_{i=1}^{l} \langle v_i, u \rangle y_i : u \in \mathbb{Z}^n\right\}.$$

Choose any vertex of P, and let $K = \langle v_{i_1}, \ldots, v_{i_d} \rangle$ be the corresponding cone of Σ. Compute normal forms of Q and of $y_{i_1} \cdots y_{i_d}$ with respect to any Gröbner basis for the ideal J. The degree d part of the quotient A_Σ / J is known to be a one-dimensional vector space. Thus these two normal forms may be divided to produce a rational number. The number of lattice points in P is then

$$|P \cap \mathbb{Z}^d| = \frac{\mathrm{nf}(Q)}{\mathrm{ind}(K)\,\mathrm{nf}(y_{i_1} \cdots y_{i_d})}.$$

SKETCH OF PROOF. The fact that the algorithm above yields the number of lattice points in P follows from a well-known application of the Riemann–Roch theorem to the toric variety X_Σ. To the polytope P, there corresponds naturally a line bundle L_P on the toric variety X_Σ. Lattice points in P are in one-to-one correspondence with a basis of global sections of this line bundle. Hence the lattice point question reduces to finding the dimension of the space of sections of L_Σ. The higher cohomology of L_Σ vanishes, and therefore the number of lattice points equals the Euler characteristic $\chi(X_\Sigma, L_P)$. The singular Hirzebruch–Riemann–Roch theorem allows us to compute this Euler characteristic by taking the intersection product of the Todd class of X_Σ, represented by T, with the Chern character of L_P, represented by C. This is expressed in step (3) above. Finally, according to Hirzebruch–Riemann–Roch, one must find the codimension d part of the product $Q = TC$. Step (4) accomplishes this by computing in the ring A_Σ/J, which is a well-known presentation for the Chow ring of X_Σ. The codimension d part of this ring is one-dimensional, consisting of multiples of the class of a point. Furthermore, for any cone

$$K = \langle v_{i_1}, \ldots, v_{i_d} \rangle,$$

as in step (4), the product of the corresponding divisors is given by

$$D_{i_1} \cdots D_{i_d} = \frac{1}{\operatorname{ind}(K)} [\mathrm{pt}].$$

Hence the formula of step (4) expresses Q as a multiple of the class of a point, and hence computes the number of lattice points in P. □

The algorithm above can also be viewed from the point of view of local formulae, in the spirit of Theorem 6.1. To see this, we give the following variation of Algorithm 8.7, which is very convenient in the case that the volumes of the faces of P are known.

Following the notation above, we begin with a simple lattice polytope P, and we compute T as in step (1). Using the linear relations of the ideal J, one may easily obtain a *squarefree* expression for T. One then replaces each squarefree monomial $y_{i_1} \cdots y_{i_k}$ occurring in T by

$$\frac{1}{\operatorname{ind}(K)} \operatorname{vol}(F),$$

where K is the k-dimensional cone

$$K = \langle v_{i_1}, \ldots, v_{i_k} \rangle,$$

and F is the corresponding $(d - k)$-dimensional face of P defined by

$$F = \{x \in P : \langle v_i, x \rangle = \beta_i \text{ for } i = i_1, \ldots, i_k\}.$$

With this replacement, one obtains a rational number, which equals the number of lattice points in P.

We note that this variant of Algorithm 8.7 may be used to compute a given coefficient in the Ehrhart polynomial of P independently of the other coefficients. If we are interested only in the top k Ehrhart coefficients, we must consider only those monomials in T of degree at most k, and only those cones $K \in \Sigma$ of dimension at most k. It follows that for fixed k, the procedure above reduces in polynomial time the computation of the top k Ehrhart coefficients of an integral polytope P to the computation of the volumes of its faces. This provides a partial answer to Problem 6.2.

QUESTION 8.8. The first polynomial-time algorithm for counting lattice points in polytopes was that of Theorem 4.4 (originally in [Barvinok 1994b]), which involved no toric varieties. Algorithm 8.7, though based on algebraic geometry, appears quite similar in flavor. In particular, both algorithms are ultimately based on subdividing cones into unimodular cones, and both are linked quite closely to the valuation $\mathfrak{F}$ introduced in Theorem 3.1. In should be noted, however, that the original algorithm is based on subdivisions of the tangent cones of the polytope, whereas Algorithm 8.7 naturally involves subdivisions of the dual cones. These observations motivate several questions: How exactly are these algorithms related? What is the precise nature of the duality involved here? If one were to implement these algorithms, what features of each algorithm would be advantageous?

The power series t also plays a key role in the lattice point formula of Brion and Vergne [1997b, Theorem 2.15], which is a generalization of the formula of Pukhlikov and Khovanskii [1992b]. In these rather remarkable formulas, the power series t_Σ is considered as an infinite-order differential operator, called the *Todd differential operator*. A deformed polytope is created, with all facets moved independently, but parallel to the original facets. The volume of this deformed polytope is calculated as a function of the displacements. Applying the Todd differential operator to this function yields the number of lattice points in the polytope. More generally, one obtains an *Euler–Maclaurin formula*, expressing the sum of any polynomial function over the lattice points in a polytope as the Todd operator applied to the integral of the polynomial function over the deformed polytope.

We now state these formulas precisely. As above, assume that P is a simple integral convex polytope. Suppose that

$$P = \{x \in \mathbb{R}^d : \langle v_i, x \rangle \geq \beta_i \text{ for } i = 1, \dots l\},$$

where the v_i are the rays of the inner normal fan Σ of P. For $h = (h_1, \dots, h_l) \in \mathbb{R}^l$, we define the deformed polytope $P(h)$ by

$$P(h) = \{x \in \mathbb{R}^d : \langle v_i, x \rangle \geq \beta_i - h_i \text{ for } i = 1, \dots l\}.$$

THEOREM 8.9. *Let P be as above, let Σ be its inner normal fan, and let ϕ be a polynomial function on $\mathbb{R}^d$. Denote by $I_\phi(h)$ the integral of ϕ over the deformed* *polytope $P(h)$:*

$$I_\phi(h) = \int_{P(h)} \phi(x)dx.$$

Then

$$\sum_{m \in P \cap \mathbb{Z}^d} \phi(m) = t_\Sigma\left(\frac{\partial}{\partial h_1}, \ldots, \frac{\partial}{\partial h_l}\right) \diamond I_\phi(h),$$

where $\diamond$ indicates that all derivatives are evaluated at $h = 0$.

In particular,

$$|P \cap \mathbb{Z}^d| = t_\Sigma\left(\frac{\partial}{\partial h_1}, \ldots, \frac{\partial}{\partial h_l}\right) \diamond \operatorname{vol}(P(h)).$$

We have thus far seen several results linking the problem of lattice point enumeration with the valuation $\mathfrak{F}$ and its relatives $\mathfrak{s}$ and $\mathfrak{t}$, which are all efficiently computable. We close this section by showing that these functions are also closely tied with the important formulas of R. Morelli.

In the late 1970's, V. I. Danilov asked for a local expression for the Todd class of a toric variety. Specifically, he asked if there exists a function μ on rational cones such that for any complete fan Σ, the Todd class of X_Σ is given by

$$\operatorname{Td} X_\Sigma = \sum_{K \in \Sigma} \mu(K)[V(K)],$$

where $V(K)$ is the closed subvariety of X_Σ corresponding to the cone $K \in \Sigma$. Morelli [1993b] showed that this assignment was indeed possible. Though this μ is far from unique, Morelli showed that a canonical μ does exist if we allow μ to take values in a certain field of rational functions (rather than simply rational or real values.) Precisely, he showed that for each $k \leq d$, there exists a function μ_k from the set of all k-dimensional rational cones in $\mathbb{R}^d$ to the field $\operatorname{Rat}(\operatorname{Gr}_{d-k+1}(\mathbb{R}^d))$ of rational functions on the Grassmannian of $d-k+1$ planes in $\mathbb{R}^d$, such that μ_k expresses the Todd class in the sense of Danilov's equation above.

The spirit of this question is quite similar to that of McMullen's Theorem, stated as Theorem 6.1 above. In fact, using the link between Todd classes and lattice point enumeration, Morelli's affirmative answer to the question above immediately settles Theorem 6.1 for integral polytopes, via the relation

$$\phi(K) = \mu(K^*).$$

The following proposition gives a relation between Morelli's $\mu_d(K)$ and the Laurent series $\mathfrak{s}_K$.

PROPOSITION 8.10 [Pommersheim 1997, Prop. 4]. *Let K be a d-dimensional rational simple cone in $\mathbb{Z}^d$, and let $u_1, \ldots, u_d$ be the primitive generators of the dual cone K^*. Let $\mathfrak{s}_K^0$ denote the degree 0 part of the Laurent series $\mathfrak{s}_K$.*

Then $\mathfrak{s}_K^0(u_1, \ldots, u_d)$, as a degree 0 rational function on $\mathbb{R}^d$, lives naturally in $\mathrm{Rat}(\mathrm{Gr}_1(\mathbb{R}^d))$. As such, it coincides with Morelli's $\mu_d(K)$:

$$\mu_d(K) = \mathfrak{s}_K^0(u_1, \ldots, u_d).$$

SKETCH OF PROOF. We can check equality of the rational functions above on unimodular cones easily. By Proposition 8.2, Part 3,

$$\mathfrak{s}_K(u_1, \ldots, u_d) = \prod_{i=1}^d \frac{1}{1 - e^{-u_i}},$$

the degree 0 part of which equals the d-th Todd polynomial $\mathrm{td}_d(u_1, \ldots, u_d)$. This equals $\mu_d(K)$, as discussed in [Morelli 1993b, following statement of Theorem 4]. Equality for all simple cones then follows from additivity under subdivisions, satisfied by both Morelli's μ, and $\mathfrak{s}$ (cf. Proposition 8.2, Part 2.) $\qquad\square$

It is also possible to relate Morelli's $\mu_k(K)$ for a simple k-dimensional cone K, with $k < d$, to the coefficients of the Laurent series $\mathfrak{s}_K$. A discussion of this connection may be found in [Pommersheim 1997].

9. Generalized Dedekind Sums and Counting Lattice Points

In Example 1.4, we saw the classical Dedekind sum appear in a formula for the number of lattice points in a certain tetrahedron. In this section we explore the important role that Dedekind sums and their higher-dimensional generalizations play in lattice point formulas. In particular, following Brion–Vergne, we show that the higher-dimensional Dedekind sums introduced by Zagier [1973] appear as coefficients in the power series t of Section 8. It then follows from the results of Section 8 that these important sums of Zagier are computable in polynomial time when the dimension is fixed.

The classical Dedekind sum first appeared long ago in Dedekind's work on the η-function, and since then has arisen in a variety other contexts. For example, Hirzebruch connected these sums with geometry by showing that they appear naturally in formulas for the signature of certain singular quotient spaces. The 1951 formula of Mordell (Example 1.4) marked the first appearance of Dedekind sums in a formula for lattice points.

We now relate the classical Dedekind sum to the power series t introduced in Section 8. This relation, less than a decade old, puts the Mordell formula in a much more general and geometric context.

Let K be a two-dimensional cone in a lattice. All such cones are simplicial, and in fact K is equivalent by a lattice isomorphism to the cone $\langle (0,1), (p,q) \rangle$ in $\mathbb{Z}^2$, with $0 \le p < q$. In this case, the cone K is said to have type (p,q). It is easily checked that q (which equals $\mathrm{ind}(K)$) is uniquely determined by K, and p is determined up to multiplicative inverses modulo q. (The cone $\langle (0,1), (p^{-1}, q) \rangle$ is equivalent to K by a lattice isomorphism that swaps the rays of K.) This allows us to make the following definition, which associates a Dedekind sum to any two-dimensional cone.

DEFINITION 9.1. Let K be a two-dimensional cone. If K has type (p,q), then we define the Dedekind sum associated to K, denoted $s(K)$ by:

$$s(K) = s(p,q).$$

This Dedekind sum appears as a coefficient in the degree two part of the power series $\mathfrak{t}_K$:

THEOREM 9.2. *For any two-dimensional cone K, the degree-two part of the power series $\mathfrak{t}_K$ is given by*

$$(\mathfrak{t}_K(y_1, y_2))_{\deg 2} = \tfrac{1}{12}(y_1^2 + y_2^2) + \mathrm{ind}(K)\big(s(K) + \tfrac{1}{4}\big)y_1 y_2.$$

SKETCH OF PROOF. Any one-dimensional cone is unimodular. Thus, by Proposition 8.2, Part 3, the coefficients of y_1^2 and y_2^2 are always $\tfrac{1}{12}$, the coefficient of z^2 in the expansion of

$$g(z) = \frac{z}{1 - e^{-z}}.$$

As for the coefficient of $y_1 y_2$, one computes directly from Proposition 8.2, Part 3. The fundamental parallelepiped Π for K^* has q elements:

$$\Pi = \left\{ \left(j, \left\lceil \frac{-pj}{q} \right\rceil \right) : j = 0, \ldots, q-1 \right\},$$

where $\lceil x \rceil$ denotes the smallest integer greater than or equal to x. The desired equality follows by expanding the expression in Proposition 8.2, and using the definition of Dedekind sum given in Example 1.4.   □

Alternatively, one could use the (dual) approach of the proof of Theorem 9.5 (see below).

Theorem 9.2, together with the ideas of Algorithm 8.7, allow one to express the degree $d - 2$ coefficient of the Ehrhart polynomial of an arbitrary d-dimensional lattice polytope. This expression involves one Dedekind sum for each face of codimension two. For each such face F, the dual to the tangent cone at F is a two-dimensional cone K, and the corresponding Dedekind sum $s(K)$ contributes to the degree $d - 2$ term in the Ehrhart polynomial.

These ideas are especially valuable for polytopes of dimensions two and three. In dimension three, the only mysterious part of the Ehrhart polynomial is its linear term. Thus, we see that the entire Ehrhart polynomial may be easily computed in terms of the Dedekind sums corresponding to the edges (1-dimensional faces) of the polytope.

For $d = 2$, the degree $d - 2$ part of the Ehrhart polynomial simply equals 1 for any polytope. Thus it would seem that a Dedekind sum expression for this coefficient would be useless. However, when we equate this Dedekind sum expression to 1, we obtain a reciprocity relation for Dedekind sums! In this way, any lattice polygon gives a relation among the Dedekind sums corresponding to its angles (see [Pommersheim 1993, Theorem 8].) One easily finds lattice triangles that demonstrate Dedekind's reciprocity relation, Rademacher's three-term law, as well as new relations for the classical Dedekind sum.

We now turn our attention to the higher-dimensional Dedekind sums introduced by Zagier. Motivated by topological considerations, Zagier made the following definition:

DEFINITION 9.3.   Let q be a positive integer, and let $p_1, \ldots, p_n$ be integers prime to q, with n even. The higher dimensional Dedekind sum $d(q; p_1, \ldots, p_n)$ is defined by

$$d(q; p_1, \ldots, p_n) = (-1)^{n/2} \sum_{k=1}^{q-1} \cot \frac{\pi k p_1}{q} \cdots \cot \frac{\pi k p_n}{q}.$$

Note that if n is odd, the sum above clearly vanishes. It should be noted that one can also define the higher-dimensional Dedekind sum in terms of the hyperbolic cotangent:

$$d(q; p_1, \ldots, p_n) = \sum_{k=1}^{q-1} \coth \frac{\pi k p_1 i}{q} \cdots \coth \frac{\pi k p_n i}{q}.$$

The name "higher-dimensional Dedekind sum" is justified in part by the equality

$$d(q; p_1, p_2) = -\tfrac{2}{3} s(p_1 p_2^{-1}, q).$$

That is, when $n = 2$, the higher-dimensional Dedekind sum reduces to a classical Dedekind sum.

Zagier showed that his sums satisfy a reciprocity relation which generalizes the reciprocity formula for the classical Dedekind sum:

THEOREM 9.4.   Let $a_0, \ldots a_n$ be coprime integers, with n even. Then

$$\sum_{j=0}^{n} \frac{1}{a_j} d(a_j; a_0, \ldots, \widehat{a_j}, \ldots, a_n) = 1 - \frac{L_n(a_0, \ldots, a_n)}{a_0 \cdots a_n}.$$

Here $L_n(a_0, \ldots, a_n)$ is the nth L-polynomial, defined as the coefficient of t^n in the power series expansion of

$$\prod_{j=0}^{n} \frac{a_j t}{\tanh a_j t}.$$

Like the classical reciprocity law, this theorem may be proved using the connection with Todd classes: see the remark following Theorem 9.5.

While Theorem 9.4 is easily seen to be a generalization of Dedekind's classical reciprocity law, there is an important difference. As we have seen, the classical reciprocity law allows for efficient computation of the classical Dedekind sum. However, in the case of the higher dimensional sums, it is not at all clear how to use Theorem 9.4 to compute these sums efficiently. In fact, as Zagier points out, it is not even clear if this reciprocity law (together with obvious symmetry and periodicity properties) *characterizes* these sums. Below (Theorem 9.6), we show that these higher-dimensional sums can be computed in polynomial time when the dimension is fixed.

We next show that higher-dimensional Dedekind sums appear as coefficients in the power series t introduced in Section 8. The following theorem is due to Brion and Vergne. The cotangent formula below also appears in the important work [Diaz and Robins 1997], which is discussed briefly at the end of this section.

THEOREM 9.5. *Let n be even, and let K be a n-dimensional simplicial cone in $\mathbb{Z}^n$. Suppose that all faces of K of dimension $n-1$ are unimodular. This means that after a change of basis, the primitive lattice points on the rays of K may be chosen as*

$$v_1 = (1, 0, \ldots, 0),$$
$$v_2 = (0, 1, \ldots, 0),$$
$$\vdots$$
$$v_n = (p_1, \ldots, p_{n-1}, q),$$

with $q = \operatorname{ind}(K) > 0$. We then have the following expression for the coefficient of $y_1 \cdots y_d$ in the power series $t(y_1, \ldots y_n)$:

$$\frac{1}{2^n q} \sum_{k=1}^{q} \left(1 + \coth \frac{\pi k p_1 i}{q}\right) \cdots \left(1 + \coth \frac{\pi k p_{n-1} i}{q}\right) \left(1 + \coth \frac{\pi k i}{q}\right)$$

SKETCH OF PROOF. Here it is useful to have a dual expression for the power series t_K, which is formally equivalent to Part 3 of Proposition 8.2. Let $u_1, \ldots, u_n$ denote the primitive elements of the dual lattice satisfying $\langle u_i, v_j \rangle = 0$ for $i \neq j$, and for $v \in \mathbb{Z}^n$, define

$$a_j(v) = \exp\left(2\pi i \frac{\langle u_j, v \rangle}{\langle u_j, v_j \rangle}\right).$$

Let Π_K denote the fundamental parallelepiped for K. (Note this differs from Π above, which is the parallelepiped for the dual cone K^*.) We then have

$$s_K(y) = \frac{1}{\operatorname{ind}(K)} \sum_{v \in \Pi_K} \prod_{j=1}^{n} \frac{1}{1 - a_j(v) e^{-y_i}} \, ;$$

see [Brion and Vergne 1997a]. The desired coefficient of $y_1 \cdots y_n$ is then easily computed, keeping in mind that

$$\frac{1}{1 - e^{-z}} = 1 + \coth \frac{z}{2}. \qquad \square$$

By multiplying out the product appearing in Theorem 9.5, one obtains an expression for this coefficient as a sum of higher-dimensional Dedekind sums. The "leading term" is the n-dimensional Dedekind sum $d(q; p_1, \ldots, p_{n-1}, 1)$.

The formula above can be used to give a simple geometric proof of Zagier's reciprocity formula (Theorem 9.4). In $\mathbb{Z}^n$, take the unimodular cone K spanned by the standard unit vectors. Subdivide K into n cones by introducing the ray through $(b_1, \ldots b_n)$, where the b_i's are pairwise coprime positive integers. Applying the additivity formula for t (Proposition 8.2) to this subdivision yields a relation on the coefficients in the power series associated to these cones. Since Theorem 9.5 identifies certain of these coefficients as Dedekind sums, we get a relation among Dedekind sums. This relation is easily seen to yield Zagier's higher reciprocity law.

We now show that the results of Section 8 provide for efficient computation of the higher-dimensional Dedekind sum, a result not obvious from the reciprocity law (Theorem 9.4).

THEOREM 9.6. *For fixed dimension n, there is a polynomial time algorithm which, given integers $q > 0$ and $p_1, \ldots p_n$ relatively prime to q, computes the higher-dimensional Dedekind sum $d(q; p_1, \ldots, p_n)$*

SKETCH OF PROOF. First notice that we can reduce to the case in which $p_n = 1$. This is because of the easily verified identity:

$$d(q; p_1, \ldots, p_n) = d(q; p_n^{-1}p_1, \ldots, p_n^{-1}p_{n-1}, 1),$$

where p_n^{-1} is a multiplicative inverse of p_n modulo q.

Applying Theorem 8.6, we see that the product of Theorem 9.5 is computable in polynomial time for fixed dimension n. This product, when expanded, can be expressed as the sum of 2^n higher-dimensional Dedekind sums (many of which are zero). Of these sums, all but the leading term, $d(q; p_1, \ldots, p_{n-1}, 1)$ are Dedekind sums of dimension less than d. By induction, these may be computed in polynomial time. Thus, $d(q; p_1, \ldots, p_{n-1}, 1)$ itself may be computed in polynomial time. $\square$

Theorem 9.5 links a certain coefficient in the power series t_K with the higher-dimensional Dedekind sum. This Dedekind sum appears in the very special case when all facets of the cone K are unimodular. For general cones K, the general coefficient in t_K represents a significant further generalization of Zagier's sums. These coefficients also admit cotangent formulas, using the ideas of the proof of Theorem 9.5.

In fact, these more general cotangent sums can be seen in the remarkable lattice point formula of Diaz and Robins [1997]. This very explicit formula expresses the number of lattice points in an arbitrary simplex in terms of cotangent sums. The sums that appear are easily seen to match the cotangent sums that appear as coefficients in the power series t. The cotangent formulas Diaz and

Robins give are expressed in an explicit and appealing form. However, as written they do not appear to be efficiently computable. This is because the size of these sums may be as large as the indices of the tangent cones to the polytope. As discussed in Section 4, these indices are not bounded by a polynomial in the input size of the cone. Nevertheless, because these sums may be expressed in terms of the coefficients of t, they are in fact computable in polynomial time in fixed dimension, by Theorem 8.6. The forthcoming paper [Pommersheim and Robins $\geq$ 1999] will explore details of this connection, as well as reciprocity relations satisfied by these interesting cotangent sums.

10. What is the Complexity of the Presburger Arithmetic?

In Section 4, we saw that the generating function for the set of integer points in a polyhedron P,

$$f(P; \boldsymbol{x}) = \sum_{m \in P \cap \mathbb{Z}^d} \boldsymbol{x}^m,$$

has a "short" (polynomial in the input size of P) representation as a rational function. We now switch gears and consider sets $S \subset \mathbb{Z}^d$ that are *given* by their rational generating function

$$f(S; \boldsymbol{x}) = \sum_{i \in I} \alpha_i \frac{\boldsymbol{x}^{p_i}}{(1 - \boldsymbol{x}^{a_{i1}}) \cdots (1 - \boldsymbol{x}^{a_{ik}})},$$

where $\alpha_i \in \mathbb{Q}$ and $p_i; a_{i1}, \ldots, a_{is} \in \mathbb{Z}^d$, which can be expanded into the Laurent series

$$f(S; \boldsymbol{x}) = \sum_{m \in S} \boldsymbol{x}^m.$$

There is an ambiguity here since $f(S; \boldsymbol{x})$ may have a number of different Laurent expansions. To fix this, we assume that the set S is finite: from the computational complexity point of view, there is not a big difference between infinite and very large finite sets.

We will be interested in the sets S that can be encoded by a "short" rational function $f(S; \boldsymbol{x})$. In particular, we will assume that both the number d of variables and the number k of binomials $(1 - \boldsymbol{x}^{a_{ij}})$ in the denominator of each fraction are fixed. Our main thesis is that the set $S \subset \mathbb{Z}^d$ is computationally tractable if and only if it can be encoded by a short rational function.

EXAMPLES 10.1. Suppose that a and b are coprime positive integers. Let $S = \{\alpha a + \beta b : \alpha, \beta \in \mathbb{Z}_+\}$ be the set of all nonnegative integer combinations of a and b. Then, for $|x| < 1$ we have

$$f(S; x) = \sum_{m \in S} x^m = \frac{1 - x^{ab}}{(1 - x^a)(1 - x^b)}.$$

The set S contains all sufficiently large positive integers, so the initial interval of S is the main interest. We can write

$$f(S; x) = p(x) + \frac{x^N}{1 - x}$$

for some polynomial $p(x)$ and some positive integer N. The polynomial $p(x)$ encodes in a compact way information regarding the initial interval of S.

Suppose that a, b, and c are coprime positive integers. Let $S = \{\alpha a + \beta b + \gamma c : \alpha, \beta, \gamma \in \mathbb{Z}_+\}$ be the set of all nonnegative integer combinations of a, b, and c. Then for $|x| < 1$, we have

$$f(S; x) = \frac{1 - x^{n_1} - x^{n_2} - x^{n_3} + x^{n_4} + x^{n_5}}{(1 - x^a)(1 - x^b)(1 - x^c)},$$

for certain positive integers n_1, n_2, n_3, n_4, and n_5. This interesting fact was discovered, apparently, by G. Denham [1996]. This result follows also from a general result from [Peeva and Sturmfels 1996]. The paper [Székely and Wormald 1986] contains an elementary proof that the number of terms in the numerator is at most 12, which is twice as many as the sharp bound.

For example, if $a = 23$, $b = 29$, and $c = 44$, then

$$f(S; x) = \frac{1 - x^{161} - x^{203} - x^{220} + x^{249} + x^{335}}{(1 - x^{23})(1 - x^{29})(1 - x^{44})}.$$

As in the previous example, the set S contains all sufficiently large positive integers, so the initial interval of S is the interesting part. It can be encoded by a short polynomial.

QUESTION. What information regarding S can be extracted from $f(S; x)$ and what operations on sets given by their generating functions $f(S; x)$ can be carried out efficiently?

For example, by substituting $x = (1, \ldots, 1)$, one can get the number of points in S (since $x = (1, \ldots, 1)$ is the pole of every fraction, one may need to compute an appropriate residue, as in Section 5).

Our next result is that one can efficiently perform Boolean operations on sets given by their functions $f(S; x)$. In fact, we prove a more general statement: the Hadamard product of short rational functions can be computed in polynomial time.

THEOREM 10.2. *Suppose that $f_1(x)$ and $f_2(x)$ are rational functions in d complex variables $x = (x_1, \ldots, x_d)$:*

$$f_1(x) = \sum_{i \in I_1} \alpha_i \frac{x^{p_i}}{(1 - x^{a_{i1}}) \cdots (1 - x^{a_{ik}})}, \quad f_2(x) = \sum_{i \in I_2} \beta_i \frac{x^{q_i}}{(1 - x^{b_{i1}}) \cdots (1 - x^{b_{ik}})},$$

where $\alpha_i, \beta_i \in \mathbb{Q}$ and $p_i, a_{i1}, \ldots, a_{ik}; q_i, b_{i1}, \ldots, b_{ik} \in \mathbb{Z}^d$.

Suppose further that

$$f_1(x) = \sum_{m \in \mathbb{Z}^d} \gamma_m x^m \quad \text{and} \quad f_2(x) = \sum_{m \in \mathbb{Z}^d} \delta_m x^m$$

are the Laurent expansions of f_1 and f_2, respectively, for $\mathbf{x}$ in a nonempty open set $U \subset \mathbb{C}^d$, such that $|x^{a_{ij}}| < 1$ and $|x^{b_{ij}}| < 1$ for all i, j and all $x \in U$.

Then there exists a rational function $f(x) = f_1(x) \star f_2(x)$, called the Hadamard product of f_1 and f_2,

$$f(x) = \sum_{i \in I} \varepsilon_i \frac{x^{r_i}}{(1 - x^{c_{i1}}) \cdots (1 - x^{c_{i(2k)}})}$$

with the Laurent expansion

$$f(x) = \sum_{m \in \mathbb{Z}^d} (\gamma_m \delta_m) x^m$$

for $x \in U$. Moreover, for fixed d and k there is a polynomial time algorithm for computing $f(x)$ from $f_1(x)$ and $f_2(x)$.

SKETCH OF PROOF. Since the Hadamard product is a bilinear operation, it suffices to compute $f_1(x) \star f_2(x)$ in the particular case of the simple fractions:

$$f_1(x) = \frac{x^p}{(1 - x^{a_1}) \cdots (1 - x^{a_k})} \quad \text{and} \quad f_2(x) = \frac{x^q}{(1 - x^{b_1}) \cdots (1 - x^{b_k})}.$$

Expanding both fractions as multiple geometric series, we get

$$f_1(x) = \sum_{\mu_1, \ldots, \mu_k \geq 0} x^{p + \mu_1 a_1 + \cdots + \mu_k a_k} \quad \text{and} \quad f_2(x) = \sum_{\nu_1, \ldots, \nu_k \geq 0} x^{q + \nu_1 b_1 + \cdots + \nu_k b_k}.$$

In the space $\mathbb{R}^{2k} = \{(\xi_1, \ldots, \xi_k; \eta_1, \ldots, \eta_k)\}$, let P be the rational polyhedron defined by the equations

$$p + \xi_1 a_1 + \cdots + \xi_k a_k = q + \eta_1 b_1 + \cdots + \eta_k b_k$$

and the inequalities

$$\xi_i, \eta_j \geq 0 \quad \text{for } i, j = 1, \ldots, k.$$

Let $\mathbb{Z}^{2k} \subset \mathbb{R}^{2k}$ be the integer lattice $\{(\mu_1, \ldots, \mu_k; \nu_1, \ldots, \nu_k) : \mu_i, \nu_j \in \mathbb{Z}\}$.
Since

$$x^m \star x^n = \begin{cases} x^m & \text{if } m = n, \\ 0 & \text{if } m \neq n, \end{cases}$$

and the Hadamard product is bilinear, we can write the series for $f(x) = f_1(\mathbf{x}) \star f_2(x)$ as a sum over the set of integer points in the polyhedron P:

$$f(x) = \sum_{P \cap \mathbb{Z}^{2k}} x^{p + \mu_1 a_1 + \cdots + \mu_k a_k} = x^p \sum_{P \cap \mathbb{Z}^{2k}} x^{\mu_1 a_1 + \cdots + \mu_k a_k}.$$

By Theorem 4.4, the series

$$\sum_{(m,n)\in P\cap\mathbb{Z}^{2k}} y^m z^n,$$

where $m = (\mu_1,\ldots,\mu_k)$, $n = (\nu_1,\ldots,\nu_k)$ and $y, z \in \mathbb{C}^k$ can be computed in polynomial time as a rational function $F(y,z)$. The function $f(x)$ is obtained by specializing $x^p F(y,z)$ for $z = (1,\ldots,1)$ and $y_i = x^{a_i}$. One may have to resolve singularities as in Section 5.   □

REMARK. If the functions f_1 and f_2 are sufficiently generic, that is, if the vectors $a_{i1},\ldots,a_{ik};b_{i1},\ldots,b_{ik}$ span $\mathbb{R}^d$, we will have dim $P = 2k - d$ and

$$f(x) = \sum_{i\in I}\varepsilon_i \frac{x^{r_i}}{(1 - x^{c_{i1}})\cdots(1 - x^{c_{i(2k-d)}})}.$$

COROLLARY 10.3. *Fix d and k. There exists a polynomial time algorithm which, for any finite sets $S_1 \subset \mathbb{Z}^d$ and $S_2 \subset \mathbb{Z}^d$, given by their generating functions*

$$f(S_1;x) = \sum_{i\in I_1}\alpha_i \frac{x^{p_i}}{(1 - x^{a_{i1}})\cdots(1 - x^{a_{ik}})}$$

and

$$f(S_2;x) = \sum_{i\in I_2}\beta_i \frac{x^{q_i}}{(1 - x^{b_{i1}})\cdots(1 - x^{b_{ik}})},$$

where $\alpha_i, \beta_i \in \mathbb{Q}$ and $p_i, a_{i1},\ldots,a_{ik}; q_i, b_{i1},\ldots,b_{ik} \in \mathbb{Z}^d$, computes the generating functions $f(S_1 \cup S_2;x)$, $f(S_1 \cap S_2;x)$, and $f(S_1 \setminus S_2;x)$.

PROOF. Choose a generic vector $c \in \mathbb{R}^d$, such that $\langle c, a_{ij}\rangle \neq 0$ and $\langle c, b_{ij}\rangle \neq 0$ for all vectors a_{ij} and b_{ij}. By multiplying, if necessary, the denominator and the numerator of each fraction by an appropriate monomial, we can always assume that $\langle c, a_{ij}\rangle < 0$ and $\langle c, b_{ij}\rangle < 0$ for all a_{ij} and b_{ij}. Then the set

$$U = \{x \in \mathbb{C}^d : |x^{a_{ij}}| < 1 \text{ for } i \in I_1,\ j = 1,\ldots,k$$

$$\text{and } |x^{b_{ij}}| < 1 \text{ for } i \in I_2,\ j = 1,\ldots,k\}$$

is a nonempty open set in $\mathbb{C}^d$, and for every $x \in U$, there are Laurent expansions:

$$f(S_1;x) = \sum_{m\in S_1} x^m \quad \text{and} \quad f(S_2;x) = \sum_{m\in S_2} x^m.$$

The corollary now follows from Theorem 10.2, since

$$f(S_1 \cap S_2;x) = f(S_1;x) \star f(S_2;x),$$
$$f(S_1 \cup S_2;x) = f(S_1;x) + f(S_2;x) - f(S_1 \cap S_2;x),$$
$$f(S_1 \setminus S_2;x) = f(S_1;x) - f(S_1 \cap S_2;x).$$ □

The most intriguing question is whether the *projection* of a set with a short generating function is a set with a short generating function.

QUESTION. Let $\pi : \mathbb{Z}^d \to \mathbb{Z}^{d-1}$ be the projection

$$\pi(\xi_1, \ldots, \xi_d) = (\xi_1, \ldots, \xi_{d-1}).$$

Let $S \subset \mathbb{Z}^d$ be a set given by its rational generating function

$$f(S; \boldsymbol{x}) = \sum_{i \in I} \alpha_i \frac{\boldsymbol{x}^{p_i}}{(1 - \boldsymbol{x}^{a_{i1}}) \cdots (1 - \boldsymbol{x}^{a_{ik}})}.$$

Assuming that d and k are fixed, is there a polynomial time algorithm for computing $f\big(\pi(S); \boldsymbol{y}\big)$, $\boldsymbol{y} \in \mathbb{C}^{d-1}$, from $f(S; \boldsymbol{x})$?

An affirmative answer to this question would lead to the solution of a long-standing open problem about the complexity of the Presburger arithmetic. Indeed, consider a formula with quantifiers and Boolean operations which involves integer variables with the usual order $<$, additive operations "+" and "−", and multiplication onto integer constants, but not variables. One can come up with an algorithm to verify the truth or falsity of such a formula, but as M. Fischer and M. Rabin proved [1974], any such algorithm will have a double exponential complexity. But what if the number of variables and Boolean operations is fixed and not a part of the input? It has long been suspected that then the problem admits a polynomial time algorithm.

Indeed, suppose that the answer to the question is Yes. If we start with the set of integer points in a given rational polytope $P \subset \mathbb{R}^d$ (whose generating function by Theorem 4.4 can be computed in polynomial time) and apply a sequence of projections and Boolean operations (see Corollary 10.3), we get a set of points described by a polynomially computable rational function, provided the dimension d and the number of Boolean operations is fixed. Generally, this way we get a set of points described by a formula of the *Presburger arithmetic*

$$x \in \mathbb{Z}^d : \quad \exists \xi_1 \in \mathbb{Z} \; \forall \xi_2 \in \mathbb{Z} \; \exists \xi_3 \in \mathbb{Z} \ldots \forall \xi_k \in \mathbb{Z} : \quad F(x, \xi_1, \ldots, \xi_k),$$

where F is a quantifier-free formula involving linear inequalities with constant rational coefficients and Boolean operations. In other words, if projection preserves "short" rational functions, the set of points described by a formula of Presburger arithmetic has a generating function whose size is bounded by a polynomial in the size of the coefficients of the formula, provided the number of variables and the number of Boolean operations is fixed. In particular, there would be a polynomial time algorithm for testing the truth/falsity of a formula of Presburger arithmetic, provided the number of variables and the number of Boolean operations is fixed. At present, a polynomial time algorithm for this problem is known if there is at most one quantifier alteration [Kannan 1990]. For example, the following question, known as the Frobenius problem, can be posed as a problem with one quantifier alteration: given coprime positive integers $a_1, \ldots, a_d$ and an integer N, is it true that any integer number $n \geq N$ can be represented as a nonnegative integer combination of $a_1, \ldots, a_d$? For a fixed

d, a polynomial time algorithm for this problem was constructed by R. Kannan [1992].

Sometimes it is easy to see that the projection indeed has a short generating function.

EXAMPLE 10.4. Let $P \subset \mathbb{Z}^d$ be a rational polytope, and let $S = P \cap \mathbb{Z}^d$ be the set of integer points in P. Then projection $\pi(S) \subset \mathbb{Z}^{d-1}$ has a generating function whose complexity is polynomial in the input size of P. (Note that $\pi(S)$ is *not* the set of integer points in a rational polytope.) Let $a = (0, \ldots, 0, 1) \in \mathbb{R}^d$ be a vector and let $Q = P \setminus (P + a)$. For $m \in \pi(S)$, the preimage $\pi^{-1}(m) \cap S$ is the set of integer points in the interval $\pi^{-1}(m) \cap P$. It then follows that $m \in \pi(S)$ if and only if there is exactly one integer point $n \in Q$ such that $\pi(n) = m$. Hence, $f(\pi(S); y) : y \in \mathbb{C}^{d-1}$ is the specialization of the generating function $f(Q \cap \mathbb{Z}^d; x) : x = (y, z) \in \mathbb{C}^d$ when $z = 1$. Using Theorem 4.4, we conclude that $f(Q \cap \mathbb{Z}^d; x)$ can be computed in polynomial time. Specialization at $z = 1$ may require resolution of singularities as in Section 5 (if the expression for $f(Q \cap \mathbb{Z}^d; x)$ contains fractions with a monomial $(1 - z^k)$ in the denominator). Hence we get a polynomial time algorithm for computing $f(\pi(S); y)$.

DEFINITION 10.5. Fix the decomposition $\mathbb{Z}^d = \mathbb{Z}^l \oplus \mathbb{Z}^{d-l}$, and let $\pi : \mathbb{Z}^d \to \mathbb{Z}^{d-l}$ be the projection on the second summand.

For a set $S \subset \mathbb{Z}^d$ and a point $m \in \mathbb{Z}^{d-l}$, we define the *fiber* $\pi^{-1}(m) \subset \mathbb{Z}^l$ as

$$\pi^{-1}(m) = \{n \in \mathbb{Z}^l : (n, m) \in S\}.$$

We have $\pi^{-1}(m) \subset \mathbb{Z}^l$, and we can consider the generating function $f(\pi^{-1}(m); z)$ for $z \in \mathbb{C}^l$.

One can prove that if $S \subset \mathbb{Z}^d$ is described by a short rational function, the fibers $\pi^{-1}(m)$ are described by a "consistent" system of short rational functions as $m \in \mathbb{Z}^{d-l}$ changes.

We use the term *open polyhedron* to mean the relative interior of a polyhedron.

THEOREM 10.6. *Let $S \subset \mathbb{Z}^d$ be a finite set with the generating function*

$$f(S; x) = \sum_{i \in I} \alpha_i \frac{x^{p_i}}{(1 - x^{a_{i1}}) \cdots (1 - x^{a_{ik}})}.$$

Then the space $\mathbb{R}^l$ can be represented as a disjoint union of open polyhedra Q_j, and for every Q_j, there exist vectors $b_{i1}, \ldots, b_{ik} \subset \mathbb{Z}^l : i \in I_j$, such that for every $m \in Q_j \cap \mathbb{Z}^l$,

$$f(\pi^{-1}(m); z) = \sum_{i \in I_j} \beta_i \frac{z^{q_i}}{(1 - z^{b_{i1}}) \cdots (1 - z^{b_{ik}})},$$

where $\beta_i = \beta_i(m)$ are rational numbers and $q_i = q_i(m) \in \mathbb{Z}^l$. In other words, as long as m stays within Q_j, the denominators of the fractions do not change.

If d and k are fixed, there is a polynomial time algorithm which, given $f(S;x)$, computes the decomposition of $\mathbb{R}^{d-l}$ into pieces Q_j and the vectors b_{ij}.

SKETCH OF PROOF. As in the proof of Corollary 10.3, we may assume that $|x^{a_{ij}}| < 1$ for all x in some open set $U \subset \mathbb{C}^d$. We may write

$$f(S;x) = \sum_{i \in I} \alpha_i \sum_{\mu_1,\dots,\mu_k \geq 0} x^{p_i + \mu_1 a_{i1} + \cdots + \mu_k a_{ik}}.$$

Let $A_i(m) \subset \mathbb{R}^k$ be the affine subspace consisting of the points $(\mu_1,\dots,\mu_k)$ such that $\pi(p_i + \mu_1 a_{i1} + \cdots + \mu_k a_{ik}) = m$. Splitting $x = (y, z)$, where $y \in \mathbb{C}^{d-l}$ and $z \in \mathbb{C}^l$, we may write the generating function $f(\pi^{-1}(m), z)$ of the fiber as a specialization of

$$\sum_{i \in I} \alpha_i \sum_{(\mu_1,\dots,\mu_k) \in \mathbb{Z}_+^k \cap A_i(m)} x^{p_i + \mu_1 b_{i1} + \cdots + \mu_k b_{ik}},$$

at $y = (1,\dots,1)$. Now for every $i \in I$, the set $\mathbb{Z}_+^k \cap A_i(m)$ is the set of integer points in a polyhedron whose facets are moved parallel to themselves as $m \in \mathbb{R}^{d-l}$ changes. Let Q_j be a partition of $\mathbb{R}^{d-l}$ such that the combinatorial type of every polytope $\mathbb{R}_+^k \cap A_i(m)$ stays the same as long as m changes within Q_j. The proof now follows from Theorem 4.4. $\qquad\square$

REMARK. From Theorem 4.4, we can deduce that $\beta_i(m)$ and $q_i(m)$ can be expressed by a polynomial size formula, involving arithmetic operations, the floor function, and Boolean functions.

An important feature of Example 10.4 is that the generating function of each fiber is *very short*. Since every fiber is an interval, we have $f(\pi^{-1}(m); z) = (z^a - z^b)/(1 - z)$. This observation can be generalized: one can show that if the function $f(\pi^{-1}(m); z)$ contains only a fixed number of terms, the complexity of the generating function $f(\pi(S); y)$ is bounded by a polynomial in the input size of the function $f(S; x)$.

EXAMPLE 10.7. Let $b_1,\dots,b_d$ be coprime positive integers, let $B = b_1,\dots,b_d$ be their product and let $a_i = B/b_i$. Let $S = \mathbb{Z}_+^d$, and let $\pi : \mathbb{Z}^d \to \mathbb{Z}$ be the projection: $\pi(\alpha_1,\dots,\alpha_d) = \alpha_1 a_1 + \cdots + \alpha_d a_d$. Then $\pi(S) \subset \mathbb{Z}$ is the set of all nonnegative integer combinations of a_i. The set S can be represented as a projection of the nonnegative integer orthant $\pi : \mathbb{Z}^d \to \mathbb{Z}$.

One can show that the fiber $\pi^{-1}(m)$ is the set of integer points in a totally unimodular simplex (all tangent cones are rational shifts of unimodular cones). Here the generating function of each fiber contains d terms; that is, it has a fixed number of terms provided d is fixed.

It turns out that the set $\pi(S)$ has a short generating function:

$$f(\pi(S); x) = \frac{(1 - x^B)^{d-1}}{(1 - x^{a_1}) \cdots (1 - x^{a_d})}.$$

Acknowledgment

Barvinok was partially supported by the Alfred P. Sloan Research Fellowship, by NSF grants DMS 9501129 and DMS 9734138, and by the Mathematical Sciences Research Institute through NSF grant DMS 9022140. Pommersheim was partially supported by NSF grant DMS 9508972. We wish to thank Michel Brion, Stavros Garoufalidis, Sinai Robins, and Bernd Sturmfels for useful conversations.

References

[Barvinok 1992]   A. Barvinok, "Computing the Ehrhart polynomial of a convex lattice polytope", preprint TRITA-MAT-1992-0036, Royal Institute of Technology, Stockholm, October 1992.

[Barvinok 1993] A. I. Barvinok, "Computing the volume, counting integral points, and exponential sums", *Discrete Comput. Geom.* **10**:2 (1993), 123–141.

[Barvinok 1994a]   A. I. Barvinok, "Computing the Ehrhart polynomial of a convex lattice polytope", *Discrete Comput. Geom.* **12**:1 (1994), 35–48.

[Barvinok 1994b]   A. I. Barvinok, "A polynomial time algorithm for counting integral points in polyhedra when the dimension is fixed", *Math. Oper. Res.* **19**:4 (1994), 769–779.

[Betke and Kneser 1985]   U. Betke and M. Kneser, "Zerlegungen und Bewertungen von Gitterpolytopen", *J. Reine Angew. Math.* **358** (1985), 202–208.

[Bollobás 1997]   B. Bollobás, "Volume estimates and rapid mixing", pp. 151–182 in *Flavors of geometry*, edited by S. Levy, Math. Sci. Res. Inst. Publications **31**, Cambridge University Press, New York, 1997.

[Brion 1988]   M. Brion, "Points entiers dans les polyèdres convexes", *Ann. Sci. École Norm. Sup.* (4) **21**:4 (1988), 653–663.

[Brion and Vergne 1997a]   M. Brion and M. Vergne, "An equivariant Riemann–Roch theorem for complete, simplicial toric varieties", *J. Reine Angew. Math.* **482** (1997), 67–92.

[Brion and Vergne 1997b]   M. Brion and M. Vergne, "Lattice points in simple polytopes", *J. Amer. Math. Soc.* **10**:2 (1997), 371–392.

[Brion and Vergne 1997c] M. Brion and M. Vergne, "Residue formulae, vector partition functions and lattice points in rational polytopes", *J. Amer. Math. Soc.* **10**:4 (1997), 797–833.

[Cappell and Shaneson 1994]   S. E. Cappell and J. L. Shaneson, "Genera of algebraic varieties and counting of lattice points", *Bull. Amer. Math. Soc.* (*N.S.*) **30**:1 (1994), 62–69.

[Cook et al. 1992]   W. Cook, M. Hartmann, R. Kannan, and C. McDiarmid, "On integer points in polyhedra", *Combinatorica* **12**:1 (1992), 27–37.

[Denham 1996]   G. Denham, "The Hilbert series of a certain module", 1996. Unpublished manuscript.

[Diaz and Robins 1997]   R. Diaz and S. Robins, "The Ehrhart polynomial of a lattice polytope", *Ann. of Math.* (2) **145**:3 (1997), 503–518. Erratum in **146**:1 (1997), 237.

[Diaz and Robins ≥ 1999] R. Diaz and S. Robins, "Solid angles and a Fourier decomposition of lattice polytopes". To appear.

[Dyer 1991] M. Dyer, "On counting lattice points in polyhedra", *SIAM J. Comput.* **20**:4 (1991), 695–707.

[Dyer and Kannan 1997] M. Dyer and R. Kannan, "On Barvinok's algorithm for counting lattice points in fixed dimension", *Math. Oper. Res.* **22**:3 (1997), 545–549.

[Dyer et al. 1991] M. Dyer, A. Frieze, and R. Kannan, "A random polynomial-time algorithm for approximating the volume of convex bodies", *J. Assoc. Comput. Mach.* **38**:1 (1991), 1–17.

[Dyer et al. 1993] M. Dyer, A. Frieze, R. Kannan, A. Kapoor, L. Perkovic, and U. Vazirani, "A mildly exponential time algorithm for approximating the number of solutions to a multidimensional knapsack problem", *Combin. Probab. Comput.* **2**:3 (1993), 271–284.

[Dyer et al. 1997] M. Dyer, R. Kannan, and J. Mount, "Sampling contingency tables", *Random Structures Algorithms* **10**:4 (1997), 487–506.

[Ehrhart 1977] E. Ehrhart, *Polynômes arithmétiques et méthode des polyèdres en combinatoire*, International Series of Numerical Mathematics **35**, Birkhäuser, Basel, 1977.

[Fischer and Rabin 1974] M. J. Fischer and M. O. Rabin, "Super-exponential complexity of Presburger arithmetic", pp. 27–41 in *Complexity of computation* (New York, 1973), edited by R. M. Karp, SIAM-AMS Proceedings **7**, Amer. Math. Soc., Providence, 1974.

[Fulton 1993] W. Fulton, *Introduction to toric varieties*, Princeton University Press, Princeton, NJ, 1993.

[Ginzburg et al. ≥ 1999] V. Ginzburg, V. Guillemin, and S. Sternberg, *Cobordism techniques in symplectic geometry*, Carus Mathematical Monographs, Mathematical Association of America, Washington, DC. To appear.

[Grötschel et al. 1993] M. Grötschel, L. Lovász, and A. Schrijver, *Geometric algorithms and combinatorial optimization*, Second ed., Springer, Berlin, 1993.

[Hirzebruch 1966] F. Hirzebruch, *Topological methods in algebraic geometry*, Third ed., Grundlehren der Mathematischen Wissenschaften **131**, Springer, New York, 1966. Reprinted 1995 in series Classics in Mathematics.

[Infirri 1992] S. Infirri, "Lefschetz fixed-point theorem and lattice points in convex polytopes", preprint, 1992.

[Kannan 1990] R. Kannan, "Test sets for integer programs, ∀∃ sentences", pp. 39–47 in *Polyhedral combinatorics* (Morristown, NJ, 1989), edited by W. Cook and P. D. Seymour, DIMACS Ser. Discrete Math. Theoret. Comput. Sci. **1**, Amer. Math. Soc., Providence, RI, 1990.

[Kannan 1992] R. Kannan, "Lattice translates of a polytope and the Frobenius problem", *Combinatorica* **12**:2 (1992), 161–177.

[Kannan et al. 1997] R. Kannan, L. Lovász, and M. Simonovits, "Random walks and an $O^*(n^5)$ volume algorithm for convex bodies", *Random Structures Algorithms* **11**:1 (1997), 1–50.

[Kantor and Khovanskii 1992] J. M. Kantor and A. G. Khovanskii, "Integral points in convex polyhedra, combinatorial Riemann–Roch Theorem and generalized Euler-Maclaurin formula", preprint 1992/24, Inst. Hautes Ét. Scient., Bures-sur-Yvette, 1992.

[Kantor and Khovanskii 1993] J.-M. Kantor and A. Khovanskii, "Une application du théorème de Riemann–Roch combinatoire au polynôme d'Ehrhart des polytopes entiers de $\mathbb{R}^{d}$", C. R. Acad. Sci. Paris Sér. I Math. **317**:5 (1993), 501–507.

[Khovanskii and Pukhlikov 1993] A. G. Khovanskii and A. V. Pukhlikov, "Integral transforms based on Euler characteristic and their applications", Integral Transform. Spec. Funct. **1**:1 (1993), 19–26.

[Lagarias 1995] J. C. Lagarias, "Point lattices", pp. 919–966 in Handbook of combinatorics, vol. 1, edited by E. by R. L. Graham et al., Elsevier, Amsterdam, 1995.

[Lawrence 1988] J. Lawrence, "Valuations and polarity", Discrete Comput. Geom. **3**:4 (1988), 307–324.

[Lawrence 1991] J. Lawrence, "Rational-function-valued valuations on polyhedra", pp. 199–208 in Discrete and computational geometry (New Brunswick, NJ, 1989/1990), edited by R. P. Jacob E. Goodman and W. Steiger, DIMACS Ser. Discrete Math. Theoret. Comput. Sci. **6**, Amer. Math. Soc., Providence, RI, 1991.

[Lenstra 1983] H. W. Lenstra Jr., "Integer programming with a fixed number of variables", Math. Oper. Res. **8**:4 (1983), 538–548.

[Lenstra et al. 1982] A. K. Lenstra, H. W. Lenstra Jr., and L. Lovász, "Factoring polynomials with rational coefficients", Math. Ann. **261**:4 (1982), 515–534.

[Lovász 1986] L. Lovász, An algorithmic theory of numbers, graphs and convexity, Soc. Industr. Appl. Mach., Philadelphia, 1986.

[McMullen 1975] P. McMullen, "Non-linear angle-sum relations for polyhedral cones and polytopes", Math. Proc. Cambridge Philos. Soc. **78**:2 (1975), 247–261.

[McMullen 1978/79] P. McMullen, "Lattice invariant valuations on rational polytopes", Arch. Math. (Basel) **31**:5 (1978/79), 509–516.

[McMullen 1989] P. McMullen, "The polytope algebra", Adv. Math. **78**:1 (1989), 76–130.

[McMullen 1993] P. McMullen, "Valuations and dissections", pp. 933–988 in Handbook of convex geometry, vol. B, edited by P. M. Gruber and J. M. Wills, North-Holland, Amsterdam, 1993.

[McMullen and Schneider 1983] P. McMullen and R. Schneider, "Valuations on convex bodies", pp. 170–247 in Convexity and its applications, edited by P. M. Gruber and J. M. Wills, Birkhäuser, Basel, 1983.

[Mordell 1951] L. J. Mordell, "Lattice points in a tetrahedron and generalized Dedekind sums", J. Indian Math. Soc. (N.S.) **15** (1951), 41–46.

[Morelli 1993a] R. Morelli, "The K-theory of a toric variety", Adv. Math. **100**:2 (1993), 154–182.

[Morelli 1993b] R. Morelli, "Pick's theorem and the Todd class of a toric variety", Adv. Math. **100**:2 (1993), 183–231.

[Morelli 1993c] R. Morelli, "A theory of polyhedra", Adv. Math. **97**:1 (1993), 1–73.

[Morelli 1993d] R. Morelli, "Translation scissors congruence", *Adv. Math.* **100**:1 (1993), 1–27.

[Peeva and Sturmfels 1996] I. Peeva and B. Sturmfels, "Syzygies of codimension 2 lattice ideals", 1996. Manuscript.

[Pick 1899] G. Pick, "Geometrisches zur Zahlentheorie", *Sitzenber. Lotos (Prague)* **19** (1899), 311–319.

[Pommersheim 1993] J. E. Pommersheim, "Toric varieties, lattice points and Dedekind sums", *Math. Ann.* **295**:1 (1993), 1–24.

[Pommersheim 1996] J. E. Pommersheim, "Products of cycles and the Todd class of a toric variety", *J. Amer. Math. Soc.* **9**:3 (1996), 813–826.

[Pommersheim 1997] J. E. Pommersheim, "Barvinok's algorithm and the Todd class of a toric variety", *J. Pure Appl. Algebra* **117/118** (1997), 519–533.

[Pommersheim and Robins ≥ 1999] J. E. Pommersheim and S. Robins, "Higher-dimensional Dedekind sums: reciprocity laws, computational complexity, and geometry". In preparation.

[Pukhlikov and Khovanskii 1992a] A. V. Pukhlikov and A. G. Khovanskiĭ, "Finitely additive measures of virtual polyhedra", *Algebra i Analiz* **4**:2 (1992), 161–185. In Russian; translation in *St. Petersburg Mathematical Journal*, **4**:2 (1993), 337–356.

[Pukhlikov and Khovanskii 1992b] A. V. Pukhlikov and A. G. Khovanskiĭ, "The Riemann–Roch theorem for integrals and sums of quasipolynomials on virtual polytopes", *Algebra i Analiz* **4**:4 (1992), 188–216. In Russian; translation in *St. Petersburg Mathematical Journal*, **4**:4 (1993), 789–812.

[Rademacher and Grosswald 1972] H. Rademacher and E. Grosswald, *Dedekind sums*, The Carus Mathematical Monographs **16**, Math. Assoc. Amer., Washington, DC, 1972.

[Schrijver 1986] A. Schrijver, *Theory of linear and integer programming*, Wiley, Chichester, 1986.

[Skriganov 1998] M. M. Skriganov, "Ergodic theory on SL(n), Diophantine approximations and anomalies in the lattice point problem", *Invent. Math.* **132**:1 (1998), 1–72.

[Stanley 1993] R. P. Stanley, "A monotonicity property of h-vectors and h^*-vectors", *European J. Combin.* **14**:3 (1993), 251–258.

[Stanley 1997] R. P. Stanley, *Enumerative combinatorics*, vol. 1, Cambridge Studies in Advanced Mathematics **49**, Cambridge University Press, Cambridge, 1997. Corrected reprint of the 1986 original.

[Székely and Wormald 1986] L. A. Székely and N. C. Wormald, "Generating functions for the Frobenius problem with 2 and 3 generators", *Math. Chronicle* **15** (1986), 49–57.

[Zagier 1973] D. Zagier, "Higher dimensional Dedekind sums", *Math. Ann.* **202** (1973), 149–172.

[Ziegler 1995] G. M. Ziegler, *Lectures on polytopes*, Graduate Texts in Mathematics **152**, Springer, New York, 1995.

ALEXANDER BARVINOK
DEPARTMENT OF MATHEMATICS
UNIVERSITY OF MICHIGAN
ANN ARBOR, MI 48109-1109
UNITED STATES
  barvinok@math.lsa.umich.edu

JAMES E. POMMERSHEIM
DEPARTMENT OF MATHEMATICAL SCIENCES
NEW MEXICO STATE UNIVERSITY
LAS CRUCES, NM 88003
UNITED STATES
  jamie@math.nmsu.edu

New Perspectives in Geometric Combinatorics
MSRI Publications
Volume **38**, 1999

Some Algebraic Properties of the Schechtman–Varchenko Bilinear Forms

GRAHAM DENHAM AND PHIL HANLON

ABSTRACT. We examine a bilinear form associated with a real arrangement of hyperplanes introduced in [Schechtman and Varchenko 1991]. Our main objective is to show that the linear algebraic properties of this bilinear form are related to the combinatorics and topology of the hyperplane arrangement. We will survey results and state a number of open problems which relate the determinant, cokernel structure and Smith normal form of the bilinear form to combinatorial and topological invariants of the arrangement including the characteristic polynomial, combinatorial structure of the intersection lattice and homology of the Milnor fibre.

1. The Varchenko B Matrices

Let $\mathcal{A} = \{H_1, \ldots, H_l\}$ be an arrangement of hyperplanes in $\mathbb{R}^n$ and let $r(\mathcal{A}) = \{R_1, \ldots, R_m\}$ denote the set of regions in the complement of the union of $\mathcal{A}$. Let $L(\mathcal{A})$ denote the collection of intersections of hyperplanes in $\mathcal{A}$. Included in $L(\mathcal{A})$ is $\mathbb{R}^n$, which we think of as the intersection of the empty set of hyperplanes. We order the elements of $L(\mathcal{A})$ by reverse inclusion thus making it into a poset. It is well known that this poset is a meet semilattice and is a geometric lattice if the arrangement is central. We will abbreviate $L(\mathcal{A})$ to L when the arrangement is clear.

For regions $S, T \in r(\mathcal{A})$, define $\mathcal{H}(S, T)$ to be the set of hyperplanes in $\mathcal{A}$ which separate S from T. Varchenko [1993] defines a matrix $B = B(\mathcal{A})$ with rows and columns indexed by the regions in $r(\mathcal{A})$ by saying that the S, T entry in B is $\prod_{H \in \mathcal{H}(S,T)} a_H$, where a_H is an indeterminate assigned to the hyperplane H. We will call $B = B(\mathcal{A})$ the *Varchenko matrix* of the arrangement $\mathcal{A}$.

EXAMPLE 1.1. As a starting example, let $F = \{H_0, H_1, H_2\}$ be the arrangement in $\mathbb{R}^2$ where H_j is the line $y = (-1)^j x$ for $j = 0, 1$ and where H_2 is the line $y = 1$. Note that $r(F)$ consists of 7 regions. Let these regions be numbered $R_1, \ldots, R_7$,

---

This research was partially supported by the National Science Foundation.

as follows:

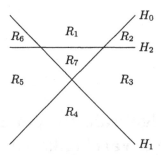

Let a_j be the weight assigned to the hyperplane H_j. Then the matrix B is given by

$$B = \begin{pmatrix} 1 & a_2 & a_0a_2 & a_0a_1a_2 & a_0a_1 & a_1 & a_0 \\ a_2 & 1 & a_0 & a_0a_1 & a_0a_1a_2 & a_1a_2 & a_0a_2 \\ a_0a_2 & a_0 & 1 & a_1 & a_1a_2 & a_0a_1a_2 & a_2 \\ a_0a_1a_2 & a_0a_1 & a_1 & 1 & a_2 & a_0a_2 & a_1a_2 \\ a_0a_1 & a_0a_1a_2 & a_1a_2 & a_2 & 1 & a_0 & a_1 \\ a_1 & a_1a_2 & a_0a_1a_2 & a_0a_2 & a_0 & 1 & a_0a_1 \\ a_0 & a_0a_2 & a_2 & a_1a_2 & a_1 & a_0a_1 & 1 \end{pmatrix}.$$

EXAMPLE 1.2. An important example that we will return to several times is the arrangement $\mathcal{A}$ consisting of the $\binom{n}{2}$ hyperplanes $H_{i,j}$ in $\mathbb{R}^n$ given by

$$H_{i,j} = \{(x_1,\ldots,x_n) : x_i = x_j\}.$$

Note that $\mathcal{A}$ consists of the reflecting hyperplanes for the root system $\mathbf{A}_{n-1}$ and so we denote this arrangement by $\mathbf{A}_{n-1}$. Two points $(x_1,\ldots,x_n)$ and $(y_1,\ldots,y_n)$ are in the same region of the complement if and only if the relative orders of their coordinates are the same. So, the permutations in S_n index the regions of the complement via the correspondence

$$\sigma \leftrightarrow \{(x_1,\ldots,x_n) : x_{\sigma 1} < x_{\sigma 2} < \cdots < x_{\sigma n}\}.$$

Let $a_{i,j} = a_{j,i}$ denote the weight assigned to the hyperplane $H_{i,j}$. For σ, τ in S_n, the σ, τ entry in B is the product of all $a_{i,j}$ such that i appears to the left of j in the one-line form of σ but to the right of j in the one-line form of τ. Another way of saying this is that $B_{\sigma,\tau}$ is the product of all $a_{i,j}$ such that $\{\sigma^{-1}(i), \sigma^{-1}(j)\}$ is an inversion of $\sigma\tau^{-1}$. In particular, if all parameters a_H are set equal to q, then the σ, τ entry of B is $q^{i(\sigma\tau^{-1})}$, where $i(\pi)$ denotes the number of inversions of π.

EXAMPLE 1.3. For each i, let O_i denote the hyperplane in $\mathbb{R}^n$ which consists of all vectors with a 0 in the i-th coordinate, and consider the arrangement $\{O_1, O_2, \ldots, O_n\}$. In this case, $r(\mathcal{A})$ has size 2^n and the individual regions can be indexed by sequences $S = (s_1, \ldots, s_n)$, where each s_i is either $+1$ or -1. The sequence S corresponds to the region R_S which contains all vectors $(x_1, \ldots, x_n)$ where $x_i < 0$ if $s_i = -1$ and $x_i > 0$ if $s_i = 1$. Given sequences S, T, the set of hyperplanes separating R_S and R_T is equal to the set of H_i such that the i-th coordinates of S and T differ. With $n = 3$ the matrix B appears below (relative to the ordering on regions which corresponds to the lexicographic order on the indexing sequences):

$$B = \begin{pmatrix} 1 & a_3 & a_2 & a_2 a_3 & a_1 & a_1 a_3 & a_1 a_2 & a_1 a_2 a_3 \\ a_3 & 1 & a_2 a_3 & a_2 & a_1 a_3 & a_1 & a_1 a_2 a_3 & a_1 a_3 \\ a_2 & a_2 a_3 & 1 & a_3 & a_1 a_2 & a_1 a_2 a_3 & a_1 & a_1 a_3 \\ a_2 a_3 & a_2 & a_3 & 1 & a_1 a_2 a_3 & a_1 a_2 & a_1 a_3 & a_1 \\ a_1 & a_1 a_3 & a_1 a_2 & a_1 a_2 a_3 & 1 & a_3 & a_2 & a_2 a_3 \\ a_1 a_3 & a_1 & a_1 a_2 a_3 & a_1 a_2 & a_3 & 1 & a_2 a_3 & a_2 \\ a_1 a_2 & a_1 a_2 a_3 & a_1 & a_1 a_3 & a_2 & a_2 a_3 & 1 & a_3 \\ a_1 a_2 a_3 & a_1 a_2 & a_1 a_3 & a_1 & a_2 a_3 & a_2 & a_3 & 1 \end{pmatrix}.$$

One of the first appearances of the B matrices was in the work of Schechtman and Varchenko [1991] on Drinfeld–Jimbo quantized Kac–Moody Lie algebras. To briefly describe this work, let H be a finite dimensional complex vector space equipped with a nondegenerate symmetric bilinear form $(\ ,\ )$. We carry the form over to the dual space H^* in the usual way. Let $\alpha_1, \ldots, \alpha_r$ be linearly independent elements in H^* and let q be a nonzero complex number. Define $U_q G$ to be the $\mathbb{C}$-algebra generated by the elements $1, e_i, f_i$ for $1 \le i \le r$ and H, subject to the relations

$$[h, e_i] = \langle \alpha_i, h \rangle e_i,$$
$$[h, f_i] = -\langle \alpha_i, h \rangle f_i,$$
$$[e_i, f_j] = (q^{h_i/2} - q^{-h_i/2}) \delta_{i,j},$$
$$[H, H] = 0.$$

Here $\delta_{i,j}$ is the Kronecker delta and $q^{uh} = \exp(u(\ln q)h)$ so that we allow convergent power series in $h \in H$ as part of $U_q G$. It is possible to define, in addition, a comultiplication Δ which maps $U_q G$ to $U_q G \otimes U_q G$ and an antipode ε on $U_q G$ so that $(U_q G, \Delta, \varepsilon)$ is a Hopf algebra. We won't need the Hopf algebra structure here so we refer the reader to [Varchenko 1995, Chapter 4] for their definition.

Let $U_q N_-$ be the subalgebra generated by $f_1, \ldots f_r$ and let L be the multilinear part of $U_q N_-$. In other words, L is spanned by words in $f_1, \ldots, f_r$ in which each f_i occurs exactly once. So L has dimension $r!$ with basis $f_\sigma := f_{\sigma 1} \ldots f_{\sigma r}$ for $\sigma \in S_r$.

Varchenko and Schechtman show that there is a natural contragredient form S on weight spaces of $U_q N_-$. Since L is one such weight space, one can ask for an explicit description of S relative to the basis $\{f_\sigma\}$ given above. An important result in [Schechtman and Varchenko 1991] is that

$$S(f_\sigma, f_\tau) = B(\sigma, \tau),$$

where B is the Varchenko matrix for the arrangement A_{r-1} (described in our Example 1.2 above) with parameters

$$a_{i,j} = (\alpha_i, \alpha_j).$$

Varchenko and Schechtman go on to show that the kernel of this contragredient form describes the Serre relations for the quantum Kac–Moody Lie algebra $U_q G$.

The main topic of this paper will be linear algebraic properties of the B matrices. We are going to report on a number of results and list a number of open questions that have to do with the nullspace of B and the Smith normal form of B. In addition, we will describe a new application of the Varchenko matrices to the problem of computing the Betti numbers of the Milnor fibre of the arrangement A.

2. The Nullspace of the B Matrices

In this section, we will consider the nullspace of the Varchenko matrices. The reader may recall that the nullspaces are important in the work of Varchenko and Schechtman as they encode the analogues of the Serre relations for quantum Kac–Moody Lie algebras in the case that the arrangement is A_{n-1}. In this section we will survey some known results and state some open problems relating to the nullspace of the Varchenko matrix.

A starting point for the study of the nullspace of B is the determinant of B. Note that $\det(B)$ is a polynomial in the a_H hence will vanish for certain choices of the a_H. The following result, due to Varchenko, gives an elegant factorization of $\det(B)$. As an immediate consequence of this theorem, one obtains a characterization of those values of the parameters for which B has a nontrivial nullspace.

THEOREM 2.1 [Varchenko 1993, Theorem 1]. *The determinant of the bilinear form of the configuration A is*

$$\det(B) = \prod_{X \in L(A)^*} (1 - a_X^2)^{l(X)},$$

where $L(A)^*$ is the set of nonempty intersections in $L(A)$, where a_X is the product of the a_H over hyperplanes H containing X and where $l(X)$ is a nonnegative integer described explicitly in [Varchenko 1993].

In fact, Varchenko gives two methods to compute the exponents $l(X)$. The first, which precedes the statement of his theorem is more geometric, the second which comes later in the paper is more combinatorial. For completeness, we will briefly describe the second method. To compute $l(X)$, first choose a hyperplane $H \in A$ which contains X. Then $l(X)$ is half the number of regions P which have the property that X is the minimal intersection containing $\bar{P} \cap H$.

EXAMPLE 2.2. Consider the B matrix for the arrangement given in Example 1.1 above. The poset of intersections consists of seven elements. Here are these seven intersections X_i, along with the exponents l_{X_i} computed according to the method of Varchenko described above.

$$
\begin{aligned}
X_0 &= \mathbb{R}^2, & l_{X_0} &= 0, \\
X_1 &= H_0, & l_{X_1} &= 3, \\
X_2 &= H_1, & l_{X_2} &= 3, \\
X_3 &= H_2, & l_{X_3} &= 3, \\
X_4 &= H_1 \cap H_2, & l_{X_4} &= 0, \\
X_5 &= H_1 \cap H_0, & l_{X_5} &= 0, \\
X_6 &= H_2 \cap H_0, & l_{X_6} &= 0.
\end{aligned}
$$

According to Varchenko's theorem, the determinant of B factors as

$$
\det(B) = (1 - a_{X_1}^2)^3 (1 - a_{X_2}^2)^3 (1 - a_{X_3}^2)^3 = (1 - a_0^2)^3 (1 - a_1^2)^3 (1 - a_2^2)^3.
$$

It is straightforward to check, using Maple or Mathematica for example, that this is a correct formula for the determinant.

Now consider this B matrix specialized so that all parameters are equal to q:

$$
B = \begin{pmatrix}
1 & q & q^2 & q^3 & q^2 & q & q \\
q & 1 & q & q^2 & q^3 & q^2 & q^2 \\
q^2 & q & 1 & q & q^2 & q^3 & q \\
q^3 & q^2 & q & 1 & q & q^2 & q^2 \\
q^2 & q^3 & q^2 & q & 1 & q & q \\
q & q^2 & q^3 & q^2 & q & 1 & q^2 \\
q & q^2 & q & q^2 & q & q^2 & 1
\end{pmatrix}.
$$

The determinant of B is

$$
\det(B) = (1 - q^2)^9.
$$

Let G be the group (isomorphic to S_3) which permutes the three hyperplanes. Any permutation of the hyperplanes induces a permutation of the seven regions and so we get a representation of S_3 on the regions which commutes with the action of B. It is straightforward to check that the vector space spanned by regions, considered as a G-module, is isomorphic to one copy of the regular representation together with one copy of the trivial representation.

There are two values of q for which the determinant vanishes. For each value we can compute the representation of G on the nullspace of B. If $q = 1$, then B is the seven by seven matrix of all ones and the nullspace is easily seen to carry the regular representation. If $q = -1$ the situation is less clear. However, one can verify that the nullspace has dimension six and consists of two copies of the defining representation.

When you study the nullspace of B for values of the hyperplane weights where $\det(B) = 0$, the first question that arises is how to determine the dimension of the nullspace. Deformation theory arguments show that dimension of the nullspace is no larger than the sum of the exponents $l(X)$ taken over intersections X with $a_X^2 = 1$. To get more information in general is difficult. However, a case of special interest in which more general statements can be made is the case where all hyperplane weights are equal to the same number q. In this case, one can define the Smith normal form of B over $\mathbb{C}[q]$ which contains the information necessary to determine the dimension of the nullspace. The Smith normal form of B over $\mathbb{C}[q]$ is the topic of the next section in this paper. As we will see in the next section, it is possible for the dimension of the nullspace of $B(\zeta)$, for ζ a complex number, to be strictly smaller than the deformation theory bound stated above which in this case is equal to the multiplicity of $q - z$ as a divisor of $\det(B(q))$.

The question of dimension can be slightly broadened to ask for the "structure" of the nullspace (or more generally the cokernel) meaning a natural choice of basis for the nullspace (cokernel) as well.

PROBLEM 1. Let $\mathcal{A}$ be an arrangement and let $\{a_H\}$ be a choice of parameters for which the determinant of B vanishes. Find a natural basis for the nullspace (or for the cokernel) of B in terms of some kind of combinatorial and geometric information about the arrangement $\mathcal{A}$ and the parameters a_H.

One could ask what sort of information might go into a solution to Problem 1. Theorem 2.1 suggests that the necessary information may include facts about the combinatorial structure of $L(\mathcal{A})$ and in particular at intersections X in $L(\mathcal{A})$ for which $a_X^2 = 1$.

The previous example suggests another interesting question that one can ask about the nullspace of the Varchenko matrices. Suppose that a finite group G acts as a group of affine linear transformations and preserves the arrangement $\mathcal{A}$. Suppose also that the weight assigned to hyperplanes is G-invariant. Then the group G acts on the regions $r(A)$ and this action commutes with B. So one can ask for the G-module structure of the nullspace of B.

PROBLEM 2. Let $\mathcal{A}$ be an arrangement and let G be a finite group of affine-linear transformations which preserve the set $\mathcal{A}$. In addition assume that the weighting on hyperplanes is invariant under the action of G. What can you say about the G-module structure of the nullspace of B?

What sort of solution might you hope to find for Problem 2? As stated above in Problem 1, you would like to determine the nullspace of B in terms of some kind of combinatorial and geometric information related to the lattice $L(\mathcal{A})$ and the choice of parameters. You would hope for an answer to Problem 2 which extends this idea to include not just combinatorial information about $L(\mathcal{A})$ but also information about how G acts on $L(\mathcal{A})$.

One further problem naturally arises in the situation where there is a group G acting on the arrangement and where all hyperplane weights are equal to a single value q. Let V denote the vector space spanned by the set of regions $r(A)$. Recall that G acts on V and that this action commutes with the matrix B.

For each irreducible representation Ψ of G, let V_Ψ denote the Ψ-isotypic component of V. Because B commutes with the action of G, B preserves the subspace V_Ψ. Varchenko's theorem gives an elegant formula for the determinant of B as a linear map of V. A natural question is whether a correspondingly elegant formula holds for B as a linear transformation of V_Ψ.

PROBLEM 3. For Ψ an irreducible representation of G, let B_Ψ be the restriction of B to V_Ψ. Find a formula for $\det(B_\Psi)$.

A couple of comments about Problem 3. First, what kind of formula could you hope to achieve in answer to this question? Following the form of Varchenko's result, we would expect that the formula for $\det(B_\Psi)$ would depend on the intersection lattice $L(A)$ and some information about the action of G on $L(A)$. This information might include, for example, the action of G on the homology groups of the intervals of $L(A)$. In the best case, the formula would be a product over elements, or perhaps over orbits of elements, in the intersection lattice $L(A)$.

Second, it should be noted that there is a variant of Problem 3 which might have a more natural solution. Let Irr denote the set of irreducible representations of G. Let ρ be a virtual representation of G written as $\rho = \sum_{\Psi \in \mathrm{Irr}} c_\Psi \Psi$. Define

$$\det{}_\rho(B) = \prod_{\Psi \in \mathrm{Irr}} (\det(B_\Psi))^{c_\Psi}.$$

PROBLEM 3'. Compute $\det(B_\rho)$, for ρ in any set of representations that span the representation space.

Varchenko [1993] solves Problem 3 for the group of order 2 induced by negation in the central hyperplane arrangement case.

We conclude this section with a summary of what is known about these problems in a particularly interesting special case. Let A be the arrangement $\mathbf{A}_{n-1}$ given in Example 1.2, with all hyperplane weights set equal to the value q. As shown in that example, the regions in $r(A)$ are indexed by permutations in S_n. Moreover, for permutations σ, τ, the σ, τ entry of B is $q^{i(\sigma\tau^{-1})}$. It follows that B is the matrix for left multiplication by

$$\Gamma = \sum_{\beta \in S_n} q^{i(\beta)} \beta.$$

The group $G = S_n$ acts on the arrangement $\mathbf{A}_{n-1}$ and hence on the set of regions in $r(A)$. When the regions are indexed by permutations, this action of S_n corresponds to the right-regular representation.

As noted earlier, Problems 1 and 2 are particularly interesting in this case because the nullspace of B has an interpretation in terms of quantum Kac–Moody Lie algebras via results of Schechtman and Varchenko. Also in this case, there is a natural group of symmetries of the arrangement, namely the group S_n. So we will assume that the weighting on hyperplanes is invariant under S_n which in this case is equivalent to their all having the same value q.

Hanlon and Stanley [1998] have studied the S_n-module structure of the nullspace of B for this arrangement with weight q on all hyperplanes. We end this section by cataloguing a number of results and open problems that appear in their work.

The first step is to understand what Theorem 2.1 says in this case about the determinant of B. The intersection lattice of $\mathcal{A}$ is the partition lattice Π_n. It turns out that the exponent $l(X)$ is 0 unless the partition X has exactly one nontrivial block. If X has one nontrivial block of size α, then $l(X) = (\alpha - 2)! \, (n - \alpha + 1)!$. So in this case, Theorem 2.1 specializes to:

COROLLARY 2.3. *Let $\mathcal{A}$ be the arrangement $\mathbf{A}_{n-1}$ and let all hyperplanes have weight q. Then*

$$\det(B) = \prod_{\alpha=2}^{n} \left(1 - q^{\alpha(\alpha-1)}\right)^{\binom{n}{\alpha}(\alpha-2)!(n-\alpha+1)!}. \qquad (2\text{--}1)$$

This factorization of the determinant in this case was first proved by Zagier [1992], who came across the B matrix in this case in an entirely different context.

In view of Corollary 2.3, we need only consider values of q that are of the form $q = e^{2\pi i s/j(j-1)}$ for $j \in \{2, 3, \ldots, n\}$. The first result that appears in [Hanlon and Stanley 1998] concerns the specialization of that form that is in some sense the most extreme. This result is stated in terms of the S_n module Lie_n which is the representation of S_n on the multilinear part of the free Lie algebra. There has been a great deal of study of this representation in part because it plays a role in many diverse mathematical situations.

THEOREM 2.4 [Hanlon and Stanley 1998, Theorem 3.3]. *Let $q = e^{2\pi i/n(n-1)}$. Then*

$$\ker\left(\Gamma_n(q)\right) = \mathrm{ind}_{S_{n-1}}^{S_n}\left(\mathrm{Lie}_{n-1}\right)/\mathrm{Lie}_n.$$

An interesting feature of Theorem 2.4 is the appearance of the module

$$W_n = \mathrm{ind}_{S_{n-1}}^{S_n}\left(\mathrm{Lie}_{n-1}\right)/\mathrm{Lie}_n.$$

It is an old result that Lie_n is contained in $\mathrm{ind}_{S_{n-1}}^{S_n}\left(\mathrm{Lie}_{n-1}\right)$. So this is a genuine (rather than virtual) module which is often called the *Whitehouse module*. This name refers to Sarah Whitehouse who came across the representation in quite a different context. The representation appears earlier in the work of Kontsevich [1993]. Whitehouse, together with Alan Robinson, investigated the homology of the space H_n of homeomorphically irreducible trees which have n labelled leaves. Here "labelled" is in the graph theory sense of the word, so the numbers $\{1, 2, \ldots, n\}$ are attached to the leaves. They say that one tree τ is a "face" of another tree ρ if τ is obtained from ρ by contracting an "internal edge", i.e., an edge which is not incident with a leaf. This notion of face gives us a complex whose boundary map commutes with the action of S_n induced by permutation of leaf labels. So one can ask for the S_n-module structure of the homology of this complex. The first result, due to Robinson and Whitehouse [1996], is that this homology vanishes except in degree $n - 2$, and that in degree $n - 2$ the dimension is $(n - 2)!$. Whitehouse went on to show that the degree $n - 2$ piece of the homology carries the S_n-module structure which we've called the Whitehouse module. Interestingly, this module has also come up in the work of Babson et al. [1999] on graphs that are not 2-connected. Recall that a graph G is "2-connected" if G is connected and remains so under the removal of any one vertex. Let N_n denote the collection of graphs with vertex set $\{1, 2, \ldots, n\}$ which are not 2-connected. If we identify a graph by its edge set, then the edge sets of the graphs in N_n form a simplicial complex which is clearly invariant under that action of S_n. An interesting result in [Babson et al. 1999] is that the simplicial homology of N_n is zero except in degree $n - 2$ and in degree $n - 2$ carries the S_n-module structure of the Whitehouse representation.

Hanlon and Stanley went on to state a conjecture which gives a description of the S_n-module structure of the kernel of B in a more general case than that covered in their Theorem 3.3. They provided a proof of this conjecture up to knowing a certain technical fact about the Smith normal form of the Varchenko matrices. This fact was subsequently proved by Denham, thereby giving a proof of the following result.

THEOREM 2.5 [Denham 1999; Hanlon and Stanley 1998]. *Suppose that q is a root of exactly one of the factors on the right-hand side of (2–1). More precisely, suppose that $q = e^{2\pi i s/j(j-1)}$ for some $j \in \{2, 3, \ldots, n\}$ and some nonnegative integer s and that $q^{k(k-1)} \neq 1$ for $k \neq j$, $2 \leq k \leq n$. Then the S_n-module structure of $\ker(\Gamma_n(q))$ is*

$$\ker(B(q)) = \operatorname{ind}_{C_{j-1}}^{S_n}(q^j)/\operatorname{ind}_{C_j}^{S_n}(q^{j-1}),$$

where C_{j-1} is the subgroup of S_n generated by the $(j-1)$-cycle $z_{j-1} = (j-1, j-2, \ldots, 2, 1)$, and where q^j denotes the linear character of C_{j-1} whose value on z_{j-1} is q^j.

It is not immediately clear why the statement of Theorem 2.5 in the special case where $j = n$ and $s = 1$ agrees with the statement of Theorem 2.4. This follows from the well-known fact that $\operatorname{Lie}_n = \operatorname{ind}_{C_n}^{S_n}(q^{n-1})$ for q a primitive $n(n-1)^{st}$ root of unity. An interesting open problem is to extend these results to the case where $1 - q^{j(j-1)}$ vanishes for more than one j. The paper [Hanlon and Stanley 1998] contains tables which give the S_n-module structure of the kernel of B in these cases for small values of n.

We next turn to Problem 3′ and see what is known here in the special case of $\mathcal{A} = \boldsymbol{A}_{n-1}$, with all hyperplane weights equal to q. Hanlon and Stanley [1998] have proved a result that gives a solution to Problem 3′, but in order to state it we need some notation.

For each partition η, let P^η be the virtual character which has value 0 on conjugacy classes not indexed by η and which has value $\frac{n!}{c_\eta}$ on the conjugacy class indexed by η where c_η is the order of the centralizer in S_n of a permutation with cycle type η.

It is well-known that we can write P^η explicitly as

$$P^\eta = \sum_\lambda \chi^\lambda(\eta)\chi^\lambda,$$

where χ^λ is the irreducible character of S_n indexed by λ (see [James and Kerber 1981] or [Sagan 1991] for more on the irreducible representations of S_n). It is easy to see that the virtual characters P^η span the representation space of S_n.

The following result, which was originally conjectured by Stanley, give an elegant solution to Problem 3′ in this case:

THEOREM 2.6 [Hanlon and Stanley 1998], Theorem 3.7. *For each $\eta \vdash n$, let $D_\eta(q)$ denote* $\det_{P^\eta}(B(q))$. *Then*

(i) $D_\eta(q) = 1$ *unless η is of the form $l^d 1^{n-ld}$ for some l, d.*

(ii) $D_{(l^d 1^s)}(q) = \left(D_{(l^d 1)}(q)\right)^{s!}$ *for all l, d and all $s \geq 1$.*

Note that (a) and (b) reduce the determination of $D_\eta(q)$ to the cases where η is of the form l^d or $l^d 1$.

(iii) $D_{(l^d)}(q) = \prod_{m|l}\left(1 - q^{dm(n-1)}\right)^{\mu(m)l^d(d-1)!/m}$, *where, in the exponent on the right-hand side, μ denotes the number-theoretic Möbius function.*

(iv) $D_{(l^d,1)}(q) = D_{(l^d)}(q)D_{(l^d)}\left(q^{n/(n-2)}\right)^{-1}$.

3. The Smith Normal Form of $B(q)$

Let notation be as in the previous section, so that $\mathcal{A}$ is an arrangement of hyperplanes and B is the Varchenko matrix of the arrangement $\mathcal{A}$. In this section we will assume that all parameters a_H are equal to a parameter q. So B is the matrix with rows and columns indexed by regions whose R, S entry is $q^{n(R,S)}$ where $n(R, S)$ is the number of hyperplanes which separate R and S.

We can specialize Varchenko's Theorem (Theorem 2.1) by setting all $a_H = q$ to get a formula for the determinant of B:

$$\det(B) = \prod_{X \in L(\mathcal{A})^*} (1 - q^{2h(X)})^{l(X)},$$

where $h(X)$ is the number of hyperplanes in $\mathcal{A}$ which contain X. A consequence of this formula is that $\det(B)$ vanishes only at q being certain roots of unity.

Because the entries of B come from $\mathbb{C}[q]$ (which is a PID), we can define the Smith normal form of B over the ring $\mathbb{C}[q]$. Recall that the Smith normal form of a matrix M over a Euclidean domain R is a normal form for left and right multiplication by unimodular matrices. In other words, the Smith normal form of M, denoted $\mathrm{SNF}(M)$, is a particular choice from the set of matrices

$$\{UMV : U \text{ and } V \text{ are unimodular matrices over } R\}.$$

The matrix $\mathrm{SNF}(M)$ is a diagonal matrix with each entry along the diagonal dividing the next one. The diagonal entries are chosen up to multiplication by units in R. In our case, the units in $\mathbb{C}[q]$ are the nonzero complex numbers so we can assume that each diagonal entry in $\mathrm{SNF}(B)$ is a monic polynomial. Note also that if M is square, the unimodular conditions on U and V imply that $\det(\mathrm{SNF}(M)) = \det(M)$.

Let Θ denote the set of roots of $\det(B)$. By the divisibility condition on successive entries of $\mathrm{SNF}(B)$, the (i,i)-entry of $\mathrm{SNF}(B)$ is of the form

$$\prod_{z \in \Theta} (q - z)^{p_z^{(i)}},$$

where the exponents $p_z^{(i)}$ satisfy

$$p_z^{(1)} \le p_z^{(2)} \le \cdots \le p_z^{(N)}$$

for each $z \in \Theta$, where N denotes the number of regions in the complement of $\mathcal{A}$. Therefore, $\mathrm{SNF}(B)$ is completely determined by the sequences $\{p_z^{(i)}\}_{i=1}^{N}$ for each $z \in \Theta$. It is sometimes more convenient to work with an equivalent sequence of numbers $\sigma_z^{(i)}$, where $\sigma_z^{(i)}$ is equal to the number of j such that $p_z^{(j)}$ is equal to i. In terms of the Smith normal form, $\sigma_z^{(i)}$ is the number of diagonal entries in $\mathrm{SNF}(B)$ that are exactly divisible by $(q - z)^j$.

EXAMPLE 3.1. Let $\mathcal{A}$ be the arrangement given in Example 2.2. The B matrix with all parameters set equal to q is given in that example, as is $\mathrm{SNF}(B)$. The sequences that determine $\mathrm{SNF}(B)$ as above are

$$\{p_1^{(i)}\} = \{p_{-1}^{(i)}\} = 0, 1, 1, 1, 2, 2, 2.$$

The σ sequences for this arrangement are

$$\{\sigma_1^{(i)}\} = \{\sigma_{-1}^{(i)}\} = 1, 3, 2.$$

This example demonstrates one of two facts about the $p_z^{(i)}$ that are straightforward to prove:

PROPOSITION 3.2. Let the sequences $p_z^{(i)}$ be those associated to the Smith normal form of the arrangement $\mathcal{A}$ in $\mathbb{R}^n$. Then

(i) If z and w are both primitive j-th roots of unity, then $p_z^{(i)} = p_w^{(i)}$ for every i.

(ii) For every root of unity z and every i, $p_z^{(i)} = p_{-z}^{(i)}$.

The computation of the Smith normal form of the Varchenko matrices arises as our next problem of interest in this paper.

PROBLEM 4. (i) Determine the Smith normal form of B in terms of some information about the arrangement $\mathcal{A}$.

(ii) Is $\mathrm{SNF}(B)$ determined just by the intersection lattice of $\mathcal{A}$?

Aside from the intrinsic interest of computing the Smith normal form of the matrices B, this problem is directly relevant to the determination of the nullspace of B in the case that all parameters a_H are set equal to q. As noted in the last chapter, facts about the Smith normal form of B were needed by Denham [1999] to prove a conjecture of Hanlon and Stanley. In the next section, we will describe another application of the Smith normal form of B (due to Denham), this time to computing the homology of the Milnor fibre of the arrangement $\mathcal{A}$.

Note that the Smith normal form of B is determined by the numbers $p_z^{(i)}$ for every $z \in \Theta$ and every i. So an alternative formulation of Problem 4 is the following.

PROBLEM 4'. Determine the numbers $p_z^{(i)}$ for $i \geq 0$ and for each $z \in \Theta$. Equivalently, determine the numbers $\sigma_z^{(i)}$ for $i \geq 0$ and for each $z \in \Theta$.

One general result is known that gives an elegant partial solution to Problem 4'.

THEOREM 3.3 [Denham and Hanlon 1997]. *For any arrangement $\mathcal{A}$ and any nonnegative integer i, $\sigma_1^{(i)} = \sigma_{-1}^{(i)}$ is the i-th Betti number of the arrangement. Equivalently,*

$$\sum_i \sigma_1^{(i)} \lambda^i = \sum_j \lambda^{p_1^{(j)}} = (-\lambda)^r \chi\left(\frac{-1}{\lambda}\right),$$

where $\chi(\lambda)$ is the characteristic polynomial of the lattice $L(\mathcal{A})$ and r is the rank of $L(\mathcal{A})$.

EXAMPLE 3.4. To illustrate an instance of Problem 4, we consider the arrangements $\mathbf{A}_{n-1}$. For each n in the range $2 \leq n \leq 5$ there is a chart below which gives the numbers $\sigma_z^{(i)}$. To understand these charts, assume that n is fixed. By Corollary 3.1 we know that $z \in \Theta$ if and only if z is a primitive d-th root of unity for some d that divides $j(j-1)$ and some $j \leq n$. These numbers d index the rows of the charts below. In view of Proposition 3.2 (b) we do not include d of the form $2d'$ where d' is odd. So the number $\sigma_z^{(j)}$ that appears in the d, j entry of the chart below is the number of entries in $\mathrm{SNF}(B)$ that are exactly divisible by $(q - e^{2\pi i/d})^j$.

With these notational conventions, the Smith normal forms of the arrangements $\mathbf{A}_{n-1}$ for small values of n are as follows:

$$
\begin{array}{cc}
d \quad j=0 \quad 1 \\
n=2: \quad 1 \quad \begin{pmatrix} 1 & 1 \end{pmatrix}
\end{array}
\qquad
\begin{array}{cc}
d \quad j=0 \quad 1 \quad 2 \\
n=3: \quad \begin{matrix} 1 \\ 3 \end{matrix} \quad \begin{pmatrix} 1 & 3 & 2 \\ 5 & 1 & 0 \end{pmatrix}
\end{array}
$$

$n = 4$:

d	$j=0$	1	2	3
1	1	6	11	6
3	17	4	3	0
4	22	2	0	0
12	22	2	0	0

$n = 5$:

d	$j=0$	1	2	3	4
1	1	10	35	50	24
3	70	20	30	0	0
4	102	10	8	0	0
5	114	6	0	0	0
12	100	20	0	0	0
20	114	6	0	0	0

Theorem 3.3 gives a description of the numbers that appear in the rows indexed by $d = 1$ in the tables above. More specifically, Theorem 3.3 equates the polynomial $Q(\lambda)$ obtained by using the numbers in the first row as coefficients to a polynomial obtained from the characteristic polynomial of the lattice of intersections of the arrangement A_{n-1}. The lattice of intersections of the arrangement A_{n-1} is the partition lattice Π_n which is well-known to have characteristic polynomial $\chi(\lambda) = \prod_{i=1, n-1}(\lambda - i)$. So Theorem 3.3 states that the polynomial $Q(\lambda)$ is equal to

$$Q(\lambda) = (-\lambda)^{n-1} \prod_{i=1}^{n-1}\left(\frac{-1}{\lambda} - i\right) = \prod_{i=1}^{n-1}(1 + i\lambda).$$

A specific case of this factorization is $n = 5$ where Theorem 3.3 predicts that

$$1 + 10\lambda + 35\lambda^2 + 50\lambda^3 + 24\lambda^4 = (1 + \lambda)(1 + 2\lambda)(1 + 3\lambda)(1 + 4\lambda).$$

It is easy to check that two sides of the above equality are in fact equal. Problem 4 asks for an extension of Theorem 3.3 to $d > 1$.

In order to refine Problem 4, we will take a different point of view on the Smith normal form. Let V be the module over $A = \mathbb{C}[q]$ spanned by the regions in the complement of $\mathcal{A}$. Note that B determines a $\mathbb{C}$-linear transformation from V to V which commutes with the action of A. Let $\zeta(q)$ be any polynomial in A. Since B commutes with the action of A on V, multiplication by $\zeta(q)$ defines a map on C, the cokernel of B.

By definition, C is $V/\operatorname{im}(B)$. As a vector space over $\mathbb{C}$, C has dimension equal to the sum of the degrees of the entries in $\operatorname{SNF}(B)$. We will want to say a bit more about the structure of C. For each $z \in \Theta$, let $K_z^{(j)}$ denote the kernel in C of multiplication by $(q - z)^j$, and let $k_z^{(j)} = \dim_{\mathbb{C}}(K_z^{(j)})$. Then $k_z^{(j)}$ equals the number of i with $p_z^{(i)} \le j$. Note that

$$K_z^{(0)} \subseteq K_z^{(1)} \subseteq K_z^{(2)} \subseteq \cdots.$$

Let $L_z^{(j)} = K_z^{(j)}/K_z^{(j-1)}$. By the comments above, we see that $\dim_{\mathbb{C}}(L_z^{(j)})$ is equal to $\sigma_z^{(j)}$, the number of $p_z^{(i)}$ which are equal to j. It is not difficult to show that

$$C = \bigoplus_{z \in Z} \bigoplus_{j \ge 0} L_z^{(j)}.$$

Now suppose that G is a group of affine transformations defined over $\mathbb{C}$ that permutes the set of hyperplanes. Then G acts on the set of regions as well and this action commutes with the actions of both B and $\mathbb{C}[q]$. So G acts on the spaces $L_z^{(j)}$ for each $z \in \Theta$ and each j. Problem 1 asks one to determine the dimensions of the $L_z^{(j)}$, or equivalently to determine the character of the identity in G acting on each $L_z^{(j)}$. One can ask more generally about their G-module structures of the $L_z^{(j)}$.

PROBLEM 5. Determine the G-module structure of the spaces $L_z^{(j)}$ in terms of some information about the action of G on the arrangement $\mathcal{A}$.

A simpler, but still interesting, variant of Problem 5 is to determine the G-module structure of the entire cokernel of B.

EXAMPLE 3.5. To illustrate these concepts and problems we do an example. Let $\mathcal{A} = \{H_0, H_1, H_2\}$ be the two-dimensional arrangement where H_0 is the line $x = 0$, H_1 is the line $y = 1$ and H_2 is the line $y = -1$. Let G be the group of order four generated by h and v, where h is the reflection of $\mathbb{R}^2$ across the line $x = 0$ and v is reflection of $\mathbb{R}^2$ across the line $y = 0$.

Index the regions in the complement of the arrangement $R_1, \ldots, R_6$ so that R_1, R_3, R_5 are those in the half-plane $x < 0$ arranged from top to bottom and R_2, R_4, R_6 are those in the half-plane $x > 0$ arranged from top to bottom. The B-matrix of this arrangement with respect to this ordering is

$$
B = \begin{pmatrix}
1 & q & q & q^2 & q^2 & q^3 \\
q & 1 & q^2 & q & q^3 & q^2 \\
q & q^2 & 1 & q & q & q^2 \\
q^2 & q & q & 1 & q^2 & q \\
q^2 & q^3 & q & q^2 & 1 & q \\
q^3 & q^2 & q^2 & q & q & 1
\end{pmatrix}.
$$

The Smith normal form of B is given by

$$
\mathrm{SNF}(B) = \mathrm{diag}\left(1,\, 1-q^2,\, 1-q^2,\, 1-q^2,\, (1-q^2)^2,\, (1-q^2)^2\right),
$$

so that the cokernel has dimension 14 (over $\mathbb{C}$.) A basis for the image of B is given by the six vectors

$$
\begin{aligned}
\mu_1 &= (1, q, q, q^2, q^2, q^3), \\
\mu_2 &= (1-q^2) \cdot (0, 1, 1, 1+2q-q^2, 1+q, 1+2q+q^2-q^3), \\
\mu_3 &= (1-q^2) \cdot (0, -2, -1, -3q, -q, 1-4q^2), \\
\mu_4 &= (1-q^2) \cdot (0, 1, 0, q, 0, 1), \\
\mu_5 &= (1-q^2)^2 \cdot (0, 0, 0, 1, 0, q-1), \\
\mu_6 &= (1-q^2)^2 \cdot (0, 0, 0, 0, 0, 1).
\end{aligned}
$$

So, the kernel of multiplication by $1 \pm q$ on $L_{\pm 1}^{(1)}$ has dimension 5 and has a basis given by

$$\nu_2 = (1 \mp q) \cdot (0, 1, 1, 1 + 2q - q^2, 1 + q, 1 + 2q + q^2 - q^3),$$
$$\nu_3 = (1 \mp q) \cdot (0, 0, 1, 2 + q - 2q^2, 2 + q, 3 + 4q - 2q^2 - 2q^3),$$
$$\nu_4 = (1 \mp q) \cdot (0, 1, 0, q, 0, 1),$$
$$\nu_5 = (1 - q^2)(1 \mp q) \cdot (0, 0, 0, 1, 0, q - 1),$$
$$\nu_6 = (1 - q^2)(1 \mp q) \cdot (0, 0, 0, 0, 0, 1).$$

The group $G = \{e, h, v, hv\}$ acts on the ordered set of regions by

$$e = (R_1)(R_2)(R_3)(R_4)(R_5)(R_6),$$
$$h = (R_1, R_2)(R_3, R_4)(R_5, R_6),$$
$$v = (R_1, R_5)(R_2, R_6)(R_3)(R_4),$$
$$hv = (R_1, R_6)(R_2, R_5)(R_3, R_4).$$

So, it is possible to explicitly calculate the action of any group element on the basis elements above. For example,

$$v \cdot \nu_2 = (1 \mp q) \cdot (1 + q, 1 + 2q + q^2 - q^3, 1, 1 + 2q - q^2, 0, 1)$$
$$= (1 \mp q) \cdot (1 + q) \cdot \mu_1 + (1 + q - q^3) \cdot \nu_2 + (-2q - q^2 + q^3) \cdot \nu_3$$
$$+ (3q + q^2 - q^3) \cdot \nu_4 + (-1 - q - q^2) \cdot \nu_5 + (4 - q)(1 - q^2) \cdot \nu_6.$$

So the contribution to the trace of v that arises from the ν_2-diagonal term is $1 + q - q^3 = 1$ since $(q - q^3)\nu_2 = 0$ in $K_z^{(1)}$.

A tedious calculation along the same lines as above shows that the characters of G acting on $L_z^{(j)}$ are

	z	$g=e$	h	v	hv		z	$g=e$	h	v	hv
$j=1$:	$+1$	$\begin{pmatrix} 5 \\ 5 \end{pmatrix}$	$\begin{matrix} -1 \\ 1 \end{matrix}$	$\begin{matrix} 1 \\ 1 \end{matrix}$	$\begin{matrix} -1 \\ 1 \end{pmatrix}$	$j=2$:	$+1$	$\begin{pmatrix} 2 \\ 2 \end{pmatrix}$	$\begin{matrix} -2 \\ 2 \end{matrix}$	$\begin{matrix} 0 \\ 0 \end{matrix}$	$\begin{matrix} 0 \\ 0 \end{pmatrix}$
	-1						-1				

4. Local Systems and the Milnor Fibre

Earlier we cited a result of Schechtman and Varchenko which describes the Serre relations for a quantum Kac–Moody Lie algebra in terms of the Varchenko matrices B. In this section, we give a second application of the Varchenko matrix of an arrangement $\mathcal{A}$ — this time to an invariant of the singularity of a hyperplane arrangement at the origin.

Let $\mathcal{A}$ be an essential arrangement in $\mathbb{C}^n$ with defining polynomial $Q : \mathbb{C}^n \to \mathbb{C}$. Let $M = Q^{-1}(\mathbb{C} \setminus \{0\})$ denote its complement, and $N = Q^{-1}(0)$ the union of the hyperplanes. A local system on M is a representation of the fundamental group $\pi_1(M)$ in a complex vector space $\mathcal{W}$.

For local systems of rank one, various authors have studied the homology $H.(M, W)$, for the most part in connection with generalized hypergeometric functions: see [Kohno 1986; Gel'fand 1986], or the references provided in [Varchenko 1995]. When the local system is trivial, one recovers the ordinary homology $H.(M, \mathbb{C})$, which is isomorphic to the Whitney homology of the lattice $L(A)$.

When the defining polynomial Q is real, there is a strong deformation retract of M onto Salvetti's CW-complex $\mathfrak{X}$ [Salvetti 1987]. In this case, there are explicit chain complexes that calculate $H_\bullet(M, W)$ and $H_\bullet(\mathbb{C}^n, N, W)$. The B matrix appears as part of an important chain map between the two complexes.

Our interest in local systems over the arrangement's complement comes from a special case. It is well known that, if A is a central arrangement, the restriction of the defining polynomial to the complement, $Q : M \to \mathbb{C}^*$, gives M the structure of a fibre bundle over $\mathbb{C}^*$ [Milnor 1968]. The typical fibre $F = Q^{-1}(1)$ is a complex manifold of dimension $n-1$, known as the Milnor fibre of the polynomial Q (or of A). F is homotopically equivalent to an infinite cyclic covering of M, so its singular homology is

$$H_\bullet(F, \mathbb{C}) = H_\bullet(M, \mathbb{C}\mathbb{Z}),$$

for a local system

$$\mathbb{C}\mathbb{Z} = \mathbb{C}[t, t^{-1}].$$

If the polynomial Q is real, then, we can exploit the complexes defined by (4–1) to study the homology of the Milnor fibre (Section 4C).

Milnor [1968] considered arbitrary polynomial maps $f : \mathbb{C}^n \to \mathbb{C}$ that vanish at 0. In particular, he showed that if the polynomial f has an isolated singularity at 0, then the Milnor fibre of f has the homotopy type of a bouquet of spheres. The Milnor fibre of defining polynomials of arrangements, then, continue to be of interest by forming a restricted family of examples in which the singularity is not isolated. Although the homotopy type of the Milnor fibre of a real arrangement is determined by the face lattice of an arrangement, an explicit description of the homology is only known in special cases; see [Cohen and Suciu 1998; Orlik and Randell 1993]. Moreover, it is not known if, in general, the intersection lattice of a complex hyperplane arrangement determines the Milnor fibre's homology.

In Section 4A, we introduce the Varchenko–Salvetti chain complexes. Section 4B indicates the role of the B matrices, and 4C specializes the construction to apply to the Milnor fibre.

4A. Salvetti's complex. In order to describe the general setup, recall that the fundamental group $\pi_1(M)$ is generated by loops $\{\alpha_H : H \in A\}$ around each hyperplane [Randell 1982]. Let W be a (complex) local coefficient system defined by a representation $\rho : \pi_1(M) \to \text{End}(W)$. For each $H \in A$, let $b_H = \rho(\alpha_H)$. We require in what follows that the image of ρ be abelian. Subject to this constraint, one can choose any set of commuting endomorphisms $b_H \in \text{End}(W)$, since then the representation ρ factors through $H_1(M, \mathbb{Z})$, and $H_1(M, \mathbb{Z})$ is freely generated by the cycles around each hyperplane.

$$\pi_1(M) \xrightarrow{\ \rho\ } \text{End}(\mathcal{W})$$

$$\text{Ab} \downarrow \qquad \qquad$$

$$H_1(M, \mathbb{Z})$$

Salvetti's complex consists of cells indexed by pairs of faces of the real arrangement corresponding to $\mathcal{A}$. Let $\mathcal{L}$ denote the face lattice of the arrangement, ordered by reverse inclusion, so that $\mathcal{L}_k$ is the set of faces of codimension k. For a face P and a hyperplane H, let $P(H) = 0$ if $P \subseteq H$, and $P(H) = \pm 1$ otherwise, depending on whether P lies on the positive or negative side of H. The face P is uniquely determined by these values. With this notation, we recall the definition of the vector product of faces with regions: for any $P \in \mathcal{L}$ and $R \in r(\mathcal{A})$, the region $PR \in r(\mathcal{A})$ is determined by

$$(PR)(H) = \begin{cases} R(H) & \text{if } P(H) = 0, \\ P(H) & \text{otherwise}, \end{cases}$$

for each $H \in \mathcal{A}$.

EXAMPLE 4.1. We return to the arrangement of Example 1.1, defined by linear forms $f_0 = x - y$, $f_1 = x + y$, and $f_2 = y - 1$. Let $P = \{(x, y) : x > 1, y = 1\}$. Then the triples $(F(H_0), F(H_1), F(H_2))$ for faces $F = P$, R_7, and PR_7 are

$$P \longleftrightarrow (+1, +1, \ 0),$$
$$R_7 \longleftrightarrow (-1, +1, -1),$$
$$PR_7 \longleftrightarrow (+1, +1, -1),$$

so $PR_7 = R_1$.

The cells in dimension k of the Salvetti complexes $\mathcal{X}$ and $\mathcal{X}'$ are indexed by pairs consisting of a face in codimension k and a region containing the face. As vector spaces, let

$$C_k(\mathcal{X}) = C_k(\mathcal{X}') = \mathbb{C}\left\{ E(P, R) : P \in \mathcal{L}_k, \ R \in \mathcal{L}_0, \ R \leq P \right\} \otimes_{\mathbb{C}} \mathcal{W}. \qquad (4\text{–}1)$$

We use the symbol $\prec$ to denote the covering relation in the lattice $\mathcal{L}$. Coorient the faces of the real arrangement, and for faces $Q \prec P$, let $\varepsilon(P, Q)$ be $+1$ or -1 according to whether or not the coorientations P and Q agree. See [Varchenko 1995, (2.4.2)]. Given a pair $P \geq R$ and a face Q such that $Q \prec P$, define

$$B(R, Q; P) = \prod_{H \in \mathcal{A} : P(H) = 0} b_H^{Q(H)R(H)/4}.$$

The boundary maps $\partial : C_k(\mathfrak{X}) \to C_{k-1}(\mathfrak{X})$ and $\partial' : C_k(\mathfrak{X}') \to C_{k-1}(\mathfrak{X}')$ are given by

$$\partial E(P,R) = \sum_{Q \prec P} \varepsilon(P,Q)B(R,Q;P)E(Q,QR) \qquad (4\text{-}2)$$

and

$$\partial' E(P,R) = \sum_{R \leq Q \prec P} \varepsilon(P,Q)E(Q,R). \qquad (4\text{-}3)$$

PROPOSITION 4.2 [Varchenko 1995]. *The chain complex with boundary map* (4–2) *computes the homology of the complement with local coefficient system* $\mathcal{W}$:

$$H_k(C_\bullet(\mathfrak{X})) \cong H_k(M, \mathcal{W}).$$

The chain complex with boundary map (4–3) *computes the relative homology of the pair* $(\mathbb{C}^n, N)$ *with local coefficient system* $\mathcal{W}$:

$$H_k(C_\bullet(\mathfrak{X}')) \cong H_k(\mathbb{C}^n, N, \mathcal{W}).$$

4B. Relation to the B matrix. For any face $P \in \mathcal{L}$, let $|P| \in L(\mathcal{A})$ denote the smallest subspace of $\mathbb{R}^n$ that contains P. For $0 \leq k \leq n$, define a map $S_k : C_k(\mathfrak{X}) \to C_k(\mathfrak{X})$ by

$$S_k(E(P,R)) = \sum_{S \leq P} B(R,S)E(P,S),$$

on the basis of $C_k(\mathfrak{X})$ given in (4–1). Here,

$$B(R,S) = \prod_{H \in \mathcal{A}:P(H)=0} b_H^{R(H)S(H)/4},$$

where R and S are regions satisfying $R, S \leq P$. Note that the subarrangement $\mathcal{A}_{|P|}$ equals $\{H \in \mathcal{A} : P(H) = 0\}$, and its regions can be identified with those regions $R \leq P$. With this in mind, one finds that

$$S_k = \bigoplus_{P \in \mathcal{L}_k} \left(\prod_{H \in \mathcal{A}_{|P|}} b_H^{-1/4} \right) B(\mathcal{A}_{|P|}),$$

where the weights of the hyperplanes of $\mathcal{A}_{|P|}$ are taken to be $a_H = b_H^{1/2}$.

One can verify that $S_{k-1}\partial_k = \partial'_k S_k$; that is, $S_\bullet : C_\bullet(\mathfrak{X}) \to C_\bullet(\mathfrak{X}')$ is a chain map [Varchenko 1995].

4C. Application to the Milnor fibre. In this section, $\mathcal{A}$ is a real, central arrangement. The (complex) homology of the Milnor fibre of $\mathcal{A}$ is isomorphic to $H_\bullet(M, \mathbb{C}[t, t^{-1}])$, where $\pi_1(M)$ acts on $\mathbb{C}[t, t^{-1}]$ by $b_H = t$, for all $H \in \mathcal{A}$. Since the determinants of the B matrices are nonzero over $\mathbb{C}[t^{\frac{1}{2}}, t^{-\frac{1}{2}}]$, the chain map $S_\bullet$ is an injection, and the short exact sequence

$$0 \to C_\bullet(\mathfrak{X}) \xrightarrow{S} C_\bullet(\mathfrak{X}') \to \operatorname{coker} S_\bullet \to 0 \qquad (4\text{-}4)$$

gives rise to a long exact sequence in homology. Since the homology of $C_\bullet(\mathfrak{X}')$ is zero for central arrangements [Varchenko 1995], one obtains an isomorphism

$$H_{k-1}(F,\mathbb{C}) \cong H_{k-1}(C_\bullet(\mathfrak{X})) \cong H_k(\operatorname{coker} S_\bullet)$$

for $1 \le k \le n$. The first isomorphism is useful for calculation, since we have an explicit description for the chain complex on the right-hand side, so calculating homology reduces to linear algebra suitable, in small enough cases, for a computer algebra program. The second isomorphism reduces the problem further to a chain complex that is finite-dimensional over $\mathbb{C}$; however, as we see in Section 3, the cokernel of $S_\bullet$ is still poorly understood.

We can write

$$\operatorname{coker} S_k = \bigoplus_{P \in \mathcal{L}_k} \operatorname{coker} B_{|P|}$$

as vector spaces, although the boundary map does not preserve this decomposition. Since multiplication by the determinant of a matrix map annihilates its cokernel, and $H_{k-1}(F,\mathbb{C})$ is isomorphic to a subquotient of $\operatorname{coker} S_k$, the determinant of S_k annihilates $H_{k-1}(F,\mathbb{C})$. However, a more precise statement can be made.

PROPOSITION 4.3. *For a real, central arrangement of l hyperplanes and k for which $1 \le k \le n$, $H_{k-1}(F,\mathbb{C})$ consists of a direct sum of modules of the form*

$$\mathbb{C}[t,t^{-1}]/(\Phi_d(t)),$$

where $d \mid \gcd(h(X),l)$ for some $X \in L_{\le k}$ that satisfies $l(X) \ne 0$. (Recall that $h(X)$ is the number of hyperplanes containing X, while $l(X)$ is introduced in Theorem 2.1.)

PROOF. A cyclic (monodromy) group $\mathbb{Z}/l\mathbb{Z}$ acts on $H_\bullet(F,\mathbb{C})$ via multiplication by t. A complete discussion can be found in [Dimca 1992]. That is, t^l acts as the identity, and $H_\bullet(F,\mathbb{C})$ contains no nilpotent elements. The proposition is proven by comparing these constraints on $H_{k-1}(F,\mathbb{C})$ with the expression for the determinant of S_k.   □

The proposition shows, for example, that multiplication by $1-t$ kills $H_{k-1}(F,\mathbb{C})$ for $0 \le k < k_0$, where k_0 is the smallest codimension of a flat X that is not, in some sense, a "general position" intersection of hyperplanes ($p(X) \ne 0$), and for which $h(X)$ divides the number of hyperplanes in the arrangement. This is of interest because the direct summand of $H_{k-1}(F,\mathbb{C})$ annihilated by $1-t$ is known [Orlik and Terao 1992; Cohen and Suciu 1995], while the higher torsion is more elusive.

We conclude this section with a remark and some numerical data. First, one can use the complex $C_\bullet(\mathfrak{X})$ to calculate group homology when our arrangement is a real, $K(\pi,1)$ arrangement. For example, the fundamental group of the arrangement A_{n-1} is the pure braid group P_n, and in this case $C_\bullet(\mathfrak{X})$ coincides with a construction from D. B. Fuks. In this case, $H_k(F, \mathbb{C}) = H_k(P'_n, \mathbb{C})$, where P'_n is the derived subgroup of the pure braid group. As in Section 2, one can attempt to make use of the S_n-module structure. Using Fuks' complex, E. V. Frenkel [1988] has found an expression for the S_n-invariant part of $H_k(P'_n, \mathbb{C})$. In a parallel vein, Cohen and Suciu [1998] construct a complex that is useful for group homology calculations for supersolvable, rather than real, arrangements.

The reader is invited to find a pattern in our computations of the homology of the Milnor fibre for the arrangement $\mathbf{A}_{n-1}$, in parallel with our Example 3.4. The table below shows the characteristic polynomial of the monodromy operator on $H_p(F, \mathbb{C})$; in particular, the Betti numbers of F are just the degrees of the polynomials. These polynomials were also calculated for $n \leq 5$ in [Cohen and Suciu 1998].

p	$n=2$	3	4	5
0	$1-t$	$1-t$	$1-t$	$1-t$
1		$(1-t)(1-t^3)$	$(1-t)^4(1-t^3)$	$(1-t)^9$
2			$(1-t)^3(1-t^3)(1-t^6)^2$	$(1-t)^{24}(1-t^2)^2$
3				$(1-t)^{16}(1-t^2)^2(1-t^{10})^6$

p	$n=$	6	7
0		$1-t$	$1-t$
1		$(1-t)^{14}$	$(1-t)^{20}$
2		$(1-t)^{70}(1-t^3)$	$(1-t)^{154}(1-t^3)$
3		$(1-t)^{134}(1-t^3)^{14}(1-t^5)^6$	$(1-t)^{560}(1-t^3)^{20}$
4		$(1-t)^{77}(1-t^3)^{13}(1-t^5)^6(1-t^{15})^{24}$	$(1-t)^{923}(1-t^3)^{121}$
5			$(1-t)^{498}(1-t^3)^{102}(1-t^{21})^{120}$

5. Factorizations of B

In this section, we will give two combinatorial recipes for writing the B matrix of an n-dimensional arrangement as a product. The first, given in Section 5B, shows that

$$B = M^t \cdot \operatorname{diag}(A_0, A_1, \ldots, A_n) \cdot M,$$

where M is invertible as a matrix of polynomials in the weights $\{a_0, \ldots, a_{l-1}\}$, and the dimension of A_k is the $n-k$th Betti number of the arrangement's complex complement.

The second factorization is only defined for central arrangements. It depends on an order of the hyperplanes, $\mathcal{A} = \{H_0, H_1, \ldots, H_{l-1}\}$. The factorization has the form

$$B = B[a_{l-1} \leftarrow 0]M_{l-1},$$

where $B[a_{l-1} \leftarrow 0]$ denotes the matrix obtained from B by specializing the weight of H_{l-1} to zero. Proceeding inductively,

$$B = M_0 M_1 \cdots M_{l-1},$$

where M_k is a matrix whose entries are polynomials in the weights $\{a_0, a_1, \ldots, a_k\}$.

Both factorizations are applications of the Möbius function of the same type of poset, studied first by Edelman [1984]. We begin with its description.

5A. Posets on the set of regions. As before, let $r(\mathcal{A})$ denote the set of regions of a real arrangement $\mathcal{A}$, and let $\mathcal{H}(R, S)$ denote the set of hyperplanes in $\mathcal{A}$ that separate regions $R, S \in r(\mathcal{A})$. For a fixed region $R \in r(\mathcal{A})$, Edelman's poset $P(\mathcal{A}, R)$ has underlying set $r(\mathcal{A})$, ordered by $S <_R T$ if and only if $\mathcal{H}(R, S) \subset \mathcal{H}(R, T)$. It is not hard to verify that R is the unique bottom element of $P(\mathcal{A}, R)$, and that the poset is graded by $|\mathcal{H}(R, S)|$, for $S \in P(\mathcal{A}, R)$. See [Edelman 1984] for details.

Let Z be any flat in $\mathbb{R}^n$ that intersects $\mathcal{A}$ transversely. By this, we mean an affine subspace Z for which, for $X \in L(\mathcal{A})$, if $X \cap Z \neq \varnothing$ then $\mathrm{codim}(X \cap Z) = \mathrm{codim}\, X + \mathrm{codim}\, Z$. Partition the regions of $\mathcal{A}$ by setting

$$r_1 = \{S \in r(\mathcal{A}) : S \cap Z \neq \varnothing\},$$

and $r_0 = r(\mathcal{A}) \backslash r_1$. For any region $R \in r(\mathcal{A})$, define a new poset $P(\mathcal{A}, Z, R)$ on $r_1 \cup \{\hat{0}\}$ using the ordering $<_R$, with an added bottom element $\hat{0}$. We shall be most interested in such posets for $R \in r_0$, in which case one simply has the subposet of $P(\mathcal{A}, R)$ restricted to $r_1 \cup \{R\}$, under the identification $\hat{0} = R$.

EXAMPLE 5.1. Let Z be the line $x = -2$ in the arrangement of Example 1.1, and let $R = R_1$. Then $r_1 = \{R_2, R_3, R_4, R_5\}$, and the posets $P(\mathcal{A}, R)$ and $P(\mathcal{A}, Z, R)$ are as follows:

$$P(\mathcal{A}, R) \qquad\qquad\qquad P(\mathcal{A}, Z, R)$$

For $a, b \in \{0, 1\}$, let B_{ab} be the submatrix of the B matrix whose rows and columns are indexed by r_a and r_b, respectively. The factorization results of this section depend on the following important observation.

LEMMA 5.2. *Under the hypotheses above, $B_{10} = B_{11}U$, where U is a $|r_1| \times |r_0|$ matrix with entries*

$$U(i, j) = -\mu_j(\hat{0}, i)B(i, j),$$

for $i \in r_1$ and $j \in r_0$. μ_j is the Möbius function of the poset $P(\mathcal{A}, Z, j)$.

PROOF. The claim is equivalent to the equation

$$\sum_{j\in r_1} -\mu_k(j)B(i,j)B(j,k) = B(i,k). \tag{5–1}$$

Observe that, for any $j \in r_1$,

$$B(i,j)B(j,k) = B(i,k)\prod_H a_H^2,$$

where the product is taken over hyperplanes $H \in \mathcal{H}(i,j) \cap \mathcal{H}(j,k)$. With this in mind, let

$$r_{ik}(S) = \{j \in r_1 : \mathcal{H}(i,j) \cap \mathcal{H}(j,k) = S\},$$

for sets $S \subseteq \mathcal{A}$. It is enough to show that, for every $S \subseteq \mathcal{A}$,

$$\sum_{j\in r_{ik}(S)} \mu_k(j) = \begin{cases} -1 & \text{if } S = \varnothing, \\ 0 & \text{otherwise.} \end{cases} \tag{5–2}$$

Let $\mathcal{A}' = \mathcal{A}\backslash\mathcal{H}(i,k)$, and consider the order-preserving surjection of posets $\pi : P(\mathcal{A}, Z, k) \to P(\mathcal{A}', Z, \pi(k))$ induced by the inclusion of regions of $\mathcal{A}$ into regions of $\mathcal{A}'$. For any region $j \in r_1$, the hyperplanes that separate j from k in $\mathcal{A}'$ equal $\mathcal{H}(i,j) \cap \mathcal{H}(j,k)$. It follows that, for $j, j' \in r_1$, $\pi(j) = \pi(j')$ if and only if $j, j' \in r_{ik}(S)$ for some $S \subseteq \mathcal{A}$.

For any $S \subseteq \mathcal{A}$, if $r_{ik}(S)$ is nonempty, let $x = \pi(r_{ik}(S))$. Let μ' denote the Möbius function of $P(\mathcal{A}', Z, \pi(k))$. A routine argument shows that the fibres $\pi^{-1}(x)$ each contain upper bounds, for all $x \in P(\mathcal{A}', Z, \pi(k))$. It follows that π induces a closure operation on $P(\mathcal{A}, Z, k)$ by letting $\bar{j}$ be the maximum element satisfying $\pi(\bar{j}) = \pi(j)$. Using a well-known property of closure operations (see [Stanley 1986]),

$$\mu'_{\pi(k)}(x) = \sum_{j\in\pi^{-1}(x)} \mu_k(j) = \sum_{j\in r_{ik}(S)} \mu_k(j).$$

Since no hyperplanes of $\mathcal{A}'$ separate regions k and i, $\pi(i)$ is the only element of $P(\mathcal{A}', Z, k)$ that covers $\hat{0}$. It follows that the Möbius function satisfies

$$\mu'_{\pi(k)}(x) = \begin{cases} -1 & \text{if } x = \pi(i), \\ 0 & \text{for any other region.} \end{cases}$$

$x = \pi(i)$ exactly when $S = \varnothing$, so equation (5–2) is proven. $\square$

The next proposition gives a more explicit description of the matrix U in a special case. Their proofs will appear in [Denham $\geq$ 1999]. Let $\mathcal{A}$ be a central arrangement and let f be a linear functional for which $H = \ker f \in \mathcal{A}$. Let $Z = \ker f + 1$, a hyperplane parallel to H. The induced arrangement $\mathcal{A}^Z$ is known as the *decone* of $\mathcal{A}$ with respect to H [Orlik and Terao 1992]. Inclusion identifies the regions of $\mathcal{A}^Z$ with regions of $\mathcal{A}$; moreover, if $R \in r(\mathcal{A})$, either R or $-R$ intersects Z.

For $R \in r(\mathcal{A})\backslash r(\mathcal{A}^Z)$ and $S \in r(\mathcal{A}^Z)$, let

$$W_R(S) = \bar{S} \cap \{H \in \mathcal{A}^Z : H \notin \mathcal{H}(-R, S)\},$$

where $\bar{S}$ denotes the topological closure of S in Z. $W_R(S)$ should be thought of as the topological space consisting of the walls around chamber S that do not separate S from $-R$, in the deconed arrangement.

LEMMA 5.3. *Let $\mathcal{A}$ be a central arrangement, and take Z as defined above, $R \in r(\mathcal{A})\backslash r(\mathcal{A}^Z)$ and $S \in r(\mathcal{A}^Z)$. Let μ_R be the Möbius function of the poset $P(\mathcal{A}, Z, R)$. Then*

$$\mu_R(\hat{0}, S) = \chi(W_R(S)),$$

the reduced Euler characteristic of $W_R(S)$.

DEFINITION 5.4. For any hyperplane $H \in \mathcal{A}$, define a map of sets $f_H : r(\mathcal{A}) \to L(\mathcal{A}^H)$ by letting $f_H(R) = |\bar{R} \cap H|$, the intersection of all hyperplanes containing $\bar{R} \cap H$.

It is shown in [Denham $\geq$ 1999] that the Möbius function of $P(\mathcal{A}, Z, R)$ has a simple description in these terms.

PROPOSITION 5.5. *For $R \in r(\mathcal{A})\backslash r(\mathcal{A}^Z)$ and $S \in r(\mathcal{A}^Z)$, $\mu_R(\hat{0}, S) = 0$ unless $f_H(R) \geq f_H(S)$. Moreover, if $f_H(R) \geq f_H(S)$, let d be the codimension of $f_H(R)$ in H. If S is bounded,*

$$\mu_R(\hat{0}, S) = \begin{cases} (-1)^{d-1} & \text{if } -R = S; \\ 0 & \text{otherwise.} \end{cases}$$

If S is unbounded,

$$\mu_R(\hat{0}, S) = \begin{cases} 0 & \text{if } -R = S; \\ (-1)^d & \text{otherwise.} \end{cases}$$

5B. Induction on dimension. Our first decomposition is not unique: it depends on an arbitrary choice of a flag of flats in general position with respect to the arrangement $\mathcal{A}$. By this, we mean affine subspaces Z_d for which

$$Z_0 \supset Z_1 \supset \cdots \supset Z_n,$$

codim $Z_d = d$, and for $X \in L(\mathcal{A})$, $X \cap Z_d = \varnothing$ if codim $X > n - d$; otherwise, codim$(X \cap Z) = \text{codim } X + d$, for $1 \leq d \leq n$.

Let

$$r_d = \{S \in r(\mathcal{A}) : S \cap Z_d \neq \varnothing,\ S \cap Z_{d+1} = \varnothing\}$$

for $0 \leq d < n$, and let r_n consist of the single region containing the point Z_n. Let $B_{a,b}$ be the submatrix of B with rows and columns indexed by r_a and r_b, respectively. For $0 \leq d < n$, B_{dd} can be identified with the B matrix of the arrangement $\mathcal{A}^{Z_d}$, and Lemma 5.2 applies. It states that $B_{d+1,d} = B_{d+1,d+1}U_d$, where U_d is the matrix with entries

$$U_d(i, j) = -\mu_j(\hat{0}, i)B(i, j)$$

for $i \in r_{d+1}$ and $j \in r_d$, and μ_j is the Möbius function of $P(A^{Z_d}, Z_{d+1}, j)$.

If one orders the basis $r(A)$ so that r_0 precedes r_1, one finds that

$$B = M_0^t \cdot \text{diag}(B_{00} - U_0^t B_{11} U_0, B_{11}) \cdot M_0,$$

where

$$M_0 = \begin{pmatrix} I_{|r_0|} & 0 \\ U_0 & I_{|r_1|} \end{pmatrix}.$$

More generally:

THEOREM 5.6. *Order the regions of a real arrangement A so that $R < S$ if $R \in r_a$ and $S \in r_b$ with $a < b$, in the notation above. Subject to this ordering, $B = M^t \cdot \text{diag}(A_0, A_1, \ldots, A_n) \cdot M$, where*

$$A_d = B_{dd} - U_d^t B_{d+1,d+1} U_d$$

is an $r_d \times r_d$ matrix for $0 \le d < n$, $A_n = (1)$, and

$$M = \overrightarrow{\prod_{0 \le d < n}} \begin{pmatrix} I_{|r_d|} & 0 \\ U_d & I_{|r_{d+1}|} \end{pmatrix}. \qquad \square$$

By the assumption that Z_1 is in general position, the intersection semilattice of A^{Z_1} is a truncation of that of A: $L(A^{Z_1}) = L_{\le n-1}(A)$. By induction, $L(A^{Z_d}) = L_{\le n-d}(A)$.

Let μ be the Möbius function of $L(A)$, and let

$$b_k = \sum_{X \in L_k(A)} (-1)^k \mu(\hat{0}, X).$$

It is well known [Orlik and Terao 1992] that $|r(A)| = \sum_{k=0}^n b_k$. Since the inclusion map from A^{Z_d} to A identifies the restriction of μ with the Möbius function of $L(A^{Z_d})$, $|r(A^{Z_d})| = \sum_{k=0}^{n-d} b_k$; consequently, $|r_d| = b_{n-d}$.

It can be shown that the matrix A_k, is equivalent to the bilinear form B^k defined in [Varchenko 1993, Section 20], under a suitable change of basis. This leads to another proof of Theorem 3.3.

5C. Induction on the number of hyperplanes.

The second factorization requires that A be a real, central arrangement. In order to give a description, let $H = \ker f$ be a hyperplane in A, and let $Z = \ker f + 1$ be a hyperplane parallel to H.

PROPOSITION 5.7. *B factors as $B = B[a_H \leftarrow 0]M_H$, where the matrix*

$$M_H(R, S) = \begin{cases} 1 & \text{if } R = S, \\ -\mu_S(\hat{0}, R)B(R, S) & \text{if } H \text{ separates } R \text{ from } S, \\ 0 & \text{otherwise,} \end{cases}$$

and μ_S is the Möbius function of the poset $P(A, \pm Z, S)$, where the sign is chosen so that $S \cap \pm Z = \varnothing$.

PROOF. For convenience, we reorder the basis of the space of regions so that $r(\mathcal{A}^Z)$ appears first in order. Let $2m = |\mathcal{A}|$, and let $\sigma : r(\mathcal{A}) \to [2m]$ be a bijection for which $\sigma(R) \leq m$ and $\sigma(-R) = \sigma(R) + m$ for $R \in r(\mathcal{A}^Z)$. Let Q be the corresponding permutation matrix, $Q(R, i) = \delta_{\sigma(R),i}$, for $R \in r(\mathcal{A})$ and $i \in [2m]$. In the notation of Lemma 5.2,

$$Q^{-1}BQ = \begin{pmatrix} B_{00} & B_{01} \\ B_{10} & B_{11} \end{pmatrix}.$$

Since $B(-R, -S) = B(R, S)$, we have $B_{00} = B_{11}$ and $B_{01} = B_{10}$. Using Lemma 5.2, it follows that

$$Q^{-1}BQ = \begin{pmatrix} B_{00} & 0 \\ 0 & B_{11} \end{pmatrix} \begin{pmatrix} I_m & U \\ U & I_m \end{pmatrix} = Q^{-1}B[a_H \leftarrow 0]M_H Q.$$

□

The proposition relates two B matrices, one with the hyperplanes weighted arbitrarily, and the other with one hyperplane given weight zero. One can apply the proposition to each hyperplane in succession to obtain the following.

THEOREM 5.8. Let $\mathcal{A} = \{H_0, H_1, \ldots, H_{l-1}\}$ be a real, central arrangement of hyperplanes. For $0 \leq d \leq l - 1$, let Z_d be a parallel translate of H_d. Then $B(\mathcal{A}) = M_0 M_1 \cdots M_{l-1}$, where M_d is a matrix over $\mathbb{Z}[a_0, \ldots, a_d]$. Explicitly,

(i) $M_d(S, S) = 1$,

(ii) $M_d(R, S) = -\mu_S(\hat{0}, R)B(R, S)$ if $d = \max\{k : H_k \in \mathcal{H}(R, S)\}$,

(iii) $M_d(R, S) = 0$ otherwise.

Here again, μ_S is the Möbius function of $P(\mathcal{A}, \pm Z_d, S)$, with the sign chosen so that $S \cap Z_d = \varnothing$.

EXAMPLE 5.9. Consider the arrangement $\mathbf{A}_2$ from Example 1.2. Order the hyperplanes H_{12}, H_{23}, H_{13}, and order the regions $123, 213, 132, 231, 312, 321$. Then $B = M_0 M_1 M_2$, where

$$M_1 = \begin{pmatrix} 1 & 0 & a_{23} & & & \\ 0 & 1 & 0 & & & \\ a_{23} & a_{12}a_{23} & 1 & & & \\ & & & 1 & a_{12}a_{23} & a_{23} \\ & & & 0 & 1 & 0 \\ & & & a_{23} & 0 & 1 \end{pmatrix}$$

and

$$M_2 = \begin{pmatrix} 1 & & & 0 & 0 & -a_{12}a_{23}a_{13} \\ & 1 & & a_{13} & 0 & a_{23}a_{13} \\ & & 1 & 0 & a_{13} & a_{12}a_{13} \\ a_{12}a_{13} & a_{13} & 0 & 1 & & \\ a_{23}a_{13} & 0 & a_{13} & & 1 & \\ -a_{12}a_{23}a_{13} & 0 & 0 & & & 1 \end{pmatrix},$$

and the off-diagonal entries of M_0 are all zero, except for a_{12} in entries $(123, 213)$, $(213, 123)$, $(312, 321)$, and $(321, 312)$.

We conclude with an example of the sort of information this theorem gives us. In the notation of Proposition 5.7,

$$Q^{-1} M_d Q = \begin{pmatrix} I & U_d \\ U_d & I \end{pmatrix},$$

so that $\det M_d = \det(I - U_d^2)$. One can use Proposition 5.5 to show that $I - U_d^2$ is upper-triangular. Then keeping track of the diagonal entries shows that

$$\det M_d = \det(I - U_d^2) = \prod_X (1 - a_X^2)^{l(X)},$$

where the product is taken over all $X \in L(\mathcal{A})$ for which $d = \max\{k : H_k \leq X\}$. This gives another proof of Varchenko's formula, Theorem 2.1.

References

[Babson et al. 1999] E. Babson, A. Björner, S. Linusson, J. Shareshian, and V. Welker, "Complexes of not i-connected graphs", *Topology* **38**:2 (1999), 271–299.

[Cohen and Suciu 1995] D. C. Cohen and A. I. Suciu, "On Milnor fibrations of arrangements", *J. London Math. Soc.* (2) **51**:1 (1995), 105–119.

[Cohen and Suciu 1998] D. C. Cohen and A. I. Suciu, "Homology of iterated semidirect products of free groups", *J. Pure Appl. Algebra* **126**:1-3 (1998), 87–120.

[Denham 1999] G. Denham, "Hanlon and Stanley's conjecture and the Milnor fibre of a braid arrangement", *J. Algebraic Combin.* (1999). To appear.

[Denham ≥ 1999] G. Denham, *Local systems on the complexification of an oriented matroid*, Ph.D. thesis, University of Michigan. In preparation.

[Denham and Hanlon 1997] G. Denham and P. Hanlon, "On the Smith normal form of the Varchenko bilinear form of a hyperplane arrangement", *Pacific J. Math.* Special issue in memoriam Olga Taussky-Todd (1997), 123–146.

[Dimca 1992] A. Dimca, *Singularities and topology of hypersurfaces*, Springer, New York, 1992.

[Edelman 1984] P. H. Edelman, "A partial order on the regions of $\mathbb{R}^n$ dissected by hyperplanes", *Trans. Amer. Math. Soc.* **283**:2 (1984), 617–631.

[Frenkel' 1988] È. V. Frenkel', "Cohomology of the commutator subgroup of the braid group", *Funktsional. Anal. i Prilozhen.* **22**:3 (1988), 91–92. In Russian; translated in *Functional Anal. Appl.* **22**:3 (1988), 248–250.

[Gel'fand 1986] I. M. Gel'fand, "General theory of hypergeometric functions", *Dokl. Akad. Nauk SSSR* **288**:1 (1986), 14–18. In Russian; translated in *Sov. Math. Doklady* **33**:3 (1986), 573–577.

[Hanlon and Stanley 1998] P. Hanlon and R. P. Stanley, "A q-deformation of a trivial symmetric group action", *Trans. Amer. Math. Soc.* **350**:11 (1998), 4445–4459.

[James and Kerber 1981] G. James and A. Kerber, *The representation theory of the symmetric group*, Addison-Wesley, Reading, MA, 1981.

[Kohno 1986] T. Kohno, "Homology of a local system on the complement of hyperplanes", *Proc. Japan Acad. Ser. A Math. Sci.* **62**:4 (1986), 144–147.

[Kontsevich 1993] M. Kontsevich, "Formal (non)commutative symplectic geometry", pp. 173–187 in *The Gel'fand Mathematical Seminars, 1990–1992,* edited by L. Corwin et al., Birkhäuser, Boston, 1993.

[Milnor 1968] J. Milnor, *Singular points of complex hypersurfaces,* Annals of Mathematics Studies **61**, Princeton Univ. Press, Princeton, NJ, 1968.

[Orlik and Randell 1993] P. Orlik and R. Randell, "The Milnor fiber of a generic arrangement", *Ark. Mat.* **31**:1 (1993), 71–81.

[Orlik and Terao 1992] P. Orlik and H. Terao, *Arrangements of hyperplanes,* Grundlehren der Mathematischen Wissenschaften **300**, Springer, Berlin, 1992.

[Randell 1982] R. Randell, "The fundamental group of the complement of a union of complex hyperplanes", *Invent. Math.* **69**:1 (1982), 103–108.

[Robinson and Whitehouse 1996] A. Robinson and S. Whitehouse, "The tree representation of Σ_{n+1}", *J. Pure Appl. Algebra* **111**:1-3 (1996), 245–253.

[Sagan 1991] B. E. Sagan, *The symmetric group: Representations, combinatorial algorithms, and symmetric functions,* Wadsworth & Brooks/Cole, Pacific Grove, CA, 1991.

[Salvetti 1987] M. Salvetti, "Topology of the complement of real hyperplanes in $\mathbb{C}^N$", *Invent. Math.* **88**:3 (1987), 603–618.

[Schechtman and Varchenko 1991] V. V. Schechtman and A. N. Varchenko, "Quantum groups and homology of local systems", pp. 182–197 in *Algebraic geometry and analytic geometry* (Tokyo, 1990), edited by A. Fujiki et al., Springer, Tokyo, 1991.

[Stanley 1986] R. P. Stanley, *Enumerative combinatorics,* Wadsworth & Brooks/Cole, Monterey, CA, 1986.

[Varchenko 1993] A. Varchenko, "Bilinear form of real configuration of hyperplanes", *Adv. Math.* **97**:1 (1993), 110–144.

[Varchenko 1995] A. Varchenko, *Multidimensional hypergeometric functions and representation theory of Lie algebras and quantum groups,* Advanced Series in Mathematical Physics **21**, World Scientific, River Edge, NJ, 1995.

[Zagier 1992] D. Zagier, "Realizability of a model in infinite statistics", *Comm. Math. Phys.* **147**:1 (1992), 199–210.

GRAHAM DENHAM
DEPARTMENT OF MATHEMATICS
UNIVERSITY OF MICHIGAN
2072 EAST HALL
ANN ARBOR MI, 48109–1109
UNITED STATES
  denham@math.lsa.umich.edu

PHIL HANLON
DEPARTMENT OF MATHEMATICS
UNIVERSITY OF MICHIGAN
2072 EAST HALL
ANN ARBOR MI, 48109–1109
UNITED STATES
  hanlon@math.lsa.umich.edu

New Perspectives in Geometric Combinatorics
MSRI Publications
Volume **38**, 1999

Combinatorial Differential Topology and Geometry

ROBIN FORMAN

ABSTRACT. A variety of questions in combinatorics lead one to the task of analyzing the topology of a simplicial complex, or a more general cell complex. However, there are few general techniques to aid in this investigation. On the other hand, the subjects of differential topology and geometry are devoted to precisely this sort of problem, except that the topological spaces in question are smooth manifolds. In this paper we show how two standard techniques from the study of smooth manifolds, Morse theory and Bochner's method, can be adapted to aid in the investigation of combinatorial spaces.

Introduction

A variety of questions in combinatorics lead one to the task of analyzing a simplicial complex, or a more general cell complex. For example, a standard approach to investigating the structure of a partially ordered set is to instead study the topology of the associated order complex. However, there are few general techniques to aid in this investigation. On the other hand, the subjects of differential topology and differential geometry are devoted to precisely this sort of problem, except that the topological spaces in question are smooth manifolds, rather than combinatorial complexes. These are classical subjects, and numerous very general and powerful techniques have been developed and studied over the recent decades.

A smooth manifold is, loosely speaking, a topological space on which one has a well-defined notion of a derivative. One can then use calculus to study the space. I have recently found ways of adapting some techniques from differential topology and differential geometry to the study of combinatorial spaces. Perhaps surprisingly, many of the standard ingredients of differential topology and differential geometry have combinatorial analogues. The combinatorial theories

This work was partially supported by the National Science Foundation and the National Security Agency.

are often simpler than the corresponding smooth theory, and in some cases imply the smooth theory. In this paper, I will discuss two new general tools to aid in the study of combinatorial spaces. One derived from Morse theory, a standard tool in differential topology, and the other derived from Bochner's method, a standard tool in differential geometry and global analysis. Most of the results in this paper have appeared in [Forman 1998d; 1998a].

Of course, many others have had the idea of "borrowing" ideas from continuous mathematics to study combinatorial objects. See for example [Björner 1995] and the numerous references therein. As we go along, I will mention those references most closely related to our topic.

1. Morse Theory

Since its introduction in [Morse 1925], Morse theory has become a fundamental technique for investigating the topology of smooth manifolds. The basic principle is that the topology of a manifold is very closely related to the critical points of a smooth function on the manifold. The simplest example of this relationship is the fact that if the manifold is compact, then any continuous function must have a maximum and a minimum. Morse theory provides a significant refinement of this observation. See [Milnor 1962] for a beautiful exposition of this subject, and [Bott 1988] for a wonderful overview of Morse theory, including some recent developments.

Many others have developed versions of Morse theory for simplicial complexes (see [Brehm and Kühnel 1987; Kühnel 1990], for example), and other types of nonsmooth topological spaces (for example, [Morse 1934; Eells and Kuiper 1962; Kuiper 1971; Goresky and MacPherson 1983; 1988]). When extending Morse theory to such spaces, one must decide what will replace smooth functions, the key ingredient in the classical theory. In the references cited, a number of choices have been made. Attention is focussed on piecewise linear functions [Brehm and Kühnel 1987; Kühnel 1990], continuous functions which behave like smooth functions near their critical points [Morse 1934; Eells and Kuiper 1962], and functions which can be extended to a larger domain to be smooth [Goresky and MacPherson 1983; 1988]). If one attempts to consider all continuous functions, one can still define the notion of a critical point, but it seems to be impossible to control the topological contribution of each critical point [Kuiper 1971; Gromov 1981].

In this paper, we take a different approach. We will present a very simple Morse theory for simplicial complexes which is entirely discrete. That is, there is no mention at all of continuous functions. Instead, our functions assign a single real number to each cell of the complex. Because we require so little structure from our functions, we need very little structure in our spaces, so the theory can be applied to very general cell complexes. After defining combinatorial Morse functions and critical points, we will show that the standard theorems of Morse theory, relating the topology of the space to the critical points of the function, are true. We also present discrete analogues of such (seemingly) intrinsically smooth notions as the gradient vector field and the corresponding gradient flow associated to a Morse function. Using these, we define a Morse complex, a differential complex built out of the critical points of our discrete Morse function, which has the same homology as the underlying space. In the smooth setting, the Morse complex has been subject to much recent study, and has played a crucial role in some recent applications (see [Klingenberg 1982; Floer 1988], for example). Most of this section will be an informal exposition of the contents of [Forman 1995; 1998d].

This combinatorial theory is not really a new theory, but rather an extraction of the combinatorial essence of the smooth theory. As evidence for this, we will indicate later that it is possible to deduce much of the smooth theory from this combinatorial theory, and, conversely, some of the results we present can be proved by "smoothing" the combinatorial Morse function and applying the corresponding smooth theory (although that is not the approach we will take in this paper).

As we mentioned above, this discrete Morse theory can be defined for any CW complex, but to avoid very minor complications we will restrict attention, in this paper, to simplicial complexes. See [Forman 1998d] for a more general treatment. Let M be any finite simplicial complex. We emphasize that M need not be a triangulated manifold, nor have any other special property. Denote by K the set of (nonempty) simplices of M, and K_p the simplices of dimension p. Write $\alpha^{(p)}$ if α has dimension p, and $\beta > \alpha$ (or $\alpha < \beta$) if α is a face of β.

A discrete Morse function on M will actually be a function on K. That is, we assign a single real number to each simplex in M. Roughly speaking, a function

$$f : K \to \mathbb{R}$$

is a Morse function if f usually assigns higher values to higher dimensional simplices, with at most one exception, locally, at each simplex. More precisely:

DEFINITION 1.1. A function

$$f : K \to \mathbb{R}$$

is a discrete Morse function if, for every $\alpha^{(p)} \in K$,

(1) $\#\{\beta^{(p+1)} > \alpha \mid f(\beta) \leq f(\alpha)\} \leq 1$ and

(2) $\#\{\gamma^{(p-1)} < \alpha \mid f(\gamma) \geq f(\alpha)\} \leq 1$.

A simple example will serve to illustrate the definition. Consider the two complexes shown in Figure 1. Here we indicate functions by writing next to each cell the value of the function on that cell. The function on the left is not a discrete Morse function as the edge $f^{-1}(0)$ violates rule (2), since it has 2 lower-dimensional "neighbors" on which f takes on higher values, and the vertex $f^{-1}(3)$

violates rule (1), since it has 2 higher dimensional "neighbors" on which f takes on lower values. The function on the right is a Morse function.

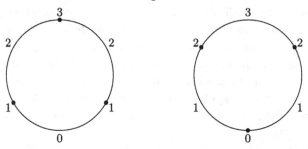

Figure 1. Left: not a discrete Morse function. Right: a discrete Morse function.

The other main ingredient in Morse theory is the notion of a critical point.

DEFINITION 1.2. A simplex $\alpha^{(p)}$ is critical (with index p) if

(1) $\#\{\beta^{(p+1)} > \alpha \mid f(\beta) \le f(\alpha)\} = 0$ and
(2) $\#\{\gamma^{(p-1)} < \alpha \mid f(\gamma) \ge f(\alpha)\} = 0$.

For example, in Figure 1, right, the vertex $f^{-1}(0)$ is critical of index 0, the edge $f^{-1}(3)$ is critical of index 1, and there are no other critical simplices. Note that if $\sigma^{(p)}$ is critical, it is necessarily critical of index p.

We will pause here to relate these definitions to the corresponding notions in the smooth category. To illustrate the main idea, we will consider a special case. Suppose x is a critical point of index 1 of a smooth Morse function F on a smooth manifold of dimension n. Then the Morse Lemma [Milnor 1962, Lemma 2.2] states that there are coordinates $(t_1, \ldots, t_n)$, with x corresponding to $(0, \ldots, 0)$, such that in these coordinates

$$F(t_1, \ldots, t_n) = F(x) - t_1^2 + \sum_{i=2}^{n} t_i^2.$$

That is, beginning at the point x, F decreases to both sides in the t_1 direction, and increases in the other coordinate directions. Now suppose e is a critical 1-simplex of a discrete Morse function f. Then $f(e)$ is greater than f at either boundary vertex, and less than f at any 2-simplex with e in its boundary (see, for example, Figure 2).

We can see that this is, in fact, a combinatorial model of the smooth situation. Moreover we see that the critical edge e can be thought of as representing a critical point of index 1 in the smooth case together with a curve pointing in the directions in which the function is decreasing. In the general case, if $\sigma^{(p)}$ is critical, σ can be thought of as representing the p-dimensional "unstable" directions at a smooth critical point of index p. If our simplicial complex is a smooth triangulation of a smooth manifold, then one can make this discussion more precise. One can think of a discrete Morse function f as assigning a number

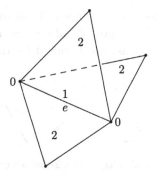

Figure 2. The function f decreases as one moves from the 1-simplex to either boundary component, and increases in each transverse direction.

to the barycenter of each simplex. One can then smoothly interpolate between these values to define a smooth function on all of M. If this is done carefully, "smoothing" a discrete Morse function f near a critical p-simplex results in a smooth Morse function F with a critical point of index p.

Our next step is to show that the topology of a general simplicial complex M is intimately related to the critical points of a discrete Morse function on M. Before stating the main theorems, we need one more definition. Suppose f is a discrete Morse function on a simplicial complex M. For any $c \in \mathbb{R}$ define the level subcomplex $M(c)$ by

$$M(c) = \bigcup_{f(\beta) \le c} \bigcup_{\alpha \le \beta} \alpha.$$

That is, $M(c)$ is the subcomplex of M consisting of all simplices β with $f(\beta) \le c$, as well as all of their faces.

THEOREM 1.3. *Suppose the interval $(a, b]$ contains no critical values of f. Then $M(a)$ is a deformation retract of $M(b)$. Moreover, $M(b)$ simplicially collapses onto $M(a)$.*

To define simplicial collapse, suppose M is a simplicial complex, $\alpha^{(p)}$ is a p-simplex of M that is not a face of any simplex, and $\gamma^{(p-1)} < \alpha$ is a $(p-1)$-face

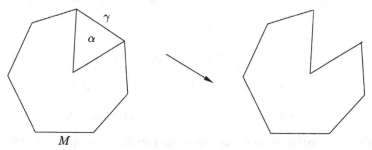

Figure 3. Simplicial collapse of M onto $M - (\alpha \cup \gamma)$.

of α that is not the face of any other simplex. Then we say that M simplicially collapses onto $M - (\alpha \cup \gamma)$; see Figure 3. More generally, a simplicial collapse is any sequence of such operations. See Figure 4 for the simplicial collapse of a 2-simplex onto a vertex.

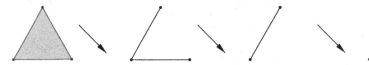

Figure 4. Simplicial collapse of a 2-simplex onto a vertex.

The equivalence relation generated by simplicial collapse is called *simple-homotopy equivalence*. This indicates that our Morse theory is particularly well suited to handling questions in this category.

THEOREM 1.4. *Suppose* $\alpha^{(p)}$ *is a critical simplex with* $f(\alpha) \in (a, b]$, *and there are no other critical simplices with values in* $(a, b]$. *Then* $M(b)$ *is (simple-)homotopy equivalent to*

$$M(a) \cup_{\dot{e}^{(p)}} e^{(p)}$$

where $e^{(p)}$ *is a p-cell, and it is glued to* $M(a)$ *along its entire boundary* $\dot{e}^{(p)}$.

Before discussing the proofs of these theorems, we pause to consider some important corollaries.

COROLLARY 1.5. *Suppose* M *is a simplicial complex with a discrete Morse function. Then* M *is (simple-)homotopy equivalent to a CW complex with exactly one cell of dimension p for each critical simplex of dimension p.*

This corollary is often most usefully applied via inequalities, known collectively as the Morse Inequalities, between numerical invariants. Let M be a simplicial complex with a discrete Morse function. Let m_p denote the number of critical simplices of dimension p. Let $\mathbb{F}$ be any field, and $b_p = dim H_p(M, \mathbb{F})$ the p-th Betti number with respect to $\mathbb{F}$. Then we have the following inequalities.

COROLLARY 1.6. I. (THE WEAK MORSE INEQUALITIES.)

(i) *For each* $p = 0, 1, 2, \ldots, n$, *where* n *is the dimension of* M, *we have*

$$m_p \geq b_p.$$

(ii) $m_0 - m_1 + m_2 - \ldots + (-1)^n m_n = \chi(M) \overset{\text{def}}{=} b_0 - b_1 + b_2 - \ldots + (-1)^n b_n.$

II. (THE STRONG MORSE INEQUALITIES.) *For each* $p = 0, 1, 2, \ldots, n, n+1$,

$$m_p - m_{p-1} + \ldots \pm m_0 \geq b_p - b_{p-1} + \ldots \pm b_0.$$

This follows immediately from the previous corollary; see [Milnor 1962]. The Strong Morse Inequalities imply the Weak Morse Inequalities.

Every simplicial complex supports a discrete Morse function. Namely, for any complex M and any simplex α of M, set $f(\alpha) = \dim(\alpha)$. Then f is a discrete Morse function, and every simplex of M is critical for f.

The results we have stated above apply to any simplicial complex. In the case that M is a combinatorial manifold [Glaser 1972, p. 19] one can often say more. For example, one can state a combinatorial analogue of the standard Sphere theorem of smooth Morse theory; see [Milnor 1962, Theorem 4.1].

COROLLARY 1.7. *Suppose M is a combinatorial n-manifold without boundary with a Morse function with exactly two critical points. Then M is a combinatorial n-sphere.*

The proof of this corollary rests heavily on Whitehead's wonderful theorem [1939] on the uniqueness of regular neighborhoods, which implies, as a special case, that a combinatorial n-manifold with boundary which simplicially collapses onto a vertex is a combinatorial n-ball. Suppose M is a combinatorial n-manifold with a discrete Morse function f with exactly two critical points. Then the maximum of f must occur at an n-simplex $\alpha^{(n)}$ which is a critical simplex of index n. Thus, f restricts on $M - \alpha$ to a Morse function with exactly one critical simplex. The minimum of f must occur at a vertex which is a critical simplex of index 0, and this is the only critical simplex of f restricted to $M - \alpha$. Theorem 1.3 implies that $M - \alpha$ simplicially collapses to that vertex. Now we apply Whitehead's theorem to conclude that $M - \alpha$ is a combinatorial n-ball. It follows that $M = (M - \alpha) \cup \alpha$ is a combinatorial n-sphere.

Rather than prove the main theorems here, we will illustrate the main ideas with an example; see Figure 5. Rigorous proofs appear in [Forman 1998d].

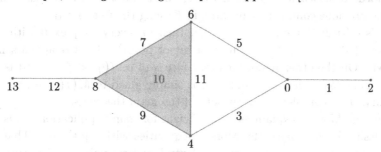

Figure 5. Example of a Morse function.

Here $f^{-1}(0)$ is a critical simplex of index 0, $f^{-1}(9)$ is a critical simplex of index 1, and there are no other critical simplices. Thus, it follows from the above corollary that M is homotopy equivalent to a CW complex built from a single 0-cell, and a single 1-cell. The only CW complex which can be built from two such cells is a circle. Therefore, we can conclude that M is homotopy equivalent to a circle, as is evident from the picture.

Consider the sequence of level subcomplexes shown in Figure 6. We can see why the theorems are true. If α is not critical, then one of two cases must occur.

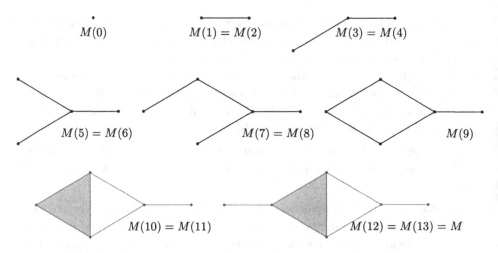

Figure 6. Level subcomplexes of the simplicial complex of Figure 5.

The first possibility is that α has a codimension one face γ with $f(\gamma) \geq f(\alpha)$ (for example $\alpha = f^{-1}(1)$, $\gamma = f^{-1}(2)$). In this case, when α is added to M, γ is added at the same time. The new subcomplex collapses onto the previous complex by pushing in α along the free face γ. (There are a few things to check. For example, we must make sure that γ is, in fact, a free face of α. That is, we must check that γ is not a face of any other simplex β with $f(\beta) \leq f(\alpha)$, but this is true.) The other possibility is that there is a simplex β with α as a codimension-one face and with $f(\beta) \leq f(\alpha)$ (for example $\alpha = f^{-1}(2), \beta = f^{-1}(1)$). In this case, α is added to M when β is added to M, and again the new complex collapses onto the previous complex by pushing in β along the free face α.

If α is critical (for example $f^{-1}(9)$), then for every simplex β with $\beta > \alpha$ we have $f(\beta) > f(\alpha)$ (this requires a bit of proof). Thus α appears first in $M(f(\alpha))$. On the other hand, for every face γ of α, $f(\gamma) < f(\alpha)$, and is added earlier. Therefore, when α is added, it is simply glued to $M(f(\alpha) - \varepsilon)$ along its boundary. This completes our "proof" of the main theorems.

While the Morse inequalities are sufficient for many applications, it is sometimes desirable to replace the Morse inequalities with equalities. That is, to understand the precise nature of any "extra" critical points which arise. This can be done, and requires the introduction of the gradient vector field. In fact, as we will see later, some of the results we have stated above are best understood in terms of the gradient vector field, rather than the original Morse function. Returning to the example in Figure 5, let us now indicate pictorially the simplicial collapse referred to in the theorem. Suppose $\alpha^{(p)}$ is a noncritical simplex with $\beta^{(p+1)} > \alpha$ satisfying $f(\beta) \leq f(\alpha)$. We then draw an arrow from α to β. The resulting diagram can be seen in Figure 7. A simplex is critical if and only if it is neither the tail nor the head of an arrow. These arrows can be viewed as the discrete analogue of the gradient vector field of the Morse function. (To

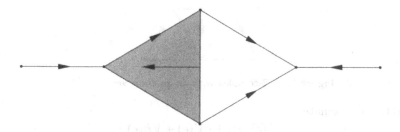

Figure 7. The gradient vector field of the Morse function.

be precise, when we say "gradient vector field" we are really referring to the negative of the gradient vector field.)

It is better to think of this gradient vector field, which we now call V, as a map of *oriented* simplices. That is, if v is a boundary vertex of an edge e with $f(e) \leq f(v)$, we want to think of $V(v)$ as a discrete tangent vector *leaving* v i.e., with e given the orientation indicated by the arrow in Figure 8.

$$v \qquad e$$

Figure 8. Edge orientation.

More generally, if $\beta^{(p+1)} > \alpha^{(p)}$ are oriented simplices and satisfy $f(\beta) \leq f(\alpha)$ then we set $V(\alpha) = \pm\beta$ with the sign chosen so that

$$\langle \alpha, \partial V(\alpha) \rangle = -1$$

where $\langle \, , \, \rangle$ is the canonical inner product on oriented chains with respect to which the simplices are orthonormal (in other words, $\langle \alpha, \partial V(\alpha) \rangle$ is the incidence number of α with respect to $V(\alpha)$). Define $V(\alpha)$ to be 0 for all simplices α for which there is no such β. Now V can be extended linearly to a map

$$V : C_p(M, \mathbb{Z}) \to C_{p+1}(M, \mathbb{Z}),$$

where, for each p, $C_p(M, \mathbb{Z})$ is the space of integer p-chains on M.

In the case of smooth manifolds, the gradient vector field defines a dynamical system, namely the flow along the vector field. Viewing the Morse function from the point of view of this dynamical system leads to important new insights [Smale 1961b]. To proceed further in our combinatorial setting, we now define a (discrete-time) flow along the gradient vector field V. Define a map

$$\Phi : C_p(M, \mathbb{Z}) \to C_p(M, \mathbb{Z}),$$

the discrete-time flow, by

$$\Phi = 1 + \partial V + V\partial.$$

We illustrate by an example. Consider the complex shown in Figure 9, with the indicated gradient vector field V. Let e be the top edge oriented from left

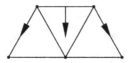

Figure 9. A 2-complex with its gradient vector field.

to right. We calculate

$$\Phi(e) = e + \partial V(e) + V\partial(e)$$

in Figure 10.

The map Φ has many of the properties that one expects from the gradient flow. For example, if $\alpha^{(p)}$ is not a critical simplex, then $\Phi(\alpha)$ is a p-chain consisting entirely of oriented p-simplices on which f is less than $f(\alpha)$, so, loosely speaking, Φ decreases f. The discrete gradient flow differs from the gradient flow in the smooth case primarily in that it stabilizes in finite time, i.e. there is an N such that

$$\Phi^N = \Phi^{N+1} = \cdots = \Phi^\infty.$$

This observation can be quite useful (for example, see the proof of the following theorem). See [Forman 1998d, Section 6] for the proofs of the main properties of Φ.

Let $C_p^\Phi \subseteq C_p(M, \mathbb{Z})$ denote the Φ-invariant p-chains, i.e., those p-chains c such that $\Phi(c) = c$. Since

$$\partial\Phi = \Phi\partial,$$

the space of Φ-invariant chains is preserved by the boundary operator ∂, so we can consider the Morse complex

$$\mathcal{C}^\Phi : 0 \to C_n^\Phi \xrightarrow{\partial} C_{n-1}^\Phi \xrightarrow{\partial} C_{n-2}^\Phi \xrightarrow{\partial} \cdots.$$

The crucial point is the following theorem.

THEOREM 1.8. $H_*(\mathcal{C}^\Phi) = H_*(M, \mathbb{Z})$.

That is, the Morse complex has the same homology as the underlying space. The proof of this theorem is not difficult. To prove that the stabilization map

$$\Phi^\infty : C_*(M, \mathbb{Z}) \to C_*^\Phi$$

induces an isomorphism on homology (with the inverse isomorphism induced by the natural injection), it is sufficient to find an algebraic homotopy operator, i.e. an operator $L : C_*(M, \mathbb{Z}) \to C_{*+1}(M, \mathbb{Z})$ such that $\Phi^\infty - 1 = \partial L + L\partial$. We now

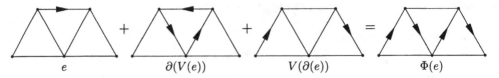

Figure 10. The discrete-time flow $\Phi = 1 + \partial V + V\partial$.

use the fact that $\Phi^\infty = \Phi^N$ for some N. If $N = 1$, then $L = V$ works. The general case is not much harder.

Our goal in introducing this complex was to further our understanding of the critical simplices. With this in mind, we now show that the Morse complex can be alternately defined in terms of the critical simplices. For each p, let $\mathcal{M}_p \subseteq C_p(M, \mathbb{Z})$ denote the span of the critical p-simplices. We then obviously have

$$\dim \mathcal{M}_p = \#\{\text{critical simplices of dimension } p\} \qquad (1.1)$$

For each p, we can apply the stabilization map Φ^∞ to get a map from $\mathcal{M}_p$ to C_p^Φ.

THEOREM 1.9. *For each p, the map*

$$\Phi^\infty : \mathcal{M}_p \to C_p^\Phi.$$

is an isomorphism.

Thus, the Morse complex can be defined equivalently as the complex

$$\mathcal{M} : 0 \to \mathcal{M}_n \xrightarrow{\tilde{\partial}} \mathcal{M}_{n-1} \xrightarrow{\tilde{\partial}} \mathcal{M}_{n-2} \xrightarrow{\tilde{\partial}} \cdots$$

where $\tilde{\partial}$ is the differential induced by the above isomorphism (that is, $\tilde{\partial} = (\Phi^\infty)^{-1} \partial \Phi^\infty$). In particular

$$H_*(\mathcal{M}) = H_*(M, \mathbb{Z}). \qquad (1.2)$$

It is interesting to note that together (1.1) and (1.2) imply the Morse inequalities, so this gives an alternate proof of Corollary 1.6. From (1.2), we see that the Morse complex $\mathcal{M}$ contains a more complete description of the relationship between the critical simplices of f, and the homology of M. All that remains is to get a explicit description of the differential $\tilde{\partial}$. This requires the introduction of the notion of a combinatorial gradient path. Let α and $\tilde{\alpha}$ be p-simplices. A *gradient path* from $\tilde{\alpha}$ to α is a sequence of simplices

$$\tilde{\alpha} = \alpha_0^{(p)}, \ \beta_0^{(p+1)}, \ \alpha_1^{(p)}, \ \beta_1^{(p+1)}, \ \alpha_2^{(p)}, \ \ldots, \ \beta_r^{(p+1)}, \ \alpha_{r+1}^{(p)} = \alpha$$

such that $f(\alpha_i) \geq f(\beta_i) > f(\alpha_{i+1})$ for each $i = 1, \ldots, r$. Equivalently, we require $V(\alpha_i) = \pm \beta_i$, and $\beta_i > \alpha_{i+1} \neq \alpha_i$. In Figure 11 we show a single gradient path from the boundary of a critical 2-simplex β to a critical edge α, where the arrows indicate the gradient vector field V.

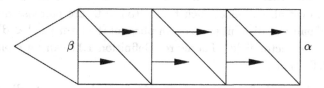

Figure 11. A gradient path from β to α.

We are now ready to state the desired formula.

THEOREM 1.10. *Choose an orientation for each simplex. Then for any critical $(p+1)$-simplex β we have*

$$\tilde{\partial}\beta = \sum_{\text{critical } \alpha^{(p)}} c_{\alpha,\beta}\alpha$$

where

$$c_{\alpha,\beta} = \sum_{\tilde{\alpha}^{(p)} < \beta} \langle \partial\beta, \tilde{\alpha} \rangle \sum_{\gamma \in \Gamma(\tilde{\alpha},\alpha)} m(\gamma)$$

for $\Gamma(\tilde{\alpha},\alpha)$ the set of gradient paths from $\tilde{\alpha}$ to α.

For any gradient path from $\tilde{\alpha} < \beta$ to α, the orientation on β induces an orientation on α. The coefficient $m(\gamma)$ is equal to ± 1, depending on whether the orientation on β induces the chosen orientation on α, or the opposite orientation.

As stated earlier, it is often easier to work with the gradient vector field, rather than the original Morse function. In fact, one need not refer to the Morse function at all, as long as one knows that the vector field under consideration is the gradient of a Morse function. Thus, it is important to characterize those vector fields which are gradient vector fields of Morse functions. In the smooth case, Smale [1961b] provided such a characterization. We can provide an analogous characterization of discrete gradient vector fields. First, we define an object we call a discrete vector field. Such objects have been previously studied, under a different name, in [Duval 1994; Stanley 1993] and some of the references therein.

DEFINITION 1.11. A discrete vector field is any map

$$U : K \to K \cup \{0\}$$

satisfying, for each $\alpha^{(p)}$, the following conditions:

(1) $U(\alpha) = 0$ or α is a codimension-one face of $U(\alpha)$.
(2) $\#\{\beta \text{ such that } U(\beta) = \alpha\} \leq 1$.
(3) If $\alpha^{(p)} \in \text{Image}(U)$ then $U(\alpha) = 0$.

A gradient vector field is not quite such an object, since gradient vector fields are functions of oriented simplices, but for now we will ignore such subtleties. We have presented the definition in this form so that the relationship to the notion of a gradient vector field is apparent. However, there is a simpler, equivalent, definition. Note that if U is a discrete vector field then the pairs $\{\alpha, U(\alpha)\}$, as α ranges over all simplices such that $U(\alpha) \neq 0$, are pairwise disjoint. Any collection of pairwise disjoint pairs of simplices of the form $\{\alpha^{(p)} < \beta^{(p+1)}\}$ arises from a discrete vector field. Therefore, Definition 1.11 can be replaced by the following definition.

DEFINITION 1.11′. A discrete vector field is a pairwise disjoint collection of pairs of simplices of the form $\{\alpha^{(p)} < \beta^{(p+1)}\}$.

Such pairings were studied in [Stanley 1993; Duval 1994] as a tool for investigating the possible f-vectors for a simplicial complex. In [Forman 1998b] we investigated the dynamical properties of the discrete-time flow corresponding to a general discrete vector field.

If U is a combinatorial vector field, we define a U-path to be any path as in Figure 11 (where now the arrows indicate the map U). More precisely, a U-path is a sequence of simplices

$$\alpha_0^{(p)}, \beta_0^{(p+1)}, \alpha_1^{(p)}, \beta_1^{(p+1)}, \alpha_2^{(p)}, \ldots, \beta_r^{(p+1)}, \alpha_{r+1}^{(p)}$$

such that for each $i = 0, \ldots, r$, $\beta_i = U(\alpha_i)$ and $\beta_i > \alpha_{i+1} \neq \alpha_i$. We say such a path is a nontrivial closed path if $r \geq 0$ and $\alpha_0 = \alpha_{r+1}$. There is a simple characterization of those discrete vector fields which are the gradient of a discrete Morse function.

THEOREM 1.12. *A discrete vector field U is the gradient vector field of a discrete Morse function if and only if there are no nontrivial closed U-paths.*

This theorem has a very nice purely combinatorial description, using which we can recast the Morse theory in an appealing form. We begin with the Hasse diagram of M, that is, the partially ordered set of simplices of M ordered by the face relation. Consider the Hasse diagram as a directed graph. The vertices of the graph are in one-to-one correspondence with the simplices of M, and there is a directed edge from β to α if and only if α is a codimension-one face of β. Now let U be a combinatorial vector field. We modify the directed graph as follows. If $U(\alpha) = \beta$ then reverse the orientation of the edge between α and β (so that it now goes from α to β). A U-path is just a directed path in this modified graph. From Theorem 1.12, U is a gradient vector field if and only if there are no closed directed paths. That is, in this combinatorial language, a discrete vector field is a partial matching of the Hasse diagram, and a discrete vector field is a gradient vector field if the partial matching is acyclic in the above sense.

We can now restate some of our earlier theorems in this language. There is a very minor complication in that one usually includes the empty set as an element of the Hasse diagram (considered as a simplex of dimension -1) while we have not considered the empty set previously.

THEOREM 1.13. *Let U be an acyclic partial matching of the Hasse diagram of M (of the sort described above). Assume that the empty set is matched with a vertex. Let u_p denote the number of unpaired p-simplices. Then M is homotopy equivalent to a CW complex with $u_0 + 1$ vertices, and, for each $p \geq 1$, u_p cells of dimension p.*

An important special case of Theorem 1.13 is when U is a complete matching, that is, every simplex is paired with another (possible empty) simplex.

THEOREM 1.14. *Let U be a complete acyclic matching of the Hasse diagram of M, then M collapses onto a vertex, so that, in particular, M is contractible.*

This language was introduced in [Chari 1996], and the reader should consult this source for a more complete presentation of the Morse theory from this point of view; see also [Shareshian 1997]. Theorem 1.14 was used to good effect in [Babson et al. 1999] to investigate the homology of the simplicial complex of "not k-connected graphs" on a fixed set of vertices. This homology plays an important role in Vassiliev's theory of invariants for links [Vassiliev 1993]. Let us be a bit more precise. Let V be a fixed set of n vertices. A graph on V is determined by specifying a subset of the $\binom{n}{2}$ possible edges. Let S be a simplex with $\binom{n}{2}$ vertices (so that $\dim S = \binom{n}{2} - 1$). A graph on V can now be determined by specifying a face of S. We say a graph is connected if every pair of vertices can be connected by a path consisting of edges. We say a graph is k-connected if it is connected after the removal of an arbitrary set of l vertices (and the edges connected to them) for any $l < k$. If a face α of S corresponds to a graph which is not k-connected, then every face β of α also corresponds to a graph which is not k-connected. Therefore, the set of "not k-connected" graphs forms a subcomplex of S. This subcomplex, which we denote by Δ_n^k, is the object of study in [Babson et al. 1999; Turchin 1997; Shareshian 1997]. (Actually, of interest is the homology of S/Δ_n^k, but since S is contractible, these questions are essentially equivalent.) We state their result in the special case $k = 2$.

THEOREM 1.15 [Babson et al. 1999; Turchin 1997; Shareshian 1997]. *The complex Δ_n^2 has the homotopy type of a wedge of $(n-2)!$ spheres of dimension $2n-5$.*

For the applications to the investigations in [Vassiliev 1993] it was necessary to find explicit generators of the homology of Δ_n^k. This was done in [Shareshian 1997], where the more general Theorem 1.13 was featured. There Shareshian constructs, by hand, an explicit acyclic partial matching on the simplicial complex of not 2-connected graphs. From the Morse theory he can then deduce the homotopy type of the complex. Moreover, the unmatched (i.e. critical) simplices provide a set of explicit generators for the homology.

In [Forman 1998c] we apply these methods to a problem in complexity theory. We will describe a special case of this problem. Suppose an oracle chooses a graph Γ on the set of n vertices V. The graph Γ is unknown to us, but we are permitted to ask questions to the oracle, one at a time, of the sort "Is the edge (v_i, v_j) in Γ?", where $v_i, v_j \in V$. We may use the answers to our previous questions when choosing our next question. The only restriction is that we choose each question according to a deterministic algorithm (which takes as its input the previous questions and answers). Our task is to decide if Γ is k-connected, using as few questions as possible. In particular, we must try to accomplish our task before asking $\binom{n}{2}$ questions. (Of course, after asking $\binom{n}{2}$ questions we have completely determined Γ.) See [Bollobás 1995, § 4.5] for a survey of the work on this and related problems.

Let A denote our algorithm for choosing questions. Say a graph Γ is an *evader* of A if, asking questions according to A, we do not determine whether or not Γ is k-connected until we ask $\binom{n}{2}$ questions. We observe that A induces a complete pairing of the set of graphs on V. Namely, pair Γ_1 and Γ_2 if we can not distinguish between them until all $\binom{n}{2}$ questions have been asked. By definition, a not-k-connected graph is an evader of A if and only if it is paired with a k-connected graph, and vice versa. The main result of [Forman 1998c] is that this is an acyclic pairing, and hence, by Theorem 1.13, the gradient vector field of a Morse function. We can restrict this vector field to Δ_n^k, to get the gradient of a Morse function on Δ_n^k. The critical simplices of this restricted gradient vector field are the simplices of Δ_n^k which were paired with simplices not in Δ_n^k. These are precisely the not-k-connected graphs which were paired with a k-connected graph, i.e., the evaders of A. We can now use the weak Morse inequalities (Corollary 1.6) along with Theorem 1.15 to conclude:

THEOREM 1.16. *Suppose our goal is to determine whether or not an unknown graph is 2-connected. For any question algorithm A, there are at least $(n-2)!$ evaders of A which are not 2-connected, and $(n-2)!$ evaders of A which are 2-connected.*

Before ending this section, we make some final remarks about the combinatorial Morse theory, and its relation to the smooth theory. One of the main problems in Morse theory is finding a Morse function, for a given space, with the fewest possible critical points. In general this is a very difficult problem, since, in particular, it contains the Poincaré conjecture — spheres can be recognized as those spaces which have a Morse function with precisely two critical points. (See [Milnor 1965] for a completely Morse theoretic presentation of Smale's proof [1961a] of the higher-dimensional Poincaré conjecture.) As an application of Theorem 1.12, we easily prove a "cancellation" theorem, which enables one, under certain conditions, to simplify a Morse function (this result is a discrete analogue of in [Milnor 1965, Theorem 5.4], the "First Cancellation Theorem"). That is, if $\alpha^{(p)}$ and $\beta^{(p+1)}$ are two critical simplices, and if there is exactly one gradient path from $\partial\beta$ to α, then α and β can be cancelled. More precisely:

THEOREM 1.17. *Suppose f is a discrete Morse function on M such that $\beta^{(p+1)}$ and $\alpha^{(p)}$ are critical, and there is exactly one gradient path from $\partial\beta$ to α. Then there is another Morse function g on M with the same critical simplices except that α and β are no longer critical. Moreover, the gradient vector field associated to g is equal to the gradient vector field associated to f except along the unique gradient path from $\partial\beta$ to α.*

In the smooth case, the proof, either as presented originally by Morse [1965] or as presented in [Milnor 1965], is rather technical. In our discrete case the proof is simple. If, in Figure 11, the indicated gradient path is the only gradient path from $\partial\beta$ to α, then we can reverse the gradient vector field along this path, replacing the figure by the vector field in Figure 12.

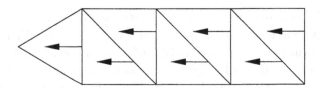

Figure 12. Modified gradient vector field.

Theorem 1.12 immediately implies that this new vector field is the gradient vector field associated to some Morse function, and α and β are no longer critical. This completes the proof.

The proof in the smooth case proceeds along the same lines. However, in addition to turning around those vectors along the unique gradient path from $\partial\beta$ to α, one must also adjust all nearby vectors so that the resulting vector field is smooth. Moreover, one must check that the new vector field is the gradient of a function, so that, in particular, modifying the vectors did not result in the creation of a closed orbit. This is an example of the sort of complications which arise in the smooth setting, but which do not make an appearance in the discrete theory.

We end this section by mentioning that much of the discrete Morse theory implies the corresponding smooth theory.

THEOREM 1.18 [Forman ≥ 1999]. *Let M be a smooth manifold with a smooth Morse function. Then there is a C^1 triangulation of M and a discrete Morse function on the resulting simplicial complex which has the same Morse data as the original function (that is, the same number of critical points, and isomorphic Morse complexes).*

2. Differential Geometry

In this section, we borrow ideas from Riemannian geometry and define local "curvature" invariants for cell complexes from which one can deduce global topological properties. In particular, given a cell complex M, we define a combinatorial Ricci curvature, which is a function on the set of edges of M. The Ricci curvature of an edge is completely determined by those cells of M which are neighbors of e. If the Ricci curvature is always positive, or nonnegative, then one can deduce strong statements about the topology of M.

Much fascinating and beautiful mathematics has resulted from earlier efforts to extend the basic notions of curvature to combinatorial spaces. In the mathematics literature, one can see, for example, [Allendoerfer and Weil 1943; Banchoff 1967; 1983; Cheeger 1983; Stone 1976]. In the mathematical physics literature such ideas appear, most notably, in the voluminous work on the so-called Regge calculus; see [Cheeger et al. 1984; Regge 1961; Williams 1992] for an introduction. Most (but not all) of this work begins with a rigid presentation of the cell complex, which we think of as a piecewise flat Riemannian manifold, so that the curvature is localized to the lower-dimensional cells. The curvature is then defined in terms of how the neighborhood of a cell differs from what one would see in a flat complex. This section is also related to efforts to derive local combinatorial formulas for the characteristic classes of combinatorial manifolds; see [Banchoff and McCrory 1979; Gabrièlov et al. 1974; Gelfand and MacPherson 1992; Halperin and Toledo 1972; Levitt 1989; Levitt and Rourke 1978; MacPherson 1993; Stone 1981; Yu 1983]. In the case of smooth manifolds, such local formulas are expressed in terms of curvature, so, presumably, local combinatorial formulas for characteristic classes can be thought of as combinatorial formulas for curvature; this connection is made explicit in [Cheeger 1983]. In the mathematical physics literature, similar questions have been studied under the name of lattice gauge theory; see [Phillips and Stone 1990], for example. The main result of [Luo and Stong 1993] is also similar in spirit to the results we will present, in that the combinatorics of the edges and faces of a combinatorial 3-manifold are related to the topology of the underlying manifold. One should also compare with [Alexandrov 1951; Gromov 1987], and the references therein, in which one defines notions of curvature on combinatorial spaces by comparing distances between points with the distances one would find on smooth Riemannian manifolds with constant curvature.

As will soon become clear, our approach is completely different from those followed in these works, and is based instead on an analysis of the combinatorial Laplace operator (this operator also appears in [Yu 1983], but plays a very different role there). The combinatorial Laplace operator has a long history. In the case of a one-dimensional simplicial complex, this operator appeared, implicitly, in Kirchoff's work on electrical networks [1847]. See [Chung 1997; Forman 1993] for more recent work on this case.

The combinatorial Laplace operator on general complexes was introduced by Eckmann [1945]. The close relationship between the combinatorial and Riemannian Laplace operators has been explored in a number of papers, most notably by Dodziuk and his collaborators; see, for example, [Dodziuk 1974; 1981; Dodziuk and Karp 1988; Dodziuk and Patodi 1976]. The work by Garland [1972; 1973; 1975], is very closely related to ours, in that he also uses the combinatorial Laplace operator to define combinatorial curvatures, which are then used to deduce global topological properties of the complex. The reader can consult [Friedman 1996; Friedman and Handron 1997; Kook et al. 1998] for some fascinating recent applications of the combinatorial Laplacian to some problems in combinatorics.

Let us begin with some examples of the theory we develop. Although the theory can be applied to very general cell complexes, in this introduction we will restrict attention to CW complexes satisfying a combinatorial condition modelled on the notion of convexity. The reader may think only of simplicial complexes, but it is insightful to allow more general convex cells.

DEFINITION 2.1. A regular CW complex M is *quasi-convex* if for each pair of open p-cells α_1, α_2, either $\bar{\alpha}_1 \cap \bar{\alpha}_2 = \varnothing$, or $\bar{\alpha}_1 \cap \bar{\alpha}_2 = \bar{\gamma}$ for some cell γ. For example, all simplicial complexes and all polyhedra are quasi-convex. From now on, by a *quasi-convex complex* we will always mean a compact regular CW complex which is quasi-convex.

Throughout this section, a primary role will be played by the notion of a *neighbor*.

DEFINITION 2.2. If α_1 and α_2 are p-cells of M, say that α_1 and α_2 are *neighbors* if at least one of these conditions is satisfied:

(1) α_1 and α_2 share a $(p+1)$-cell, that is, there is a $(p+1)$-cell β with $\beta > \alpha_1$ and $\beta > \alpha_2$.
(2) α_1 and α_2 share a $(p-1)$-cell, that is, there is a $(p-1)$-cell γ with $\gamma < \alpha_1$ and $\gamma < \alpha_2$.

We say that α_1 and α_2 are *parallel neighbors* if either condition is true but not both. If both are true, we say α_1 and α_2 are *transverse neighbors*.

Figure 13 illustrates the origin of these terms.

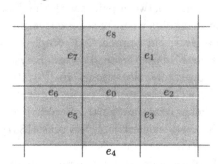

Figure 13. An illustration of neighbors in the case $p = 1$. The edge e_0 has four parallel neighbors (e_2, e_4, e_6, e_8) and four transverse neighbors (e_1, e_3, e_5, e_7).

We are now ready to state the main new definition of this section.

DEFINITION 2.3. For each p, define the p-th curvature function

$$\mathcal{F}_p : K_p \to \mathbb{R}$$

as follows. For any p-cell α, set

$$\mathcal{F}(\alpha) = \#\{(p+1)\text{-cells } \beta > a\} + \#\{(p-1)\text{-cells } \gamma < a\} - \#\{\text{parallel neighbors of } \alpha\}.$$

The significance of this definition is made apparent in the following combinatorial analogue of Bochner's Theorem [Bochner 1946; Bochner 1948; Bochner 1949].

THEOREM 2.4. *Let M be a quasi-convex complex. Suppose, for some p, that $\mathcal{F}_p(\alpha) > 0$ for each p-cell α. Then*

$$H_p(M, \mathbb{R}) = 0.$$

Of particular interest is the case $p = 1$. In this case we give the curvature function a special name. For any edge (1-cell) e of M, define the *Ricci curvature* of e by

$$\mathrm{Ric}(e) = \mathcal{F}_1(e).$$

More explicitly,

$$\mathrm{Ric}(e) = \#\{2\text{-cells } f > e\} + 2 - \#\{\text{parallel neighbors of } e\}.$$

For example, in Figure 13

$$\mathrm{Ric}(e_0) = 2 + 2 - 4 = 0.$$

Thus, the standard lattice cell decomposition of the plane is a "Ricci flat" cell complex. In fact, the reader can easily check that for all p, $\mathcal{F}_p$ vanishes on the standard lattice cell decomposition of Euclidean space of any dimension. At the right we show a 2-dimensional sphere with "cubic" cell decomposition. In this case the edge e_0 has 2 parallel neighbors (the edges e_1 and e_2) so that

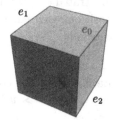

$$\mathrm{Ric}(e_0) = 2 + 2 - 2 = 2.$$

The purpose of Figure 14 is to demonstrate that the definitions and theorems discussed in this paper can, with a few exceptions, be applied to very general complexes. The complex need not be a manifold, or have any other special structure.

Figure 14. The Ricci curvature — indicated here for each edge — can be computed for arbitrary complexes.

In the Riemannian case Bochner showed that one does not need the p-th curvature function to be strictly positive, as we required in Theorem 2.4, in order to draw some topological conclusions. In fact, nonnegativity also has strong implications. In the combinatorial setting, we can prove a corresponding result in the case p=1.

THEOREM 2.5. *Let M be a connected quasi-convex complex.*

(i) *Suppose that $\mathrm{Ric}(e) \geq 0$ for all edges e of M. If there is a vertex v such that $\mathrm{Ric}(e) > 0$ for all $e > v$ then*

$$H_1(M, \mathbb{R}) = 0.$$

(ii) *Suppose that M is a combinatorial n-manifold, $n \leq 3$, and $\mathrm{Ric}(e) \geq 0$ for all edges of M. Then*

$$\dim H_1(M, \mathbb{R}) \leq n.$$

(*This theorem is true for all $n \geq 0$ with an additional technical hypothesis.*)

We will later indicate the proof of these results. We conjecture that, just as in the Riemannian setting, Theorem 2.5 holds for general p (with the right hand side of the inequality in (ii) replaced by $\binom{n}{p}$). However, we have run into interesting obstacles in trying to prove Theorem 2.5 in the desired generality. We will later give an idea as to how $p = 1$ differs from the general case. For a more complete discussion of this point, see [Forman 1998a, § 4].

It is interesting to see how these theorems apply to the very simple complexes illustrated on page 14. In the case of the 2-sphere S^2 given the cubic cell decomposition, each edge has Ricci curvature 2. Thus we may apply Theorem 2.4 to conclude

$$H_1(S^2, \mathbb{R}) = 0.$$

Each edge of the complex M in Figure 14 has Ricci curvature ≥ 0. Moreover, every edge $e > v$ has positive Ricci curvature. In this case, Theorem 2.5(i) implies

$$H_1(M, \mathbb{R}) = 0.$$

The 2-dimensional torus T^2 can be given a cell decomposition which looks everywhere like the cell complex shown in Figure 13. With this cell structure T^2 is a connected combinatorial 2-manifold. Since every edge has Ricci curvature $= 0$, we may apply Theorem 2.5(ii) to learn that

$$\dim H_1(T^2, \mathbb{R}) \leq 2.$$

More generally, for every n, the n-dimensional torus T^2 can be given such a cubic cell decomposition. In this case, T^n will be a connected combinatorial n-manifold, with every edge having Ricci curvature 0. Moreover, T^n satisfies the (unstated) technical hypothesis of the above theorem, so we may conclude

$$\dim H_1(T^n, \mathbb{R}) \leq n.$$

Since, in fact, $\dim H_1(T^n, \mathbb{R}) = n$, we see that the conclusion of this combinatorial Bochner theorem, just as the conclusion of the classical Bochner theorem, is sharp.

In [Myers 1941], a precursor to Bochner's work, Myers proves a theorem (in the Riemannian setting) which is a strengthening of Theorem 2.4 in the case $p = 1$. The analogous result holds in the combinatorial setting.

THEOREM 2.6. *Let M be a quasi-convex complex. Suppose that $\mathrm{Ric}(e) > 0$ for every edge e. Then $\pi_1(M)$ is finite.*

Stone [Stone 1976] defines a different notion of curvature, in a somewhat more restricted setting, from which a version of Myers' theorem is deduced. Even though the proof of Theorem 2.6 is very similar to Stone's proof (and the proof in [Myers 1941]), the relationship between the two notions of combinatorial curvature is not clear to this author at present.

These theorems show that there are strong topological restrictions for a topological space to have a quasi-convex cell decomposition with $\mathrm{Ric}(e) > 0$ (or ≥ 0) for each edge e. It is natural to consider the other extreme, and to investigate the possibility of decompositions with $\mathrm{Ric}(e) < 0$ for every edge e. In [Gao 1985; Gao and Yau 1986; Lohkamp 1994a; 1994b] it was shown that every smooth manifold of dimension ≥ 3 has a Riemannian metric with everywhere negative Ricci curvature. The same, and more, is true in the combinatorial setting.

THEOREM 2.7. *Let M be a (not necessarily compact) combinatorial n-manifold with $n \geq 2$. Then there is a finite subdivision M^* of M such that $\mathrm{Ric}(e) < 0$ for all edges e of M^*.*

It is interesting to note that the corresponding Riemannian theorem is only true for manifolds of dimension ≥ 3. The Gauss–Bonnet Theorem provides an obstruction for a surface to have a metric with negative Ricci curvature (which is to say Gaussian curvature if $n = 2$). Theorem 2.7 implies that there is no simple Gauss–Bonnet theorem for combinatorial Ricci curvature. See [Forman 1998a, § 5] for a way to recover a version of Gauss–Bonnet in this setting.

We will now give an idea of the proofs of Theorems 2.4 and 2.5. Readers familar with Bochner's original proof will notice that we are simply following his proof (now known as "Bochner's method") in a combinatorial setting. Let M be a finite CW complex. Consider the cellular chain complex

$$0 \longrightarrow C_n(M,\mathbb{R}) \xrightarrow{\partial} C_{n-1}(M,\mathbb{R}) \xrightarrow{\partial} \cdots \xrightarrow{\partial} C_0(M,\mathbb{R}) \longrightarrow 0$$

Endow each $C_p(M,\mathbb{R})$ with a positive definite inner product such that the cells of M are orthogonal. This involves choosing a positive "weight" w_α for each cell α and setting

$$\langle \alpha, \alpha \rangle = w_\alpha.$$

(The formulas presented in this paper result from setting $w_\alpha = 1$ for every cell α.) Let $\partial^* : C_p(M,\mathbb{R}) \to C_{p+1}(M,\mathbb{R})$ denote the adjoint of ∂ and

$$\square_p = \partial\partial^* + \partial^*\partial : C_p(M,\mathbb{R}) \to C_p(M,\mathbb{R})$$

the corresponding Laplacian. The combinatorial Hodge theorem

$$\mathrm{Ker}\,\square_p \cong H_p(M,\mathbb{R})$$

follows from basic linear algebra.

To motivate what follows, we now present a very brief overview of Bochner's original proof in the smooth category. We will freely use standard notation and terminology of Riemannian geometry. The reader may feel free to skip this rather dense paragraph. The next step in Bochner's original proof is the Bochner–Weitzenböck formula

$$\Box_p = \nabla_p^* \nabla_p + F_p$$

where now $\Box_p$ is the Riemannian Laplace operator on p-forms on a compact Riemannian manifold, ∇_p is the (Levi-Civita) covariant derivative operator, and F_p is a 0th order operator whose value at $m \in M$ depends only on derivatives of the Riemannian metric at m. The operator $\nabla_p^* \nabla_p$ is called the Bochner Laplacian. The main point is that this is a nonnegative operator, in the sense that for every p-form η on M

$$\langle \nabla_p^* \nabla_p \eta, \eta \rangle = \langle \nabla_p \eta, \nabla_p \eta \rangle \geq 0$$

where $\langle \cdot, \cdot \rangle$ is the L^2 inner product on p-forms induced by the Riemannian metric. It immediately follows that if F_p is a positive operator, in the sense that $\mathcal{F}_p(\eta) := (F_p(m)\eta, \eta) > 0$ for all $m \in M$ and $\eta \in \bigwedge^p T_m^* M - \{0\}$, then $\Box_p$ is a positive operator, and hence the kernel is trivial. This implies that $H_p(M, \mathbb{R}) = 0$, the desired conclusion of Theorem 2.4. For $p > 1$, the function $\mathcal{F}_p$ is rather mysterious. On the other hand, it can be shown that $\mathcal{F}_1$ is equal to the Ricci curvature. That is, for all $\omega \in T^* M$,

$$\mathrm{Ric}(\omega) = \mathcal{F}_1(\omega). \tag{2.1}$$

Now we go back to the combinatorial setting. Following Bochner's proof, our next step is to develop a Bochner–Weitzenböck formula for $\Box_p$, the combinatorial Laplace operator. This is rather mysterious, since we begin by knowing neither a combinatorial analogue for $(\nabla_p^*)\nabla_p$ nor for F_p. However, we show that there is, in fact, a canonical decomposition

$$\Box_p = B_p + F_p,$$

where B_p is a nonnegative operator and, when expressed as a matrix with respect to a basis of $C_p(M, \mathbb{R})$ consisting of the p-cells of M, F_p is diagonal.

To describe this decomposition, we begin by defining a decomposition of any $n \times n$ symmetric matrix. Rather than writing a general definition, we will illustrate the decomposition in the case of a symmetric 3×3 matrix:

$$\begin{pmatrix} a & b & c \\ b & d & e \\ c & e & f \end{pmatrix} = \begin{pmatrix} |b| + |c| & b & c \\ b & |b| + |e| & e \\ c & e & |c| + |e| \end{pmatrix}$$

$$+ \begin{pmatrix} a - (|b| + |c|) & 0 & 0 \\ 0 & d - (|b| + |e|) & 0 \\ 0 & 0 & f - (|c| + |e|) \end{pmatrix}$$

$$= B + F. \tag{2.2}$$

The linear map defined by the matrix B is nonnegative (with respect to the standard inner product on $\mathbb{R}^3$). This is a special case of the fact that any symmetric matrix $(a_{i,j})$ such that $a_{i,i} \geq \sum_{j \neq i} |a_{i,j}|$ for each i is nonnegative. This is one of those very useful facts in matrix theory which is constantly being rediscovered. For the history of this result (not stated in quite the same way) and extensions, see [Taussky 1949]. The desired conclusion is also implied by the classical eigenvalue estimate known as "the Gershgorin circle method" [Gershgorin 1931], which says that the eigenvalues lie in circles centered at $a_{i,i}$ with radius $\sum_{j \neq i} |a_{i,j}|$.

The nonnegativity of B implies that in this simple example we can already make the Bochner-like statement that if each of the 3 numbers $a - (|b| + |c|)$, $d - (|b| + |e|)$, $f - (|c| + |e|)$ is positive then the original 3×3 matrix has a trivial kernel.

To apply this decomposition to the combinatorial Laplace operator, we must first represent the Laplace operator as a symmetric matrix, and this requires choosing an ordered orthonormal basis for the space of chains. We can define such a basis by choosing an orientation for each cell of M, and then choosing an ordering of the cells. Making a different choice for the orientation of a cell has the effect of multiplying the corresponding row and column of the symmetric matrix by -1. Permuting the ordering of the cells has the effect of applying the same permutation to both the rows and the columns of the matrix. Therefore, for a decomposition of symmetric matrices to induce a well-defined decomposition of the combinatorial Laplace operator, the decomposition must be equivariant under multiplying a row and column by -1, and applying a single permutation to both the rows and the columns. It is easy to see that the decomposition shown in (2.2) has this property, and hence induces a decomposition $\Box_p = B_p + F_p$. In analogy with the Bochner–Weitzenböck formula, we call B_p the combinatorial Bochner Laplacian, and F_p the p-th combinatorial curvature function.

It follows immediately that if $F_p > 0$ then $\Box_p > 0$, so that $H_p(M, \mathbb{R}) = 0$. For any p-cell α, we define

$$\mathcal{F}_p(\alpha) = \langle F_p(\alpha), \alpha \rangle.$$

Our definition of Ricci curvature, $\mathrm{Ric}(e) = \mathcal{F}_1(e)$ was motivated by formula (2.1). We note that (2.1) is a theorem in the Riemannian setting, since Ricci curvature was originally defined in a completely different manner, but it is a definition in the combinatorial case. Proving Theorem 2.4 is now reduced to explicitly calculating $\mathcal{F}_p$ as defined above, and showing that this is equivalent to the formula given in Definition 2.3. See [Forman 1998a] for the details of this straight-forward calculation.

In the Riemannian case (I hope the reader is not getting dizzy from all of this bouncing back and forth between categories), Theorem 2.5 follows from the Bochner–Weitzenböck formula with just a bit more work. Namely, if F_p is a nonnegative operator, then since $\nabla_p^* \nabla_p$ is also a nonnegative operator, we have

$$\operatorname{Ker} \square_p = \operatorname{Ker}(\nabla_p^* \nabla_p) \cap \operatorname{Ker} F_p.$$

We note that

$$\operatorname{Ker}(\nabla_p^* \nabla_p) = \operatorname{Ker} \nabla_p.$$

Any $\omega \in \operatorname{Ker} \nabla_p$ is said to be *parallel*, and is completely determined by its value at any one point. This is essentially all that is required to complete the proof.

The proofs of Theorem 2.5(i) and (ii) in the combinatorial case, just as in the Riemannian case, would follow if we knew that a 1-chain $c \in \operatorname{Ker} B_1$ is completely determined by its values in some neighborhood in M, that is, if there was a unique continuation theorem for 1-chains in $\operatorname{Ker} B_1$. This is not quite true. However, if $c \in \operatorname{Ker} \square_1 \cap \operatorname{Ker} B_1$, then c is, in fact, determined by its value in a neighborhood of any vertex. This is sufficient for the proof of Theorem 2.5(i) and (ii). It can also be shown that for $p > 1$, p-chains $c \in \operatorname{Ker} \square_p \cap \operatorname{Ker} B_p$, in drastic contradiction to the Riemannian case, are not determined by local information. This explains the difficulty in extending Theorem 2.5 to the case of general p (see § 4 of [Forman 1998a]).

We briefly indicate the proof of the unique continuation theorem for $c \in \operatorname{Ker} \square_1 \cap \operatorname{Ker} B_1$. Consider, for example, a vector $x = (x_1, x_2, x_3)$ in the kernel of the matrix B in (4). It is not too hard to see that if $b \neq 0$ then x_1 determines x_2. Namely,

$$x_2 = -(\operatorname{sign}(b)) x_1.$$

More generally, if $x \in \operatorname{Ker} B$, and the (i,j)-th entry of B is nonzero, then x_i determines x_j (so that x_i is zero if and only if x_j is zero).

In the case of the Laplace operator, if α_1 and α_2 are p-cells, then the entry of B_p corresponding to α_1 and α_2 is nonzero if α_1 and α_2 are parallel neighbors. Therefore, if $c = \sum_{\alpha^{(p)}} c_\alpha \alpha \in \operatorname{Ker} B_p$, and α_1 and α_2 are parallel neighbors, then c_{α_1} determines c_{α_2}. Consider the polygon shown in Figure 15, left. Every side of the polygon is a parallel neighbor of either e_1 of e_2. This is the case whenever the polygon has at least 4 sides. Therefore, if $c = \sum_e c_e e \in \operatorname{Ker} B_1$, then the value of c on the boundary of the polygon is completely determined by c_{e_1} and c_{e_2}. This argument doesn't quite work if the polygon has 3 sides, as in Figure 15, right. Here, e_3 is a transverse neighbor of both e_1 and e_2. However, if

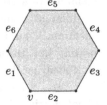

Figure 15. For a polygon with more than three sides, every edge is a parallel neighbor of one of the edges incident on v.

we knew that $c \in \operatorname{Ker} \partial^*$, then c_{e_3} would be determined by c_{e_1} and c_{e_2}. Putting these ideas together, it is not too hard to get to the general result that if M is connected, and $c \in \operatorname{Ker} \partial^* \cap \operatorname{Ker} B_1$, then for any vertex v, c is completely determined by c_e for $e > v$. Since

$$\operatorname{Ker} \square_p = \operatorname{Ker} \partial \cap \operatorname{Ker} \partial^* \subset \operatorname{Ker} \partial^*$$

the desired theorem follows.

To see how Theorem 2.5(i) follows, suppose $\operatorname{Ric}(e) \geq 0$ for each edge e. Then, since B_1 is also a nonnegative operator,

$$\operatorname{Ker} \square_1 = \operatorname{Ker} B_1 \cap \operatorname{Ker} \operatorname{Ric} = (\operatorname{Ker} B_1 \cap \operatorname{Ker} \square_1) \cap \operatorname{Ker} \operatorname{Ric}$$
$$= (\operatorname{Ker} B_1 \cap \operatorname{Ker} \partial^*) \cap \operatorname{Ker} \operatorname{Ric}.$$

Now suppose $c = \sum_e c_e e \in \operatorname{Ker} \square_p$. Then $c \in \operatorname{Ker}(\operatorname{Ric})$. Since $\operatorname{Ric}(e) > 0$ for each edge $e > v$, we must have $c_e = 0$ for each edge $e > v$. Now we observe that since $c \in \operatorname{Ker} B_1 \cap \operatorname{Ker} \partial^*$, c is completely determined by c_e for $e > v$, and hence we must have $c = 0$. This demonstrates that $H_1(M, \mathbb{R}) = 0$.

The results we have presented here, as well as those in the references cited earlier, indicate that there should exist more complete theories of combinatorial differential topology and geometry. However, at this point the combinatorial theories capture only a small portion of these beautiful and far-reaching classical subjects. There remains much to be done.

Acknowledgements

The author thanks the staff of the Mathematical Sciences Research Institute, as well as the organizers of the special year in Combinatorics and Topology, for their hospitality while some of this work was completed. The author also thanks the referees and editor, whose thoughtful comments have drastically improved this paper.

References

[Alexandrov 1951] A. D. Alexandrov, "A theorem on triangles in a metric space and some of its applications", *Trudy Mat. Inst. Steklov* **38** (1951), 5–23. In Russian.

[Allendoerfer and Weil 1943] C. B. Allendoerfer and A. Weil, "The Gauss–Bonnet theorem for Riemannian polyhedra", *Trans. Amer. Math. Soc.* **53** (1943), 101–129.

[Babson et al. 1999] E. Babson, A. Björner, S. Linusson, J. Shareshian, and V. Welker, "Complexes of not i-connected graphs", *Topology* **38**:2 (1999), 271–299.

[Banchoff 1967] T. Banchoff, "Critical points and curvature for embedded polyhedra", *J. Differential Geometry* **1** (1967), 245–256.

[Banchoff 1983] T. F. Banchoff, "Critical points and curvature for embedded polyhedra, II", pp. 34–55 in *Differential geometry* (College Park, MD, 1981/1982), edited by R. Brooks et al., Progr. Math. **32**, Birkhäuser, Boston, 1983.

[Banchoff and McCrory 1979] T. Banchoff and C. McCrory, "A combinatorial formula for normal Stiefel-Whitney classes", *Proc. Amer. Math. Soc.* **76**:1 (1979), 171–177.

[Björner 1995] A. Björner, "Topological methods", pp. 1819–1872 in *Handbook of combinatorics, v. 2*, edited by R. Graham et al., North-Holland, Amsterdam, 1995.

[Bochner 1946] S. Bochner, "Vector fields and Ricci curvature", *Bull. Amer. Math. Soc.* **52** (1946), 776–797.

[Bochner 1948] S. Bochner, "Curvature and Betti numbers", *Ann. of Math.* (2) **49** (1948), 379–390.

[Bochner 1949] S. Bochner, "Curvature and Betti numbers, II", *Ann. of Math.* (2) **50** (1949), 77–93.

[Bollobás 1995] B. Bollobás, "Extremal graph theory", pp. 1231–1292 in *Handbook of combinatorics, vol. 2*, edited by E. by R. L. Graham et al., Elsevier, Amsterdam, 1995.

[Bott 1988] R. Bott, "Morse theory indomitable", *Inst. Hautes Études Sci. Publ. Math.* **68** (1988), 99–114.

[Brehm and Kühnel 1987] U. Brehm and W. Kühnel, "Combinatorial manifolds with few vertices", *Topology* **26**:4 (1987), 465–473.

[Chari 1996] M. Chari, "On discrete Morse functions and combinatorial decompositions", preprint, 1996.

[Cheeger 1983] J. Cheeger, "Spectral geometry of singular Riemannian spaces", *J. Differential Geom.* **18**:4 (1983), 575–657.

[Cheeger et al. 1984] J. Cheeger, W. Müller, and R. Schrader, "On the curvature of piecewise flat spaces", *Comm. Math. Phys.* **92**:3 (1984), 405–454.

[Chung 1997] F. R. K. Chung, *Spectral graph theory*, CBMS Regional conference series in mathematics **92**, Amer. Math. Soc., Providence, 1997.

[Dodziuk 1974] J. Dodziuk, "Combinatorial and continuous Hodge theories", *Bull. Amer. Math. Soc.* **80** (1974), 1014–1016.

[Dodziuk 1981] J. Dodziuk, "Sobolev spaces of differential forms and de Rham–Hodge isomorphism", *J. Differential Geom.* **16**:1 (1981), 63–73.

[Dodziuk and Karp 1988] J. Dodziuk and L. Karp, "Spectral and function theory for combinatorial Laplacians", pp. 25–40 in *Geometry of random motion* (Ithaca, NY, 1987), edited by R. Durrett and M. A. Pinsky, Contemporary mathematics **73**, Amer. Math. Soc., Providence, RI, 1988.

[Dodziuk and Patodi 1976] J. Dodziuk and V. K. Patodi, "Riemannian structures and triangulations of manifolds", *J. Indian Math. Soc.* (*N.S.*) **40** (1976), 1–52.

[Duval 1994] A. M. Duval, "A combinatorial decomposition of simplicial complexes", *Israel J. Math.* **87**:1-3 (1994), 77–87.

[Eckmann 1945] B. Eckmann, "Harmonische Funktionen und Randwertaufgaben in einem Komplex", *Comment. Math. Helv.* **17** (1945), 240–255.

[Eells and Kuiper 1962] J. Eells, Jr. and N. H. Kuiper, "Manifolds which are like projective planes", *Publ. Math. Inst. Hautes Études Sci.* **14** (1962), 5–46 (181–222 in year).

[Floer 1988]   A. Floer, "Morse theory for Lagrangian intersections", *J. Differential Geom.* **28**:3 (1988), 513–547.

[Forman 1993]   R. Forman, "Determinants of Laplacians on graphs", *Topology* **32**:1 (1993), 35–46.

[Forman 1995]   R. Forman, "A discrete Morse theory for cell complexes", pp. 112–125 in *Geometry, topology, and physics for Raoul Bott*, edited by S.-T. Yau, Internat. Press, Cambridge, MA, 1995.

[Forman 1998a]   R. Forman, "Combinatorial Ricci curvature", preprint, 1998. Available at http://math.rice.edu/~forman/ricci.tex.

[Forman 1998b]   R. Forman, "Combinatorial vector fields and dynamical systems", *Math. Z.* **228**:4 (1998), 629–681.

[Forman 1998c]   R. Forman, "Morse theory and evasiveness", preprint, 1998. Available at http://math.rice.edu/~forman/evade.tex.

[Forman 1998d]   R. Forman, "Morse theory for cell complexes", *Adv. Math.* **134**:1 (1998), 90–145.

[Forman ≥ 1999]   R. Forman, "On triangulations of vector fields". In preparation.

[Friedman 1996]   J. Friedman, "Computing Betti numbers via combinatorial Laplacians", pp. 386–391 in *Proceedings of the Twenty-eighth Annual ACM Symposium on the Theory of Computing* (Philadelphia, 1996), ACM, New York, 1996.

[Friedman and Handron 1997]   J. Friedman and P. Handron, "On the Betti numbers of chessboard compexes", preprint, 1997.

[Gabrièlov et al. 1974]   A. M. Gabrièlov, I. M. Gel'fand, and M. V. Losik, "The combinatorial computation of characteristic classes", *Funktsional. Anal. i Prilozhen.* **9**:1 (1974), 54–55. In Russian; translated in *Func. Anal. and Appl.* **10** (1976), 12–15.

[Gao 1985]   L. Z. Gao, "The construction of negatively Ricci curved manifolds", *Math. Ann.* **271**:2 (1985), 185–208.

[Gao and Yau 1986]   L. Z. Gao and S.-T. Yau, "The existence of negatively Ricci curved metrics on three-manifolds", *Invent. Math.* **85**:3 (1986), 637–652.

[Garland 1972]   H. Garland, "*p*-adic curvature and a conjecture of Serre", *Bull. Amer. Math. Soc.* **78** (1972), 259–261.

[Garland 1973]   H. Garland, "*p*-adic curvature and the cohomology of discrete subgroups of *p*-adic groups", *Ann. of Math.* (2) **97** (1973), 375–423.

[Garland 1975]   H. Garland, "On the cohomology of discrete subgroups of semi-simple Lie groups", pp. 21–30 in *Discrete subgroups of Lie groups and applications to moduli* (Bombay, 1973), Tata Institute Studies in Mathematics **7**, Oxford Univ. Press, Bombay, 1975.

[Gelfand and MacPherson 1992]   I. M. Gelfand and R. D. MacPherson, "A combinatorial formula for the Pontrjagin classes", *Bull. Amer. Math. Soc.* (*N.S.*) **26**:2 (1992), 304–309.

[Gershgorin 1931]   S. Gershgorin, "Über die Abgrenzung der Eigenwerte einer Matrix", *Izv. Akad. Nauk. USSR Otd. Fiz.-Mat. Nauk* **7** (1931), 749–754.

[Glaser 1972]   L. C. Glaser, *Geometrical combinatorial topology*, v. 1, Van Nostrand Reinhold Mathematical Studies **28**, Van Nostrand Reinhold, New York, 1972.

[Goresky and MacPherson 1983] M. Goresky and R. MacPherson, "Stratified Morse theory", pp. 517–533 in *Singularities,* v. 1 (Arcata, CA, 1981), edited by P. Orlik, Proceedings of symposia in pure mathematics **40**:1, Amer. Math. Soc., Providence, 1983.

[Goresky and MacPherson 1988] M. Goresky and R. MacPherson, *Stratified Morse theory,* Ergebnisse der Mathematik (3) **14**, Springer, Berlin, 1988.

[Gromov 1981] M. Gromov, "Curvature, diameter and Betti numbers", *Comment. Math. Helv.* **56**:2 (1981), 179–195.

[Gromov 1987] M. Gromov, "Hyperbolic groups", pp. 75–263 in *Essays in group theory,* edited by S. M. Gersten, Math. Sci. Res. Inst. Publ. **8**, Springer, New York, 1987.

[Halperin and Toledo 1972] S. Halperin and D. Toledo, "Stiefel-Whitney homology classes", *Ann. of Math.* (2) **96** (1972), 511–525.

[Kirchhoff 1847] G. Kirchhoff, "Über die Auflösung der Gleichungen auf welche man bei der Untersuchen der linearen Vertheilung galvanischer Ströme gefüht wird", *Ann. der Phys. und Chem.* **72** (1847), 495–508.

[Klingenberg 1982] W. Klingenberg, "The Morse complex", pp. 117–122 in *Symposia Mathematica,* v. 26 (Rome, 1980), Academic Press, London, 1982.

[Kook et al. 1998] W. Kook, V. Reiner, and D. Stanton, "Combinatorial Laplacians of matroid complexes", preprint, 1998.

[Kühnel 1990] W. Kühnel, "Triangulations of manifolds with few vertices", pp. 59–114 in *Advances in differential geometry and topology,* edited by F. Tricerri, World Sci., Teaneck, NJ, 1990.

[Kuiper 1971] N. H. Kuiper, "Morse relations for curvature and tightness", pp. 77–89 in *Proceedings of Liverpool Singularities Symposium* (Liverpool, 1969/1970), vol. 2, edited by C. T. C. Wall, Lecture Notes in Math. **209**, Springer, Berlin, 1971.

[Levitt 1989] N. Levitt, *Grassmannians and Gauss maps in piecewise-linear topology,* Springer, Berlin, 1989.

[Levitt and Rourke 1978] N. Levitt and C. Rourke, "The existence of combinatorial formulae for characteristic classes", *Trans. Amer. Math. Soc.* **239** (1978), 391–397.

[Lohkamp 1994a] J. Lohkamp, "Metrics of negative Ricci curvature", *Ann. of Math.* (2) **140**:3 (1994), 655–683.

[Lohkamp 1994b] J. Lohkamp, "Negative bending of open manifolds", *J. Differential Geom.* **40**:3 (1994), 461–474.

[Luo and Stong 1993] F. Luo and R. Stong, "Combinatorics of triangulations of 3-manifolds", *Trans. Amer. Math. Soc.* **337**:2 (1993), 891–906.

[MacPherson 1993] R. D. MacPherson, "Combinatorial differential manifolds: a symposium in honor of John Milnor's sixtieth birthday", pp. 203–221 in *Topological methods in modern mathematics* (Stony Brook, NY, 1991), edited by L. R. Goldberg and A. Phillips, Publish or Perish, Houston, 1993.

[Milnor 1962] J. Milnor, *Morse theory,* Annals of Mathematics Study **51**, Princeton Univ. Press, Princeton, NJ, 1962.

[Milnor 1965] J. Milnor, *Lectures on the h-cobordism theorem,* Princeton Mathematical Notes, Princeton Univ. Press, Princeton, NJ, 1965.

[Morse 1925] M. Morse, "Relations between the critical points of a real function of n independent variables", *Trans. Amer. Math. Soc.* **27** (1925), 345–396.

[Morse 1934] M. Morse, *The calculus of variations in the large*, Colloquium Publications **18**, Amer. Math. Soc., Providence, 1934.

[Morse 1965] M. Morse, "Bowls of a non-degenerate function on a compact differentiable manifold", pp. 81–103 in *Differential and Combinatorial Topology*, edited by S. S. Cairns, Princeton Univ. Press, Princeton, NJ, 1965.

[Myers 1941] S. B. Myers, "Riemannian manifolds with positive mean curvature", *Duke Math. J.* **8** (1941), 401–404.

[Phillips and Stone 1990] A. V. Phillips and D. A. Stone, "The computation of characteristic classes of lattice gauge fields", *Comm. Math. Phys.* **131**:2 (1990), 255–282.

[Regge 1961] T. Regge, "General relativity without coordinates", *Nuovo Cimento* (10) **19** (1961), 558–571.

[Shareshian 1997] J. Shareshian, "Discrete Morse theory for complexes of 2-connected graphs", preprint, 1997.

[Smale 1961a] S. Smale, "Generalized Poincaré's conjecture in dimensions greater than four", *Ann. of Math.* (2) **74** (1961), 391–406.

[Smale 1961b] S. Smale, "On gradient dynamical systems", *Ann. of Math.* (2) **74** (1961), 199–206.

[Stanley 1993] R. P. Stanley, "A combinatorial decomposition of acyclic simplicial complexes", *Discrete Math.* **120**:1-3 (1993), 175–182.

[Stone 1976] D. A. Stone, "A combinatorial analogue of a theorem of Myers", *Illinois J. Math.* **20**:1 (1976), 12–21. Correction in **20**:3 (1976), 551–554.

[Stone 1981] D. A. Stone, "On the combinatorial Gauss map for C^1 submanifolds of Euclidean space", *Topology* **20**:3 (1981), 247–272.

[Taussky 1949] O. Taussky, "A recurring theorem on determinants", *Amer. Math. Monthly* **56** (1949), 672–676.

[Turchin 1997] V. È. Turchin, "Homology of complexes of biconnected graphs", *Uspekhi Mat. Nauk* **52**:2 (issue 314) (1997), 189–190. In Russian; translated in *Russian Math. Surveys* **52**:2 (1997), 426–427.

[Vassiliev 1993] V. A. Vassiliev, "Complexes of connected graphs", pp. 223–235 in *The Gel'fand Mathematical Seminars, 1990–1992*, edited by L. Corwin et al., Birkhäuser, Boston, 1993.

[Whitehead 1939] J. H. C. Whitehead, "Simplicial spaces, nuclei, and m-groups", *Proc. London Math. Soc.* **45** (1939), 243–327.

[Williams 1992] R. M. Williams, "Discrete quantum gravity: the Regge calculus approach", *Internat. J. Modern Phys. B* **6**:11-12 (1992), 2097–2108.

[Yu 1983] Y. L. Yu, "Combinatorial Gauss–Bonnet–Chern formula", *Topology* **22**:2 (1983), 153–163.

ROBIN FORMAN
DEPARTMENT OF MATHEMATICS
RICE UNIVERSITY
HOUSTON, TX 77251
UNITED STATES
  forman@math.rice.edu

New Perspectives in Geometric Combinatorics
MSRI Publications
Volume **38**, 1999

Macdonald Polynomials and Geometry

MARK HAIMAN

ABSTRACT. We explain some remarkable connections between the two-parameter symmetric polynomials discovered in 1988 by Macdonald, and the geometry of certain algebraic varieties, notably the Hilbert scheme $\mathrm{Hilb}^n(\mathbb{C}^2)$ of points in the plane, and the variety C_n of pairs of commuting $n \times n$ matrices.

CONTENTS

1. Introduction

This article is an explication of some remarkable connections between the two-parameter symmetric polynomials discovered by Macdonald [1988] and the geometry of certain algebraic varieties, notably the Hilbert scheme $\mathrm{Hilb}^n(\mathbb{C}^2)$ of points in the plane and the variety C_n of pairs of commuting $n \times n$ matrices ("commuting variety", for short). The conjectures on diagonal harmonics introduced in [Haiman 1994; Garsia and Haiman 1996a] also relate to this geometric setting.

1991 *Mathematics Subject Classification.* Primary 05-02; Secondary 14-02, 05E05, 13H10, 14M05.

Key words and phrases. Macdonald polynomials, Hilbert scheme, commuting variety, sheaf cohomology, Cohen–Macaulay, Gorenstein.

Supported in part by NSF Mathematical Sciences grant DMS-9400934.

I have sought to give a reasonably self-contained treatment of these topics, by providing an introduction to the theory of Macdonald polynomials, to the "plethystic substitution" notation for symmetric functions (which is invaluable in dealing with them), and to the conjectures and related phenomena that we aim to explain geometrically. The geometric discussion is less self-contained, it being unavoidable to use scheme-theoretic language, constructions such as blowups, and some sheaf cohomological arguments. I do however give geometric descriptions in elementary terms of the various algebraic varieties encountered, and review whatever of their special features we might use, so as to orient the reader not previously familiar with them.

The linchpin of the geometric connections we consider is the so-called "$n!$ conjecture" [Garsia and Haiman 1993; 1996b], which remains unproved at present. The $n!$ conjecture proposes a combinatorial interpretation of the famous Kostka–Macdonald coefficients $K_{\lambda\mu}(q,t)$, which relate the Macdonald polynomials to Schur functions and which were conjectured by Macdonald to be polynomials with nonnegative integer coefficients in the parameters q, t. (Part of the conjecture is that the $K_{\lambda\mu}(q,t)$ are polynomials at all, which is not obvious from their definition and was only proved recently, in five independent papers [Garsia and Remmel 1998; Garsia and Tesler 1996; Kirillov and Noumi 1998; Knop 1997; Sahi 1996].)

The $n!$ conjecture is really two conjectures: first, that certain simply defined spaces, quotient rings of the polynomial ring

$$\mathbb{C}[\boldsymbol{x}, \boldsymbol{y}] = \mathbb{C}[x_1, y_1, \ldots, x_n, y_n],$$

have dimension $n!$; and second, that these spaces, viewed as doubly graded representations of the symmetric group S_n, have Hilbert polynomials that are essentially the Kostka–Macdonald coefficients. It turns out, as we shall show, that the first (apparently weaker) part of the conjecture is equivalent to the Cohen–Macaulay property of a certain "isospectral" variety X_n over $\mathrm{Hilb}^n(\mathbb{C}^2)$. Using this fact we can prove that the first part of the conjecture actually implies the second part, and with it the Macdonald positivity conjecture for the $K_{\lambda\mu}(q,t)$.

As we shall see, the variety X_n is the blowup of $(\mathbb{C}^2)^n$ at the ideal J generated by those elements of its coordinate ring $\mathbb{C}[\boldsymbol{x}, \boldsymbol{y}]$ that alternate in sign under the action of the symmetric group S_n. An obvious conjecture is that J is the ideal of the locus in $(\mathbb{C}^2)^n$ where two or more of the n points (x_i, y_i) coincide, that is,

$$J = \bigcap_{i \neq j} (x_i - x_j, y_i - y_j). \qquad (1.1)$$

It is easy to see that J defines the coincidence locus set-theoretically, which is to say, the radical of J is the intersection on the right hand side above, but it is not obvious, and indeed it remains an open question, that J is a radical ideal.

More generally, the geometry of the blowup X_n depends on module-theoretic properties of the powers J^m. We are led to extend (1.1) and conjecture that

$$J^m = \bigcap_{i \neq j} (x_i - x_j, \, y_i - y_j)^m, \tag{1.2}$$

that is, the powers of J are the symbolic powers of the ideal of the coincidence locus. In fact, we conjecture that the variables $x_1, \ldots, x_n$ form a regular sequence for the $\mathbb{C}[x, y]$-module J^m, for all m. As we show, this conjecture implies (1.2). It further implies that the x_i's form a regular sequence on X_n (that is, on its structure sheaf $\mathcal{O}$, viewed as a sheaf of $\mathbb{C}[x, y]$-algebras). Assuming this regular-sequence conjecture, we are able to give an inductive sheaf-cohomological argument to show that X_n is Cohen–Macaulay, and thus the $n!$ and Macdonald positivity conjectures follow.

An important point to remark on here is that the $n!$ conjecture and many of the related geometric conjectures have evident analogous statements in more than two sets of variables $X, Y, Z, \ldots$. For the most part, these analogs fail to hold.[1] However, the above conjectures on the ideals J^m are an exception, as we expect them to hold in any number of sets of variables (the last conjecture then being that any one of the sets of variables forms a regular sequence). Of course the reasoning leading from there to the $n!$ conjecture makes essential use of having only two sets.

The isospectral Hilbert scheme X_n also provides the geometric setting for the study of *diagonal harmonics*, the subject of a series of conjectures [Garsia and Haiman 1996a; Haiman 1994]. The space of diagonal harmonics may be identified with the quotient ring R_n of $\mathbb{C}[x, y]$ by the ideal I generated by all S_n invariant polynomials with zero constant term. It is conjectured, among other things, that the dimension of R_n as a vector space is $(n+1)^{n-1}$. Further conjectures in [Haiman 1994] describe aspects of its structure as a graded S_n module in combinatorial terms. In [Garsia and Haiman 1996a] we conjectured a complete formula for the doubly graded character of R_n, in terms of Macdonald polynomials, and proved that this master formula implies all the earlier combinatorial conjectures.

In geometric terms R_n is the coordinate ring of the scheme-theoretic fiber over the origin under the natural map $(\mathbb{C}^2)^n \to S^n\mathbb{C}^2$ from ordered n-tuples of points in the plane to unordered n-tuples. Now, there is a fiber square

$$\begin{array}{ccc} X_n & \longrightarrow & (\mathbb{C}^2)^n \\ {\scriptstyle \sigma} \downarrow & & \downarrow \\ \mathrm{Hilb}^n(\mathbb{C}^2) & \overset{\tau}{\longrightarrow} & S^n\mathbb{C}^2, \end{array}$$

---

[1] The $n!$ conjecture has acquired a minor history of exciting but unsuccessful ideas for simple proofs, by the author and others. A good reality check on a contemplated proof is to ask where the argument breaks down — as it must — in three sets of variables.

giving rise to a natural homomorphism from R_n to the global sections of the structure sheaf on the fiber $(\tau\sigma)^{-1}(0) \subseteq X_n$. If X_n is Cohen–Macaulay, that is, if the $n!$ conjecture is true, these sections may be identified with global sections of a vector bundle on the zero-fiber $\tau^{-1}(0)$ in $\mathrm{Hilb}^n(\mathbb{C}^2)$. Under a suitable cohomology vanishing hypothesis, the homomorphism from R_n to this space of global sections will be an isomorphism. Moreover, we can give its character explicitly, using a variant of the Atiyah–Bott Lefschetz formula. This yields the master formula for diagonal harmonics, on the assumption that the $n!$ conjecture and vanishing hypotheses hold. The agreement of this master formula with computational results for $n \leq 7$ is in my view striking evidence for the probable validity of the geometric conjectures that give rise to it.

Finally, the commuting variety C_n enters the picture because it contains a natural nonsingular open set C_n^0 with a smooth map to $\mathrm{Hilb}^n(\mathbb{C}^2)$. The analog in this context of the isospectral Hilbert scheme X_n is the "isospectral commuting variety" IC_n of pairs (X, Y) of commuting matrices, together with $2n$-tuples $(a_1, b_1 \ldots, a_n, b_n)$ for which the (a_i, b_i) are the joint eigenvalues of the matrices X, Y, that is, they satisfy the equations

$$\det(I + rX + sY) = \prod_{i=1}^{n}(1 + ra_i + sb_i), \tag{1.3}$$

where r, s are indeterminates. The open set IC_n^0 of IC_n lying over C_n^0 is the fiber product of X_n with C_n^0 over $\mathrm{Hilb}^n(\mathbb{C}^2)$:

$$
\begin{array}{ccc}
IC_n^0 & \longrightarrow & X_n \\
\downarrow & & \downarrow \\
C_n^0 & \longrightarrow & \mathrm{Hilb}^n(\mathbb{C}^2).
\end{array}
\tag{1.4}
$$

Thus IC_n^0 is smooth over X_n, and hence X_n is Cohen–Macaulay if and only if IC_n^0 is. In fact, as we shall see, if X_n is Cohen–Macaulay it is even Gorenstein, and thus the same is true of IC_n^0. We are led to conjecture that the *whole* isospectral commuting variety IC_n is Gorenstein, and not just the open subset IC_n^0. We have been able to verify this for small values of n. This conjecture not only implies the $n!$ conjecture, it also implies that the ordinary commuting variety C_n is Cohen–Macaulay, which is an open problem of long standing.

2. Symmetric Functions and Macdonald Polynomials

The general reference for material in this section is [Macdonald 1995], whose notation and terminology we follow, except as to the plethystic substitution, and as to the transformed Macdonald polynomials $\tilde{H}_\mu$ defined below. We give some definitions and derive some properties that are also in the same reference, both for completeness and to illustrate the utility of the plethystic notation. We also derive some additional facts that will be needed later.

We work throughout with symmetric functions in infinitely many indeterminates $x_1, x_2, \ldots$, with coefficients in the field $\mathbb{Q}(q, t)$ of rational functions of two variables q and t. The various classical bases of the ring of symmetric functions are indexed by integer partitions μ, and denoted as follows: the monomial symmetric functions by m_μ, the power-sums by p_μ, the elementary symmetric functions by e_μ, the complete homogeneous symmetric functions by h_μ, and the Schur functions by s_μ. In each basis, as μ ranges over partitions of a given integer d, we obtain a basis for the symmetric polynomials homogeneous of degree d.

The standard partial ordering on partitions of d is the *dominance order*, defined by $\lambda \leq \mu$ if $\lambda_1 + \cdots + \lambda_k \leq \mu_1 + \cdots + \mu_k$ for all k. The triangularity of transition matrices between certain bases of symmetric functions with respect to dominance plays a crucial role in the definition and development of Macdonald polynomials, as well as in the reasoning we will use later to deduce the Macdonald positivity conjecture from the $n!$ conjecture.

We now turn to the important device of plethystic substitution. The fact that the power-sums p_μ form a basis means, equivalently, that the ring of symmetric functions can be identified with the ring of polynomials in the power-sums $p_1, p_2, \ldots$. In particular, the p_k's may be specialized arbitrarily to elements of any algebra over the coefficient field, and the specialization extends uniquely to an algebra homomorphism on all symmetric functions.

Now let A be a formal Laurent series with rational coefficients in indeterminates $a_1, a_2, \ldots$, which may include our parameters q and t. We define $p_k[A]$ to be the result of replacing each indeterminate a_i in A by a_i^k. Extending the specialization $p_k \mapsto p_k[A]$ to arbitrary symmetric functions f, we obtain the *plethystic substitution* of A into f, denoted $f[A]$.

If A is merely a sum of indeterminates, $A = a_1 + \cdots + a_n$, then we see that $p_k[A] = p_k(a_1, a_2, \ldots, a_n)$, and hence for every f we have $f[A] = f(a_1, a_2, \ldots, a_n)$. This is why we view the operation as a kind of substitution. Similarly, if A has a series expansion as a sum of monomials, $f[A]$ is f evaluated on these monomials; for example,

$$f[1/(1-t)] = f(1, t, t^2, \ldots).$$

Our convention will be that in a plethystic expression X stands for the sum of the original indeterminates $x_1 + x_2 + \cdots$, so that $f[X]$ is the same as $f(X)$,

$$f[X/(1-t)] = f(x_1, x_2, \ldots, tx_1, tx_2, \ldots, t^2 x_1, t^2 x_2, \ldots),$$

and so forth. Among the virtues of this notation is that the substitution of $X/(1-t)$ for X as above has an explicit inverse, namely the plethystic substitution of $X(1-t)$ for X.

The one caution that must be observed with plethystic notation is that indeterminates must always be treated as formal symbols, never as variable numeric quantities. For instance, if f is homogeneous of degree d then it is true (and easy to see) that

$$f[tX] = t^d f[X],$$

but it is *false* that $f[-X] = (-1)^d f[X]$; that is, we cannot set $t = -1$ in the equation above. In fact $f[-X]$ is a very interesting quantity: it equals $(-1)^d \omega f(X)$, where ω is the classical involution on symmetric functions defined by $\omega p_k = (-1)^{k+1} p_k$, which interchanges the elementary and complete symmetric functions e_λ and h_λ, and more generally exchanges the Schur function s_λ with $s_{\lambda'}$, where λ' is the conjugate partition.

It is convenient when using plethystic notation to define

$$\Omega(X) = \exp\left(\sum_{k=1}^{\infty} p_k(X)/k \right). \tag{2.1}$$

Then since $p_k[A + B] = p_k[A] + p_k[B]$ and $p_k[-A] = -p_k[A]$ we have

$$\Omega[A + B] = \Omega[A]\Omega[B], \quad \Omega[-A] = 1/\Omega[A]. \tag{2.2}$$

From this and the single-variable evaluation $\Omega[x] = \exp\left(\sum_{k\geq 1} x^k/k\right) = 1/(1-x)$ we obtain

$$\Omega[X] = \prod_i \frac{1}{1-x_i} = \sum_{n=0}^{\infty} h_n(X) \tag{2.3}$$

and

$$\Omega[-X] = \prod_i (1-x_i) = \sum_{n=0}^{\infty} (-1)^n e_n(X). \tag{2.4}$$

Recall that the standard Hall inner product $\langle \cdot, \cdot \rangle$ is defined so that the Schur functions s_μ are an orthonormal basis, and the complete symmetric functions h_μ are dual to the monomials m_μ. The *Cauchy identity* is that for Hall-dual bases $\{u_\mu\}, \{v_\mu\}$ we have

$$\prod_{i,j} \frac{1}{1-x_i y_j} = \sum_\mu u_\mu(X) u_\nu(Y),$$

or, plethystically,

$$\Omega[XY] = \sum_\mu u_\mu[X] v_\mu[Y]. \tag{2.5}$$

This may be written in a basis-free way as

$$\langle \Omega[AX], f(X) \rangle = f[A], \tag{2.6}$$

which follows from (2.5) by taking $f = v_\mu$, and extending to arbitrary f by linearity. In particular,

$$\langle \Omega[B(AX)], \Omega[CX] \rangle = \langle \Omega[BX], \Omega[C(AX)] \rangle = \Omega[ABC].$$

But since B and C are arbitrary, we may set $f(X) = \Omega[BX]$, $g(X) = \Omega[CX]$, to obtain the identity

$$\langle f[AX], g(X) \rangle = \langle f(X), g[AX] \rangle, \tag{2.7}$$

valid for all f, g. In other words, the plethystic substitution of AX for X is self-adjoint.

Macdonald defines his polynomials by first introducing a q, t-analog of the Hall inner product $\langle \cdot, \cdot \rangle$, which in plethystic notation is simply

$$\langle f, g \rangle_{q,t} = \left\langle f(X), g\left[X\frac{1-q}{1-t}\right]\right\rangle.$$

In view of (2.7), this definition is symmetric in f and g. If $\{u_\mu\}$ and $\{v_\mu\}$ are $\langle \cdot, \cdot \rangle_{q,t}$-dual bases, then $\{u_\mu\}$ and $\{v_\mu[\frac{1-q}{1-t}]\}$ are Hall dual, so the Cauchy identity gives

$$\Omega[XY] = \sum_\mu u_\mu[X] v_\mu\left[X\frac{1-q}{1-t}\right], \quad \text{or} \quad \Omega\left[XY\frac{1-t}{1-q}\right] = \sum_\mu u_\mu[X] v_\mu[Y]. \quad (2.8)$$

Note that the nonplethystic expression for $\Omega[XY\frac{1-t}{1-q}]$, as in [Macdonald 1995], is the rather mysterious product

$$\prod_{i,j} \frac{(tx_iy_j; q)_\infty}{(x_iy_j; q)_\infty}, \quad \text{where} \quad (a; q)_\infty = \prod_{k=0}^{\infty}(1 - aq^k).$$

As a particular case of (2.8), we see that the $\langle \cdot, \cdot \rangle_{q,t}$-dual basis to the monomials m_μ is the basis of transformed complete symmetric functions $h_\mu[X\frac{1-t}{1-q}]$ (denoted g_μ in [Macdonald 1995]).

The Macdonald polynomials $P_\mu(X; q, t)$ may be defined by requiring that they are orthogonal with respect to $\langle \cdot, \cdot \rangle_{q,t}$, and lower unitriangular with respect to the monomials, that is,

$$P_\mu(X; q, t) = m_\mu(X) + \sum_{\lambda < \mu} c_{\lambda\mu}(q, t) m_\lambda, \quad (2.9)$$

for some coefficients $c_{\lambda\mu}$. Here, however, we shall define them directly as eigenfunctions of the plethystic operator

$$\Delta' f(X) = f[X - (1-q)/z]\Omega[zX(1-t^{-1})]\big|_{z^0}, \quad (2.10)$$

where the vertical bar indicates we are to take the constant term with respect to z.

Before further examining the operator Δ' we define an important quantity B_μ that appears in the eigenvalues of the operator, and will turn out to have geometric significance later on. (The theory of Macdonald polynomials is rife with numerology. Quantities such as B_μ and various q, t-hook products crop up again and again; see [Garsia and Haiman 1998; 1996a] for many examples. In our geometric context, these quantities will turn out to have natural interpretations.)

First recall that the *diagram* of a partition μ is the array of lattice points

$$D(\mu) = \{(i, j) \in \mathbb{N} \times \mathbb{N} : j < \mu_{i+1}\}. \quad (2.11)$$

As is customary, we regard i as indexing rows and j as indexing columns, so that the rows of $D(\mu)$ have lengths equal to the parts of μ. We now set

$$B_\mu(q,t) = \sum_{(i,j)\in D(\mu)} t^i q^j, \qquad (2.12)$$

a kind of generating function describing $D(\mu)$, with a term for each cell in the diagram, as illustrated here.

$$\mu : (4,2,1) \qquad D_\mu : \begin{matrix} \bullet \\ \bullet \;\; \bullet \\ \bullet \;\; \bullet \;\; \bullet \;\; \bullet \end{matrix} \qquad B_\mu : \begin{matrix} t^2 + \\ t + qt + \\ 1 + q + q^2 + q^3 \end{matrix} \qquad (2.13)$$

We now return to the study of the operator Δ'.

PROPOSITION 2.1. *A symmetric function $f(X;q,t)$ is an eigenfunction of Δ' with eigenvalue $\alpha(q,t^{-1})$ if and only if $f[X/(1-t^{-1});q,t^{-1}]$ is an eigenfunction of the operator*

$$\Delta f = f\big[X + (1-q)(1-t)/z\big]\,\Omega[-zX]\big|_{z^0},$$

with eigenvalue $\alpha(q,t)$.

PROOF. We verify directly from the definitions that

$$\Delta\big(f[X/(1-t^{-1});q,t^{-1}]\big) = (\Delta'f)\big[X/(1-t^{-1});q,t^{-1}\big],$$

which implies the result.   □

PROPOSITION 2.2. *The operator Δ' is lower-triangular with respect to the basis of monomial symmetric functions. More precisely,*

$$\Delta'm_\mu = \big(1 - (1-q)(1-t^{-1})B_\mu(q,t^{-1})\big)\,m_\mu + \sum_{\lambda < \mu} b_{\lambda\mu}m_\lambda,$$

for some coefficients $b_{\lambda\mu}$.

PROOF. Since the Schur functions are lower unitriangular with respect to the basis of monomials, it will do equally well to prove

$$\Delta'm_\mu = \sum_{\lambda \leq \mu} a_{\lambda\mu}s_\lambda,$$

for some coefficients $a_{\lambda\mu}$, with $a_{\mu\mu} = 1 - (1-q)(1-t^{-1})B_\mu(q,t^{-1})$. It is also sufficient to restrict to a finite set of variables $X = x_1 + \cdots + x_n$. Then $\Omega[zX(1-t^{-1})]$ has the partial fraction expansion

$$\Omega[zX(1-t^{-1})] = \prod_{i=1}^{n} \frac{1-t^{-1}zx_i}{1-zx_i} = t^{-n} + \sum_{i=1}^{n} \frac{1}{1-zx_i}\frac{\prod_{j=1}^{n}(1-x_j/tx_i)}{\prod_{j\neq i}(1-x_j/x_i)}$$

$$= t^{-n} + t^{1-n}(1-t^{-1})\sum_i \frac{1}{1-zx_i}\frac{v(X)_{(x_i\mapsto tx_i)}}{v(X)}, \qquad (2.14)$$

where $v(X) = \prod_{i<j}(x_i - x_j)$ is the Vandermonde determinant.

From (2.14), using the identity $f(1/z)/(1-zx)|_{z^0} = f(x)$, we see that for any function f,

$$f(1/z)\Omega[zX(1-t^{-1})]\big|_{z^0} = t^{-n}f(1/z)\big|_{z^0} + t^{1-n}(1-t^{-1})\sum_i f(x_i)\frac{v(X)_{(x_i\mapsto tx_i)}}{v(X)},$$

and therefore, since $m_\mu[X - x_i(1-q)] = m_\mu(X)_{(x_i\mapsto qx_i)}$,

$$\Delta'm_\mu(X) = t^{-n}m_\mu(X) + t^{1-n}(1-t^{-1})\sum_i m_\mu(X)_{(x_i\mapsto qx_i)}\frac{v(X)_{(x_i\mapsto tx_i)}}{v(X)}.$$

Note that the substitution of x_i for z^{-1} inside the plethysm is permissible, since we are substituting one indeterminate for another.

Recall the Jacobi formula

$$s_\lambda(X)v(X) = \det\big[x_i^{\lambda_j+n-j}\big]_{i,j=1}^n.$$

From this we see that the coefficient of s_λ in the Schur function expansion of any symmetric function $f(X)$ is the coefficient of $x^{\lambda+\delta}$ in $f(X)v(X)$, where $\delta = (n-1, n-2, \ldots, 1, 0)$. In particular the coefficient $a_{\lambda\mu}$ of s_λ in $\Delta'm_\mu$ is given by

$$t^{-n}k_{\lambda\mu} + t^{1-n}(1-t^{-1})\sum_i m_\mu(X)_{(x_i\mapsto qx_i)}v(X)_{(x_i\mapsto tx_i)}\bigg|_{x^{\lambda+\delta}}, \qquad (2.15)$$

where $k_{\lambda\mu}$ is the coefficient of s_λ in m_μ, so $k_{\mu\mu} = 1$. Now in each summand above, the leading term in dominance order is clearly $x^{\mu+\delta}$, establishing the triangularity. This term arises from the term x^μ in m_μ, multiplied by the term x^δ in $v(X)$. In the i-th summand the indicated substitutions multiply it by $q^{\mu_i}t^{n-i}$. Thus we find

$$a_{\mu\mu} = t^{-n} + t^{1-n}(1-t^{-1})\sum_{i=1}^n q^{\mu_i}t^{n-i}.$$

With the understanding that μ_i is zero for i exceeding the number of parts $l(\mu)$ of μ, we readily verify that the above expression is independent of n for $n > l(\mu)$, as it must be, and reduces to

$$(1-t^{-1})\sum_{i=1}^\infty q^{\mu_i}t^{1-i} = 1 - (1-q)(1-t^{-1})B_\mu(q,t^{-1}). \qquad \square$$

COROLLARY 2.3. *The operator Δ' has distinct eigenvalues $1 - (1-q)(1-t^{-1}) \times B_\mu(q,t^{-1})$ and its corresponding eigenfunction is a linear combination of the monomial symmetric functions $m_\lambda : \lambda \leq \mu$, with nonzero coefficient of m_μ.*

DEFINITION. The *Macdonald polynomial* $P_\mu(X; q, t)$ is the eigenfunction of the operator Δ',

$$\Delta'P_\mu = \big(1 - (1-q)(1-t^{-1})B_\mu(q,t^{-1})\big)P_\mu,$$

normalized so that

$$P_\mu = m_\mu + \sum_{\lambda < \mu} c_{\lambda\mu} m_\lambda.$$

The coefficients $c_{\lambda\mu}(q,t)$ are rational functions with nontrivial denominators. Macdonald proposed an alternate normalization, called the *integral form*

$$J_\mu = \prod_{s \in D(\mu)} (1 - q^{a(s)} t^{1+l(s)}) P_\mu, \qquad (2.16)$$

and conjectured that its coefficients are *polynomials* in q and t, i.e., the above product clears the denominators. Here $a(s)$ and $l(s)$ are the *arm* and *leg* of the cell s in the diagram of μ, defined to be the number of cells strictly east and north of s, respectively.

This *integrality conjecture* has recently been proved [Garsia and Remmel 1998; Garsia and Tesler 1996; Kirillov and Noumi 1998; Knop 1997; Sahi 1996]. Macdonald made a further remarkable conjecture, that if we write the integral forms as

$$J_\mu(X; q, t) = \sum_\lambda K_{\lambda\mu}(q,t) s_\lambda[X(1-t)] \qquad (2.17)$$

then the coefficients $K_{\lambda\mu}(q,t)$ are not only polynomials in q and t, but they have *nonnegative integer coefficients*. The search for an algebraic-combinatorial proof of this *Macdonald positivity conjecture*, which remains open, has been the moving force behind the work described in this article. It is pertinent to mention here that $J_\mu(X; 0, t)$ specializes to the Hall–Littlewood function $Q_\mu(X; t)$, and therefore the coefficients $K_{\lambda\mu}(0, t)$ specialize to the famous *t-Kostka coefficients* $K_{\lambda\mu}(t)$ that have been central to much beautiful work in combinatorics, geometry and representation theory. This is one of many reasons for the great interest in Macdonald polynomials in the decade since their discovery.

For our purposes it is convenient to work with the following variant. In all that follows we fix $|\mu| = n$ (not to be confused with $n(\mu)$).

DEFINITION. The *transformed Macdonald polynomials* are

$$\tilde{H}_\mu(X; q, t) = t^{n(\mu)} J_\mu \left[\frac{X}{1 - t^{-1}}; q, t^{-1} \right], \qquad (2.18)$$

where $n(\mu) = \sum_i (i - 1)\mu_i = \sum_{s \in D(\mu)} l(s)$.

From Proposition 2.1 and equation (2.17) we immediately obtain this result:

PROPOSITION 2.4. *The transformed polynomial $\tilde{H}_\mu$ is an eigenfunction of the operator Δ,*

$$\Delta \tilde{H}_\mu = \left(1 - (1-q)(1-t) B_\mu\right) \tilde{H}_\mu,$$

and its Schur function expansion is

$$\tilde{H}_\mu = \sum_\lambda \tilde{K}_{\lambda\mu}(q,t) s_\lambda,$$

where $\tilde{K}_{\lambda\mu}(q,t) = t^{n(\mu)}K_{\lambda\mu}(q,t^{-1})$. In particular (since it is known that $K_{(n),\mu} = t^{n(\mu)}$ [Macdonald 1995]), $\tilde{H}_{\mu}$ is normalized so that its coefficient of $s_{(n)}$ is 1.

Now the operator Δ is symmetric in q and t, while $B_{\mu}(t,q) = B_{\mu'}(q,t)$. Hence we obtain

PROPOSITION 2.5. For all μ we have $\tilde{H}_{\mu'}(X;q,t) = \tilde{H}_{\mu}(X;t,q)$ and, consequently, $\tilde{K}_{\lambda\mu'}(q,t) = \tilde{K}_{\lambda\mu}(t,q)$.

As mentioned earlier, the functions $\tilde{H}_{\mu}$ can be characterized by certain triangularity relations.

PROPOSITION 2.6. The transformed Macdonald polynomials $\tilde{H}_{\mu}$ satisfy, and are uniquely characterized by, these conditions:

(1) $\tilde{H}_{\mu}[X(1-q);q,t] \in \mathbb{Q}(q,t)\{s_{\lambda} : \lambda \geq \mu\}$;
(2) $\tilde{H}_{\mu}[X(1-t);q,t] \in \mathbb{Q}(q,t)\{s_{\lambda} : \lambda \geq \mu'\}$;
(3) $\langle \tilde{H}_{\mu}, s_{(n)} \rangle = 1$.

PROOF. Note that $\tilde{H}_{\mu}[X(1-t);q,t] = t^{|\mu|}\tilde{H}_{\mu}[-X(1-t^{-1});q,t]$ is a scalar multiple of $P_{\mu}[-X;q,t^{-1}]$ and thus of $\omega P_{\mu}(X;q,t^{-1})$. Since P_{μ} belongs to the space $\mathbb{Q}(q,t)\{s_{\lambda} : \lambda \leq \mu\}$, and conjugation reverses the dominance order, ωP_{μ} belongs to $\mathbb{Q}(q,t)\{s_{\lambda} : \lambda \geq \mu'\}$, which is (2) above. From the symmetry given by Proposition 2.5 we then obtain (1). The normalization from Proposition 2.4 is (3).

For uniqueness, suppose $H_{\mu}'(X)$ is another solution of (1) and (2). Then (1) implies that $H_{\mu}'[X(1-q)] \in \mathbb{Q}(q,t)\{\tilde{H}_{\lambda}[X(1-q)] : \lambda \geq \mu\}$ and hence that $H_{\mu}' \in \mathbb{Q}(q,t)\{\tilde{H}_{\lambda} : \lambda \geq \mu\}$. Similarly, (2) implies that $H_{\mu}' \in \mathbb{Q}(q,t)\{\tilde{H}_{\lambda} : \lambda \leq \mu\}$. Together these mean that H_{μ}' is a scalar multiple of $\tilde{H}_{\mu}$, and (3) fixes the scalar factor as 1.   □

COROLLARY 2.7. For all μ we have $\omega\tilde{H}_{\mu}(X;q,t) = t^{n(\mu)}q^{n(\mu')}\tilde{H}_{\mu}(X;q^{-1},t^{-1})$ and, consequently, $\tilde{K}_{\lambda'\mu}(q,t) = t^{n(\mu)}q^{n(\mu')}\tilde{K}_{\lambda\mu}(q^{-1},t^{-1})$

PROOF. One verifies easily that $\omega t^{n(\mu)}q^{n(\mu')}\tilde{H}_{\mu}(X;q^{-1},t^{-1})$ satisfies (1) and (2) of Proposition 2.6, and hence is a scalar multiple of $\tilde{H}_{\mu}$. To fix the scalar as 1 requires that $\tilde{K}_{(1^n),\mu} = t^{n(\mu)}q^{n(\mu')}$. But this is known [Macdonald 1995], as it is equivalent to $K_{(1^n),\mu} = q^{n(\mu')}$.   □

To conclude, we will recover the orthogonality of the P_{μ}'s with respect to $\langle \cdot, \cdot \rangle_{q,t}$, as in Macdonald's original definition. Replacing t by t^{-1}, we are to show that

$$\left\langle P_{\mu}(X;q,t^{-1}), P_{\nu}\left(X\frac{1-q}{1-t^{-1}};q,t^{-1}\right)\right\rangle = 0$$

for $\mu \neq \nu$, or equivalently that

$$\left\langle P_{\mu}(X;q,t^{-1}), P_{\nu}\left(-X\frac{1-q}{1-t};q,t^{-1}\right)\right\rangle = 0.$$

Since $P_\mu(X; q, t^{-1})$ is a scalar multiple of $\tilde{H}_\mu[X(1-t)]$, we are to show that

$$\langle \tilde{H}_\mu[X(1-t)], \tilde{H}_\nu[-X(1-q)] \rangle = 0.$$

Now from Proposition 2.6 and the orthogonality of Schur functions it is clear that this last inner product vanishes unless $\nu \le \mu$. But then by symmetry it also vanishes unless $\mu \le \nu$, that is, unless $\mu = \nu$.

3. The $n!$ Conjecture

Let $D = \{(p_1, q_1), \ldots, (p_n, q_n)\}$ be an n-element subset of $\mathbb{N} \times \mathbb{N}$. We define a polynomial in $2n$ variables $x_1, y_1, \ldots, x_n, y_n$ as follows:

$$\Delta_D(\boldsymbol{x}, \boldsymbol{y}) = \det \left[x_i^{p_j} y_i^{q_j} \right]_{i,j=1}^n. \tag{3.1}$$

Note that Δ_D is well defined, up to a change of sign, independent of the ordering chosen for the elements of D, and that it alternates in sign under the action of the symmetric group S_n permuting the $\boldsymbol{x}$ and the $\boldsymbol{y}$ variables simultaneously. That is,

$$w\Delta_\mu = \varepsilon(w)\Delta_\mu \quad \text{for all} \quad w \in S_n,$$

where $\varepsilon(w)$ is the sign of the permutation w. For a partition diagram, we set

$$\Delta_\mu = \Delta_{D(\mu)}.$$

Then Δ_μ is doubly homogeneous, of degree $n(\mu)$ in the $\boldsymbol{x}$ variables and $n(\mu')$ in the $\boldsymbol{y}$ variables. In the cases $\mu = (1^n)$ and $\mu = (n)$, Δ_μ is the usual Vandermonde determinant in the $\boldsymbol{x}$ and $\boldsymbol{y}$ variables, respectively.

Our conjectures concern the space of all derivatives of Δ_μ,

$$D_\mu = \{ p(\partial x_1, \partial y_1, \ldots, \partial x_n, \partial y_n) \Delta_\mu : p \in \mathbb{Q}[\boldsymbol{x}, \boldsymbol{y}] \}.$$

Since Δ_μ is doubly homogeneous, this space is doubly graded:

$$D_\mu = \bigoplus_{r,s} (D_\mu)_{r,s},$$

where $(D_\mu)_{r,s}$ consists of those elements of D_μ that are doubly homogeneous of degree r in the $\boldsymbol{x}$ variables and s in the $\boldsymbol{y}$ variables. Since Δ_μ is alternating, the space D_μ is stable under the action of S_n. Thus it affords a doubly graded representation of the symmetric group S_n. We denote by $\text{mult}(\chi^\lambda, \chi)$ the multiplicity of the irreducible S_n-character χ^λ in a given character χ.

CONJECTURE 3.1 (THE $n!$ CONJECTURE). *The space D_μ has dimension $n!$.*

CONJECTURE 3.2. *The bivariate character multiplicity Hilbert series*

$$\sum_{r,s} t^r q^s \, \text{mult}(\chi^\lambda, \text{ch}(D_\mu)_{r,s}) \tag{3.2}$$

is equal to $\tilde{K}_{\lambda\mu}(q, t)$. In particular, the latter is a polynomial with nonnegative integer coefficients.

It is known [Macdonald 1995] that $\tilde{K}_{\lambda\mu}(1,1) = \chi^\lambda(1)$, the degree of the character χ^λ or the number of standard Young tableaux of shape λ. Hence, according to Conjecture 3.2, we must have $\text{mult}(\chi^\lambda, \text{ch}(D_\mu)) = \chi^\lambda(1)$, so that when we ignore the grading, D_μ affords the regular representation of S_n, and hence has dimension $n!$. Thus Conjecture 3.2 implies Conjecture 3.1. One of the chief things we will achieve in the geometric setting of Sections 4 and 5 is to prove the converse implication.

It will be helpful to reformulate the conjectures in two ways. The first is to introduce a more convenient notation for (3.2). Recall that the *Frobenius map* from S_n characters to symmetric functions homogeneous of degree n is defined by

$$\Phi(\chi) = \frac{1}{n!} \sum_{w \in S_n} \chi(w) p_{\tau(w)}(X), \qquad (3.3)$$

where $\tau(w)$ is the partition whose parts are the lengths of the cycles of the permutation w, and where the indeterminates X are not to be confused with the coordinates x. For the irreducible characters we have the symmetric function identity $\Phi(\chi^\lambda) = s_\lambda(X)$, from which follows, for any character,

$$\Phi(\chi) = \sum_\lambda \text{mult}(\chi^\lambda, \chi) s_\lambda.$$

Now, by analogy to the Hilbert series, we define the *Frobenius series* of a doubly graded S_n representation D to be

$$\mathcal{F}_D(X; q, t) = \sum_{r,s} t^r q^s \Phi \, \text{ch}(D)_{r,s}.$$

This given, Conjecture 3.2 takes the simple form

$$\mathcal{F}_{D_\mu} = \tilde{H}_\mu. \qquad (3.4)$$

In Section 5 we extend the notion of Frobenius series to S_n actions on modules over a geometric regular local ring with an equivariant two-dimensional torus action, providing the basic tool to link Conjecture 3.2 with the geometry.

The second reformulation we need is of the definition of D_μ itself—the definition in terms of derivatives is simple, but geometrically misleading, and we need a derivative-free version. This is given by the next propositions. We will need the following definition here and later.

DEFINITION. The *alternation* operator over the symmetric group is

$$\text{Alt} \, f = \sum_{w \in S_n} \varepsilon(w) w(f).$$

Since the polynomials Δ_D form a basis of all the S_n-alternating polynomials in $\mathbb{Q}[x, y]$, it makes sense to speak of the coefficient of Δ_D in $\text{Alt} \, f$. Indeed, it is merely the coefficient of the monomial $x_1^{p_1} y_1^{q_1} \cdots x_n^{p_n} y_n^{q_n}$.

PROPOSITION 3.3. *The ideal J_μ of polynomials $p(x, y) \in \mathbb{Q}[x, y]$ for which the differential operator $p(\partial x, \partial y)$ annihilates Δ_μ can be characterized as follows: $p \in J_\mu$ if and only if for all $g \in \mathbb{Q}[x, y]$, the coefficient of Δ_μ in $\mathrm{Alt}\, gp$ is zero.*

PROOF. Observe that the constant term of $g(\partial x, \partial y)p(\partial x, \partial y)\Delta_\mu$ is, apart from a constant factor, the coefficient of Δ_μ in $\mathrm{Alt}\, gp$. Hence if $p(\partial x, \partial y)\Delta_\mu = 0$, the characterization certainly holds. Conversely, if the characterization holds, then $p(\partial x, \partial y)\Delta_\mu$ has the property that it and all its partial derivatives of all orders have zero constant term. By Taylor's theorem this implies that $p(\partial x, \partial y)\Delta_\mu = 0$. $\qquad\square$

PROPOSITION 3.4. *The quotient ring $\mathbb{Q}[x, y]/J_\mu$ is isomorphic as a doubly graded S_n representation to D_μ.*

PROOF. Define an inner product $(\,\cdot\,, \cdot\,)$ on $\mathbb{Q}[x, y]$ by

$$(f, g) = f(\partial x, \partial y)g(x, y)\big|_{x, y = 0}. \tag{3.5}$$

One sees immediately that monomials are mutually orthogonal under $(\,\cdot\,, \cdot\,)$. Hence the form is symmetric and nondegenerate, separately on each doubly homogeneous subspace $(\mathbb{Q}[x, y])_{r,s}$. It follows that $\mathbb{Q}[x, y]/J_\mu$ is isomorphic as a doubly graded S_n representation to the orthogonal complement $J_\mu^\perp$.

Now I claim that $J_\mu^\perp = D_\mu$. Taking $g = \Delta_\mu$ and $f \in J_\mu$ in (3.5), we see that $\Delta_\mu \in J_\mu^\perp$. From the definition we have $(p(x, y)f, g) = (f, p(\partial x, \partial y)g)$ for all $p(x, y)$, *i.e.*, multiplication is adjoint to differentiation. Since J_μ is an ideal it follows that $J_\mu^\perp$ is closed under differentiation and hence contains D_μ.

Now both $J_\mu^\perp$ and D_μ are finite-dimensional, and the map $p \mapsto p(\partial x, \partial y)\Delta_\mu$ is an isomorphism of vector spaces from $\mathbb{Q}[x, y]/J_\mu$ onto D_μ. Therefore $J_\mu^\perp$ and D_μ have the same dimension and hence are equal. $\qquad\square$

For geometric purposes we will replace $\mathbb{Q}$ by $\mathbb{C}$, extending J_μ to $J_\mu \otimes_\mathbb{Q} \mathbb{C}$, which we again denote J_μ, and for which the characterization in Proposition 3.3 still holds.

DEFINITION. The ring R_μ is $\mathbb{C}[x, y]/J_\mu$.

Of course R_μ has the same Frobenius series as $\mathbb{Q}[x, y]/J_\mu$, and also, by Proposition 3.4, as D_μ.

The evidence for Conjecture 3.2 includes the fact that various known symmetries and specializations of $\tilde{H}_\mu$ can also be established for $\mathcal{F}_{D_\mu}$. We conclude this section by demonstrating a few of these.

PROPOSITION 3.5. *We have the identity $\mathcal{F}_{D_{\mu'}}(X; q, t) = \mathcal{F}_{D_\mu}(X; t, q)$ (compare Proposition 2.5).*

PROOF. Obvious, since $\Delta_{\mu'}(x, y) = \Delta_\mu(y, x)$. $\qquad\square$

PROPOSITION 3.6. *We have the identity*

$$\omega\mathcal{F}_{D_\mu}(X; q, t) = t^{n(\mu)}q^{n(\mu')}\mathcal{F}_{D_\mu}(X; q^{-1}, t^{-1})$$

(compare Corollary 2.7).

PROOF. In the proof of Proposition 3.4 there are two isomorphisms of $\mathbb{Q}[x,y]/J_\mu$ with D_μ. The first is the inclusion of D_μ in $\mathbb{Q}[x,y]$, followed by projection mod J_μ, which is an isomorphism of graded S_n representations. The second is the map $p \mapsto p(\partial x, \partial y)\Delta_\mu$. As Δ_μ is S_n-alternating and homogeneous of degrees $(n(\mu), n(\mu'))$, this map reverses degrees and tensors S_n characters by the sign character. Recalling that the Frobenius map satisfies $\Phi(\varepsilon \otimes \chi) = \omega\Phi(\chi)$, we see that reversing degrees and tensoring with the sign character on D_μ yields a space with Frobenius series $\omega t^{n(\mu)} q^{n(\mu')} \mathcal{F}_{D_\mu}(X; q^{-1}, t^{-1})$. Combining the two isomorphisms shows this is equal to $\mathcal{F}_{D_\mu}$.   □

Remark: The algebraic correlate of this symmetry of D_μ is that the ring R_μ is *Gorenstein.* More generally, for any homogeneous ideal $J \subseteq \mathbb{Q}[x]$, the quotient $\mathbb{Q}[x]/J$ is finite-dimensional (as a vector space) and Gorenstein if and only if J is the ideal of differential operators annihilating some homogeneous polynomial.

PROPOSITION 3.7. *We have the identity $\mathcal{F}_{D_\mu}(X; 0, t) = \tilde{H}_\mu(X; 0, t)$.*

PROOF. The left-hand side is the Frobenius series of the subspace $\mathbb{Q}[x] \cap D_\mu$. The *Garnir polynomial* $g_\mu(x)$ is defined as the product of the Vandermonde determinants in the first μ'_1 variables, the next μ'_2 variables, and so on. It is not hard to see that the space $\mathbb{Q}[x] \cap D_\mu$ is spanned by all derivatives of g_μ and its images under permutation of the variables.

This space has been well studied [Bergeron and Garsia 1992; De Concini and Procesi 1981; Garsia and Procesi 1992; Hotta and Springer 1977; Kraft 1981; Springer 1978], and its Frobenius series (in one variable t) is known to be the transformed Hall–Littlewood polynomial $t^{n(\mu)} Q_\mu[X/(1-t^{-1}); t^{-1}]$. Since $J_\mu(X; 0, t) = Q_\mu(X; t)$, the result follows.   □

It is also possible to give an elementary proof (see [Garsia and Haiman 1996b]) that the $n!$ conjecture implies $\mathcal{F}_{D_\mu}(X; 1, t) = \tilde{H}_\mu(X; 1, t)$. Since $\mathcal{F}_{D_\mu}(X; 1, 1)$ determines the character and thus the dimension of D_μ, one must of course assume the $n!$ conjecture for this. But as we are going to prove that the $n!$ conjecture implies $\mathcal{F}_{D_\mu} = \tilde{H}_\mu$, the proof for the $q = 1$ specialization would be redundant here.

4. The Hilbert Scheme and X_n

Let $\mathbb{C}^2 = \operatorname{Spec}\mathbb{C}[x,y]$ be the affine plane over $\mathbb{C}$. By definition, closed subschemes $S \subseteq \mathbb{C}^2$ correspond to ideals $I \subseteq \mathbb{C}[x,y]$. The subscheme S is 0-*dimensional*, of *length* n, if $\mathbb{C}[x,y]/I$ has dimension n as a vector space. The generic example of such a subscheme S is a set of n points in $\mathbb{C}^2$, with the reduced subscheme structure. In this case I is a radical ideal and $\mathbb{C}[x,y]/I$ can be identified with the ring of complex-valued functions on the finite set S, which is clearly n-dimensional.

Other such subschemes S have fewer than n points, with nonreduced subscheme structures at these points. If we define the multiplicity of each point $P \in S$ as the length of the local ring $\mathcal{O}_{S,P}$ then the total length, n, is the sum of the multiplicities. Associated to each partition μ of n is a nonreduced subscheme S_μ whose ideal I_μ is spanned by the monomials $\{x^r y^s : (r,s) \notin D(\mu)\}$. Note that this is indeed an ideal. Since the remaining monomials

$$\mathcal{B}_\mu = \{x^r y^s : (r,s) \in D(\mu)\} \tag{4.1}$$

form a basis of $\mathbb{C}[x,y]/I_\mu$, S_μ has length n, and the Hilbert series of $\mathbb{C}[x,y]/I_\mu$ as a doubly graded algebra is the quantity $B_\mu(q,t)$ defined in (2.12). We readily see that in fact, the S_μ's are the only 0-dimensional length-n subschemes whose ideals are spanned by monomials. As a set, S_μ has only one point, the origin, with multiplicity n.

The set of all 0-dimensional length-n subschemes of $\mathbb{C}^2$, or equivalently, the set of ideals I such that $\dim_\mathbb{C} \mathbb{C}[x,y]/I = n$, has the structure of an algebraic variety, the *Hilbert scheme of n points in the plane*, or $\mathrm{Hilb}^n(\mathbb{C}^2)$. We will prefer the point of view that the (closed) points of $\mathrm{Hilb}^n(\mathbb{C}^2)$ are the ideals I. Its variety structure may be described in either of two ways.

First, for every $I \in \mathrm{Hilb}^n(\mathbb{C}^2)$, at least one of the sets $\mathcal{B}_\mu$ spans modulo I [Gordan 1900]. In particular, the set $M_N = \{x^r y^s : r + s < N\}$ always spans mod I, for $N \geq n$. Thus $\mathbb{C}[x,y]/I$ may be identified with an element of the Grassmann variety $G_n(W)$ of n-dimensional quotients of the space $W = \mathbb{C}M_N$, giving a map from $\mathrm{Hilb}^n(\mathbb{C}^2)$ into $G_n(W)$. For N sufficiently large — in fact, for $N \geq n + 1$ [Gotzmann 1978] — this map is injective, its image is a locally closed subvariety of $G_n(W)$, and the induced variety structure on $\mathrm{Hilb}^n(\mathbb{C}^2)$ is independent of N.

The second description is via Grothendieck's universal property [Grothendieck 1961], as follows. There exists a subscheme $U \subseteq \mathrm{Hilb}^n(\mathbb{C}^2) \times \mathbb{C}^2$,

$$\begin{array}{ccc} U & \longrightarrow & \mathbb{C}^2 \\ {\scriptstyle \pi}\downarrow & & \\ \mathrm{Hilb}^n(\mathbb{C}^2), & & \end{array} \tag{4.2}$$

called the *universal family*, whose scheme-theoretic fiber over a point I of the scheme $\mathrm{Hilb}^n(\mathbb{C}^2)$ is the corresponding subscheme $S \subseteq \mathbb{C}^2$. In particular, U is flat and finite of degree n over $\mathrm{Hilb}^n(\mathbb{C}^2)$. The universal property is that for any scheme T and any family $F \subseteq T \times \mathbb{C}^2$ of subschemes of $\mathbb{C}^2$,

$$\begin{array}{ccc} F & \longrightarrow & \mathbb{C}^2 \\ \downarrow & & \\ T, & & \end{array} \tag{4.3}$$

flat and finite of degree n over T, there is a unique morphism $\phi : T \to \mathrm{Hilb}^n(\mathbb{C}^2)$, such that F is the fiber product $F = T \times_{\mathrm{Hilb}^n(\mathbb{C}^2)} U$. The universal property, as usual, characterizes $\mathrm{Hilb}^n(\mathbb{C}^2)$ and the universal family U up to canonical isomorphism.

The following result, which is unique to $\mathbb{C}^2$ and fails in more than two dimensions, is one of the true miracles of the mathematical universe.

THEOREM 4.1 [Fogarty 1968]. *The Hilbert scheme* $\mathrm{Hilb}^n(\mathbb{C}^2)$ *is irreducible and nonsingular, of dimension* $2n$.

Given $I \in \mathrm{Hilb}^n(\mathbb{C}^2)$, the operators X and Y of multiplication by x and y, respectively, are commuting endomorphisms of the vector space

$$\mathbb{C}[x,y]/I \cong \prod_{P \in S} \mathcal{O}_{S,P}$$

which preserve the direct factors $\mathcal{O}_{S,P}$. For $P = (x_0, y_0)$, $X - x_0$ and $Y - y_0$ are nilpotent on $\mathcal{O}_{S,P}$, so that $\mathcal{O}_{S,P}$ is the characteristic subspace associated with the joint eigenvalue (x_0, y_0) of (X, Y). Thus the points $(x_1, y_1), \ldots, (x_n, y_n)$ of S, each included with its multiplicity in S, are the joint spectrum of (X, Y). In particular, the *polarized power sums* $p_{h,k}(x, y) = \sum_{i=1}^n x_i^h y_i^k$ satisfy the identity

$$p_{h,k}(x, y) = \mathrm{tr}\, X^h Y^k. \tag{4.4}$$

Note that trace map associated to the finite morphism $\pi : U \to \mathrm{Hilb}^n(\mathbb{C}^2)$ sends the regular function $x^h y^k$ on U (coming from $U \to \mathbb{C}^2$) to $\mathrm{tr}\, X^h Y^k$, so the latter is a regular function on $\mathrm{Hilb}^n(\mathbb{C}^2)$.

Now the symmetric group S_n acts on the variety

$$(\mathbb{C}^2)^n = \mathrm{Spec}\, \mathbb{C}[x_1, y_1, \ldots, x_n, y_n]$$

of ordered n-tuples of points in the plane, and we may identify Spec of the ring of invariants as the variety of unordered n-tuples, or n-element multisets, $S^n \mathbb{C}^2 = \mathrm{Spec}\, \mathbb{C}[x, y]^{S_n}$. By a theorem of Weyl [1946], the polarized power-sums $p_{h,k}$ generate this ring of invariants. It follows from this and the preceding paragraph that the map

$$\tau : \mathrm{Hilb}^n(\mathbb{C}^2) \to S^n \mathbb{C}^2$$

sending I to the multiset of points of the corresponding subscheme S, with their multiplicities in S, is a morphism. It extends to a morphism $\hat{\tau} : \mathrm{Hilb}^n(\mathbb{P}^2) \to S^n \mathbb{P}^2$, with τ being the restriction to $\hat{\tau}^{-1}(S^n \mathbb{C}^2)$. Since $\mathrm{Hilb}^n(\mathbb{P}^2)$ is a projective variety, τ is a projective morphism, called the *Chow morphism*. Note that τ restricts to an isomorphism of the open sets where the n points are distinct, and so is birational.

DEFINITION. The *isospectral Hilbert scheme* X_n is the reduced fiber product

$$
\begin{array}{ccc}
X_n & \longrightarrow & (\mathbb{C}^2)^n \\
\sigma \downarrow & & \downarrow \\
\mathrm{Hilb}^n(\mathbb{C}^2) & \overset{\tau}{\longrightarrow} & S^n\mathbb{C}^2.
\end{array}
\tag{4.5}
$$

In other words, a point of X_n is a point I of the Hilbert scheme, together with an ordered n-tuple of points $(x_1, y_1), \ldots, (x_n, y_n)$ whose underlying unordered multiset is $\tau(I)$, that is, the joint spectrum of the operators (X, Y) on $\mathbb{C}[x,y]/I$. We should stress here that the scheme-theoretic fiber product indicated by the above diagram is *not* reduced, but our definition is that X_n is the underlying reduced subscheme. What this reflects is that the equations $p_{h,k}(x_1, y_1, \ldots, x_n, y_n) = \mathrm{tr}\, X^h Y^k$, which define X_n set-theoretically, do not generate its ideal. Indeed we do not know a fully explicit set of generators for the ideal of X_n. Such a set of generators will necessarily be complicated, since it must specialize to give generators of all the ideals J_μ of Proposition 3.3.

It is possible to describe the ideal of X_n implicitly, as we shall do next.

Let $U^{\times n}$ denote the n-fold fiber product of the universal family U over the Hilbert scheme $\mathrm{Hilb}^n(\mathbb{C}^2)$. It is a subscheme of $\mathrm{Hilb}^n(\mathbb{C}^2) \times (\mathbb{C}^2)^n$, flat and finite of degree n^n, and generically reduced, since it has reduced fibers whenever S consists of n distinct points. As the Hilbert scheme is irreducible, this implies that $U^{\times n}$ is reduced.[2] Thus $U^{\times n}$ is just the set of tuples $I, (x_1, y_1), \ldots, (x_n, y_n)$ with all $(x_i, y_i) \in S = V(I)$, irrespective of multiplicities. X_n is the subscheme of $U^{\times n}$ consisting of tuples for which the points occur with their correct multiplicities in S. Generically, when S has n distinct points, this means the points (x_i, y_i) are a permutation of S.

Now let $B = \pi_* \mathcal{O}_U$, the sheaf of $\mathcal{O}$-algebras on $\mathrm{Hilb}^n(\mathbb{C}^2)$ such that $U = \mathrm{Spec}\, B$ as a scheme over $\mathrm{Hilb}^n(\mathbb{C}^2)$. Since π is flat of degree n, B is locally free of rank n, that is, it is the sheaf of sections of a rank-n vector bundle over the Hilbert scheme. Indeed, B is the *tautological bundle* whose fiber over a point $I \in \mathrm{Hilb}^n(\mathbb{C}^2)$ is $\mathbb{C}[x,y]/I$. We have $U^{\times n} = \mathrm{Spec}\, B^{\otimes n}$, and we seek to identify the ideal sheaf of X_n in $U^{\times n}$ as a submodule sheaf of the sheaf of $\mathcal{O}$-algebras $B^{\otimes n}$. Note that the S_n action here permutes the tensor factors.

PROPOSITION 4.2. *Let*

$$
B^{\otimes n} \otimes B^{\otimes n} \to B^{\otimes n} \to \textstyle\bigwedge^n B
\tag{4.6}
$$

be the map given by multiplication, followed by the operator Alt. *Then the ideal sheaf of X_n is the kernel of the map*

$$
\phi : B^{\otimes n} \to (B^{\otimes n})^* \otimes \textstyle\bigwedge^n B
\tag{4.7}
$$

---

[2] By the same reasoning, the universal family U itself is reduced. This fails in higher dimension, when the Hilbert scheme is not irreducible.

induced by (4.6).

PROOF. Since X_n is reduced, a section of $B^{\otimes n}$ belongs to the ideal sheaf of X_n if and only if, regarded as a regular function on X_n, it vanishes on any dense open subset. Hence it is enough to check the proposition generically, over the locus where S consists of n distinct points. Suppose s is a section that vanishes on X_n. Then so do gs and $\operatorname{Alt} gs$, for any section g of $B^{\otimes n}$. But since it is alternating, $\operatorname{Alt} gs$ also vanishes at any point $I, P_1, \ldots, P_n$ of $U^{\times n}$ for which two of the points P_i, P_j coincide. Hence it vanishes on $U^{\times n} \setminus X_n$ and thus on all of $U^{\times n}$, which means it is zero in $B^{\otimes n}$. The condition that s belongs to the kernel of (4.7) is precisely that $\operatorname{Alt} gs = 0$ for all g.

Conversely, suppose s does not vanish on X_n, and choose a point $Q = (I, P_1, \ldots, P_n) \in X_n$ outside the vanishing locus $V(s)$, with the P_i all distinct. After multiplying by a suitable g we can arrange that gs vanishes at all points of the S_n orbit of Q, except Q. Then $\operatorname{Alt} gs$ does not vanish at Q, so $\operatorname{Alt} gs \neq 0$. $\square$

Now consider the situation over one of the distinguished points I_μ. The fiber $B(I_\mu)$ of the vector bundle B is $\mathbb{C}[x, y]/I_\mu$, and that of $B^{\otimes n}$ is

$$\mathbb{C}[\boldsymbol{x}, \boldsymbol{y}]/\big(I_\mu(x_1, y_1) + \cdots + I_\mu(x_n, y_n)\big).$$

Notice that every alternating polynomial Δ_D as in (3.1) vanishes modulo

$$(I_\mu(x_1, y_1) + \cdots + I_\mu(x_n, y_n)),$$

except for Δ_μ. In other words, the 1-dimensional space $\bigwedge^n B(I_\mu)$ is spanned by the image of Δ_μ, and the linear functional $\operatorname{Alt} : B^{\otimes n}(I_\mu) \to \bigwedge^n B(I_\mu) \cong \mathbb{C}\{\Delta_\mu\}$, composed with the natural projection $\mathbb{C}[\boldsymbol{x}, \boldsymbol{y}] \to B^{\otimes n}(I_\mu)$, is just the map sending f to the coefficient of Δ_μ in $\operatorname{Alt} f$. Together with Proposition 3.3, this proves the next result:

PROPOSITION 4.3. *The ideal J_μ of polynomials that as differential operators annihilate Δ_μ is the kernel of the map*

$$\mathbb{C}[\boldsymbol{x}, \boldsymbol{y}] \to B^{\otimes n}(I_\mu) \to (B^{\otimes n})^*(I_\mu) \otimes \textstyle\bigwedge^n B(I_\mu)$$

induced by the map ϕ in (4.7).

Comparing this with Proposition 4.2 we might well expect the fiber of the ideal sheaf of X_n over I_μ to be J_μ, and the scheme-theoretic fiber of X_n over I_μ to be $\operatorname{Spec} R_\mu$. We cannot yet draw this conclusion, however, since the map ϕ is only a sheaf homomorphism, not necessarily a homomorphism of vector bundles, and thus the fiber of its kernel at I_μ need not equal the kernel of its fiber. What we can say is is that the fiber of the kernel factors through the kernel of the fiber, which gives the following result.

PROPOSITION 4.4. *The image of the ideal J_μ in $B^{\otimes n}(I_\mu)$ contains the image of the fiber map $\mathcal{J}(I_\mu) \to B^{\otimes n}(I_\mu)$, where $\mathcal{J}(I_\mu)$ is the fiber of the ideal sheaf $\mathcal{J}$ of*

X_n at I_μ. *As a consequence* $\dim R_\mu \leq n!$, *and* R_μ *is isomorphic to a submodule of the regular representation of* S_n.

PROOF. The first part is clear, by the previous propositions. It only remains to prove the consequence.

By Proposition 4.2, we have an exact sequence of sheaves on $\text{Hilb}^n(\mathbb{C}^2)$,

$$0 \to \mathcal{J} \to B^{\otimes n} \to \sigma_* \mathcal{O}_{X_n} \to 0, \tag{4.8}$$

in which $\sigma_* \mathcal{O}_{X_n}$ is the image of the sheaf homomorphism ϕ given by (4.7). At generic ideals $I \in \text{Hilb}^n(\mathbb{C}^2)$ corresponding to sets of n distinct points $S \subseteq \mathbb{C}^2$, the fibers $\sigma_* \mathcal{O}_{X_n}(I)$ have constant dimension $n!$. This implies that the rank of the fiber map

$$\phi(I): B^{\otimes n}(I) \to (B^{\otimes n})^*(I) \otimes \textstyle\bigwedge^n B(I)$$

is generically $n!$.

At the special ideal I_μ the rank of $\phi(I_\mu)$ cannot exceed the generic rank, and by Proposition 4.3 this implies $\dim R_\mu \leq n!$. Since S_n acts equivariantly on everything, the same considerations apply to the isotypic components corresponding to each irreducible character of S_n, to show that the character multiplicities in R_μ cannot exceed those in a generic fiber $\sigma_* \mathcal{O}_{X_n}(I)$. When $I = I(S)$, this fiber is the coordinate ring of the set of all permutations of S, and thus affords the regular representation of S_n. $\qquad\square$

Continuing with this argument, we see that if the $n!$ conjecture holds for μ then the rank of $\phi(I)$ does not decrease at I_μ, and so is constant on a neighborhood of I_μ. This is the criterion for ϕ to be locally a homomorphism of vector bundles on a neighborhood of I_μ. When this holds, the sheaves in (4.8) are locally free, and the scheme X_n is flat of degree $n!$ over $\text{Hilb}^n(\mathbb{C}^2)$ at I_μ.

Conversely, if X_n is flat over $\text{Hilb}^n(\mathbb{C}^2)$ at I_μ, then $B^{\otimes n}$ and $B^{\otimes n}/\mathcal{J}$ are both locally free, which implies that $\mathcal{J}$ is locally free and ϕ is a homomorphism of vector bundles. By Propositions 4.2 and 4.3 it then follows that $\mathcal{J}(I_\mu) = J_\mu$ and $\dim R_\mu = n!$. Recall that a finite, surjective morphism $X \to H$ with H nonsingular is flat if and only if X is Cohen–Macaulay. We have proved:

THEOREM 4.5. *The following statements are equivalent:*

(1) *The $n!$ conjecture holds for the partition μ.*
(2) *The map $X_n \to \text{Hilb}^n(\mathbb{C}^2)$ is flat over a neighborhood of I_μ.*
(3) *The isospectral Hilbert scheme X_n is Cohen–Macaulay in a neighborhood of the point $Q_\mu = (I_\mu, 0, \ldots, 0)$.*

This theorem has an interesting connection with the Hilbert scheme of regular S_n orbits in $(\mathbb{C}^2)^n$, which we shall discuss briefly. Given a set S of n distinct points in the plane, its $n!$ permutations describe a regular orbit of S_n in $(\mathbb{C}^2)^n$. The ideal J of the orbit is a point of $\text{Hilb}^{n!}((\mathbb{C}^2)^n)$. Let Z_n be the closure in $\text{Hilb}^{n!}((\mathbb{C}^2)^n)$ of the set of such points. It is not hard to show that the set of ideals J which are S_n stable and for which $\mathbb{C}[\boldsymbol{x}, \boldsymbol{y}]/J$ has a given character is closed in $\mathrm{Hilb}^{n!}((\mathbb{C}^2)^n)$, so for every $J \in Z_n$, $\mathbb{C}[\boldsymbol{x}, \boldsymbol{y}]/J$ affords the regular representation of S_n.

Now there is a natural map from Z_n to $\mathrm{Hilb}^n(\mathbb{C}^2)$, which may be described as follows. Since $\mathbb{C}[\boldsymbol{x}, \boldsymbol{y}]/J$ affords the regular representation, its only invariants are the constants. This means that modulo J we have $p_{h,k}(\boldsymbol{x}, \boldsymbol{y}) = c_{h,k}$ for some constant $c_{h,k}$, for all h, k. By Weyl's theorem, the S_{n-1}-invariants in $\mathbb{C}[\boldsymbol{x}, \boldsymbol{y}]$, for the action of S_{n-1} on x_2, y_2 through x_n, y_n, are generated by x_1, y_1, and the polarized power-sums $p_{h,k}(x_2, y_2, \ldots, x_n, y_n)$. Modulo J, the latter are congruent to $c_{h,k} - x_1^h y_1^k$, so the S_{n-1} invariants of $\mathbb{C}[\boldsymbol{x}, \boldsymbol{y}]/J$ are generated by x_1 and y_1. In other words, $\mathbb{C}[x_1, y_1]/(J \cap \mathbb{C}[x_1, y_1]) = (\mathbb{C}[\boldsymbol{x}, \boldsymbol{y}]/J)^{S_{n-1}}$. It follows that $J \cap \mathbb{C}[x_1, y_1]$ belongs to $\mathrm{Hilb}^n(\mathbb{C}^2)$, after identifying x_1, y_1 with x, y.

The above construction defines the map $Z_n \to \mathrm{Hilb}^n(\mathbb{C}^2)$, which also has the follow geometric description. Let W be the universal family over Z_n. It has a natural S_n-action, in which every fiber affords the regular representation. Then W/S_{n-1} is flat and finite of degree n over Z_n, and by the above calculation can be identified with a family of subschemes of $\mathbb{C}^2$. The map $Z_n \to \mathrm{Hilb}^n(\mathbb{C}^2)$ is the one given by the universal property of $\mathrm{Hilb}^n(\mathbb{C}^2)$, for the family W/S_{n-1}.

If the equivalent conditions of Theorem 4.5 hold, then X_n is a flat family of subschemes of $(\mathbb{C}^2)^n$, of degree $n!$ over $\mathrm{Hilb}^n(\mathbb{C}^2)$. The universal property of $\mathrm{Hilb}^{n!}((\mathbb{C}^2)^n)$ then yields a map $\mathrm{Hilb}^n(\mathbb{C}^2) \to \mathrm{Hilb}^{n!}((\mathbb{C}^2)^n)$, whose image lies in Z_n. Generically, for sets S of n distinct points in $\mathbb{C}^2$ and their corresponding regular orbits in $(\mathbb{C}^2)^n$, these two maps are mutually inverse. Hence, assuming the $n!$ conjecture, they are inverse everywhere, and the natural map $Z_n \to \mathrm{Hilb}^n(\mathbb{C}^2)$ is an isomorphism.

Conversely, if $Z_n \to \mathrm{Hilb}^n(\mathbb{C}^2)$ is an isomorphism, then its inverse defines a family $X' \subseteq \mathrm{Hilb}^n(\mathbb{C}^2) \times (\mathbb{C}^2)^n$ that is flat over $\mathrm{Hilb}^n(\mathbb{C}^2)$ and coincides with X_n generically. But then X' is reduced and hence equal to X_n, so X_n is flat, and the $n!$ conjecture holds.

It would even suffice to show that $Z_n \to \mathrm{Hilb}^n(\mathbb{C}^2)$ is injective. For it is proper and birational, hence surjective, and a bijective morphism onto a nonsingular variety (or any normal variety) is an isomorphism, by Zariski's theorem. To summarize, we have

PROPOSITION 4.6. *The equivalent conditions of Theorem 4.5, for all partitions μ of n, are also equivalent to:*

(4) *The natural map $Z_n \to \mathrm{Hilb}^n(\mathbb{C}^2)$ is injective.*

(5) *This map is an isomorphism.*

Proposition 4.6 implies that the $n!$ conjecture is equivalent to an instance of a conjecture of Nakamura [1996], cited in [Reid 1997], connected with the *McKay correspondence*, as we now explain.

Let G be a finite subgroup of $SL(V)$, where $V = \mathbb{C}^n$. For any finite linear group action on V, $V/G = \mathrm{Spec}\, \mathcal{O}(V)^G$ is Cohen–Macaulay, and its canonical module is the isotypic component $\mathcal{O}(V)^{\delta}$ of $\mathcal{O}(V)$, where δ is the determinant representation of G on $\bigwedge^n(V)$. For $G \subseteq SL(V)$ the determinant is trivial and the canonical sheaf $\omega_{V/G}$ is equal to $\mathcal{O}_{V/G}$, that is, V/G is *Gorenstein*.

A resolution of singularities $H \rightarrow V/G$ is said to be *crepant* (from "not discrepant") if $\omega_H = \mathcal{O}_H$. The Gorenstein condition on V/G is necessary, but not sufficient, for a crepant resolution to exist. When they do exist, crepant resolutions need not be unique, but they do enjoy a good minimality property: given

$$Z \xrightarrow{f} H \longrightarrow V/G,$$

if Z and H are both crepant resolutions, then f is an isomorphism.

The McKay correspondence is a conjecture to the effect that if H is a crepant resolution of V/G, then the dimension of the cohomology ring of the complex analytic manifold H is equal to the number of conjugacy classes (or irreducible representations) of G. A refinement of this given in [Batyrev and Dais 1996] specifies the dimension of each cohomology group separately. For $V = (\mathbb{C}^2)^n$ and $G = S_n$, the Hilbert scheme $\mathrm{Hilb}^n(\mathbb{C}^2)$ is a crepant resolution, by Lemma 6.16, and the computation of its cohomology in [Ellingsrud and Strømme 1987] (see Lemma 6.7) verifies that the McKay correspondence holds in this case.

Returning to the general G, if $x \in V$ is chosen generically, then its G-orbit has $N = |G|$ elements, and thereby defines a point of the Hilbert scheme $\mathrm{Hilb}^N(V)$. Nakamura defines the *G-Hilbert scheme* $\mathrm{Hilb}^G(V)$ to be the closure of the locus in $\mathrm{Hilb}^N(V)$ consisting of such orbits. There is a canonical morphism $\mathrm{Hilb}^G(V) \rightarrow V/G$. Nakamura's conjecture is as follows.

CONJECTURE 4.7. $\mathrm{Hilb}^G(V)$ *is a crepant resolution of* V/G *whenever one exists.*

The relevance of this to the McKay correspondence is that there are vector bundles on $\mathrm{Hilb}^G(V)$ canonically associated to the irreducible representations of G. It is expected that when the conjecture applies, they will form a basis of the Grothendieck group, and the latter will be isomorphic to the cohomology ring, establishing the McKay correspondence for the resolution $\mathrm{Hilb}^G(V)$.

In our situation, Nakamura's $\mathrm{Hilb}^G(V)$ is our Z_n, and we have already established the factorization

$$Z_n \rightarrow \mathrm{Hilb}^n(\mathbb{C}^2) \rightarrow S^n\mathbb{C}^2 = V/G.$$

If Nakamura's conjecture holds in this case, then both Z_n and $\mathrm{Hilb}^n(\mathbb{C}^2)$ are crepant resolutions, hence they are isomorphic. Conversely if $Z_n \cong \mathrm{Hilb}^n(\mathbb{C}^2)$, then obviously Z_n is a crepant resolution. The equivalent conditions of Theorem 4.5 and Proposition 4.6 are therefore also equivalent to Conjecture 4.7 in the case $V = (\mathbb{C}^2)^n$, $G = S_n$.

5. Frobenius Series

Let $\mathbb{T}$ denote the two dimensional algebraic torus group, $i.e.$, the multiplicative group $\mathbb{C}^* \times \mathbb{C}^*$. It acts algebraically on $\mathbb{C}^2$ by the rule

$$(t, q) \cdot (x, y) = (tx, qy), \quad \text{for } (t, q) \in \mathbb{T}, \; (x, y) \in \mathbb{C}^2.$$

The universal construction of the Hilbert scheme is functorial with respect to automorphisms of $\mathbb{C}^2$, so $\mathbb{T}$ also acts on $\mathrm{Hilb}^n(\mathbb{C}^2)$. This action sends a subscheme $S \subseteq \mathbb{C}^2$ to $(t, q) \cdot S$, so on ideals $I \subseteq \mathbb{C}[x, y]$ it is given by the pullback through the ring endomorphism (t, q), that is,

$$(t, q) \cdot I = I(x/t, y/q).$$

Similarly $\mathbb{T}$ acts on the the schemes U, $U^{\times n}$, $(\mathbb{C}^2)^n$ and X_n, and the various maps between these schemes are $\mathbb{T}$-equivariant.

Observe that a polynomial $p(x, y)$ is doubly homogeneous of degree (r, s) if and only if it is an eigenfunction for the $\mathbb{T}$ action with eigenvalue $t^r q^s$. Hence the bivariate Hilbert series of a finite-dimensional doubly graded space of polynomials D is simply the character of the $\mathbb{T}$-action, as a function of q and t. Similar considerations apply to the Frobenius series when S_n acts, as it is just a generating function for the Hilbert series of the the various S_n-isotypic subspaces.

In the geometric situation we have to deal with $\mathbb{T}$ actions on local rings and sheaves of modules that may be neither graded nor finite-dimensional. For this we need recourse to a "formal" Hilbert series that captures the naive $\mathbb{T}$ character in the finite dimensional case, and extends to the more general setting.

DEFINITION. Let R be the local ring of a scheme X of finite type over $\mathbb{C}$ at a closed point x with maximal ideal $\mathfrak{m}$. Assume X is nonsingular at x and that x is an isolated fixed point for an algebraic action of $\mathbb{T}$ on X. Let M be a finitely generated R-module with an equivariant $\mathbb{T}$ action. Then the *formal Hilbert series* of M is given by

$$\mathcal{H}_M(q, t) = \frac{\sum_i (-1)^i \operatorname{tr}(\operatorname{Tor}_i^R(M, \mathbb{C}), \lambda)}{\det(\mathfrak{m}/\mathfrak{m}^2, 1 - \lambda)}, \quad \text{with } \lambda = (t, q) \in \mathbb{T}. \tag{5.1}$$

To see that the definition is sound, observe first that the modules $\operatorname{Tor}_i^R(M, \mathbb{C})$ and the cotangent space $\mathfrak{m}/\mathfrak{m}^2$ are finite dimensional representations of $\mathbb{T}$, so the trace and determinant in the formula make sense. Since R is regular, the syzygy theorem implies that the sum in the numerator is finite. Since x is an isolated fixed point, the $\mathbb{T}$ action on the cotangent space does not have 1 as an eigenvalue, so the denominator does not vanish identically, but is a product of factors of the form $(1 - t^r q^s)$ with r, s not both zero. Note that $\mathcal{H}_M(q, t)$ is a rational function of q and t.

PROPOSITION 5.1. (1) *If* $0 \to M \to N \to P \to 0$ *is an exact sequence, then*
$$\mathcal{H}_N = \mathcal{H}_M + \mathcal{H}_P.$$

(2) *If M has finite length then $\mathcal{H}_M$ is its ordinary bivariate Hilbert series as a doubly graded space.*

PROOF. (1) follows from the long exact sequence for Tor and the additivity of the trace on exact sequences. In view of (1), (2) reduces to the case $M = \mathbb{C}$. Since R is regular, we have in this case from the Koszul resolution of $\mathbb{C}$ that $\mathrm{Tor}_i(M, \mathbb{C}) \cong M \otimes \wedge^i T_x^*$, where $T_x^* = \mathfrak{m}/\mathfrak{m}^2$ is the cotangent space. Here we have kept the one-dimensional factor M explicit, since $\mathbb{T}$ might not act trivially on it. Let $t^{r_1} q^{s_1}, \ldots, t^{r_d} q^{s_d}$ be the eigenvalues of $\mathbb{T}$ on T_x^*, repeated according to their multiplicities. Then the denominator is

$$\prod_j (1 - t^{r_j} q^{s_j}) = \sum_i (-1)^i e_i(t^{r_1} q^{s_1}, \ldots, t^{r_d} q^{s_d}),$$

while the numerator is the same thing multiplied by the $\mathbb{T}$ character of M. $\square$

When M has an S_n action commuting with the $\mathbb{T}$ action, we can define a formal Frobenius series in an analogous manner. The character of a finite dimensional $S_n \times \mathbb{T}$ module V is now a function $\mathrm{tr}(V, (w, \lambda))$ of both $w \in S_n$ and $\lambda = (t, q) \in \mathbb{T}$, which we can regard as a $\mathbb{Q}(q, t)$-valued character on S_n. With the same definition of the Frobenius map Φ as in (3.3), $\Phi \, \mathrm{ch} \, V$ is now a symmetric function with coefficients in $\mathbb{Q}(q, t)$. Indeed if we identify the $\mathbb{T}$ action with a double grading of V then $\Phi \, \mathrm{ch} \, V$ in this sense is our earlier Frobenius series $\mathcal{F}_V(X; q, t)$.

DEFINITION. With R, X, x, and M as in the definition of formal Hilbert series, and an action of S_n by R-module automorphisms of M commuting with the $\mathbb{T}$ action, the *formal Frobenius series* of M is given by

$$\mathcal{F}_M(X; q, t) = \frac{\sum_i (-1)^i \Phi \, \mathrm{ch}(\mathrm{Tor}_i^R(M, \mathbb{C}))}{\det(\mathfrak{m}/\mathfrak{m}^2, 1 - \lambda)}, \quad \text{with } \lambda = (t, q) \in \mathbb{T}.$$

The formula makes sense for the same reasons as in the Hilbert series case and the analog of Proposition 5.1 holds.

PROPOSITION 5.2. (1) *If $0 \to M \to N \to P \to 0$ is an exact sequence, then $\mathcal{F}_N = \mathcal{F}_M + \mathcal{F}_P$.*

(2) *If M has finite length then $\mathcal{F}_M$ is its ordinary Frobenius series as a doubly graded S_n representation.*

PROOF. The same as the previous Proposition, except that for (2) we reduce to the case that M is an irreducible $S_n \times \mathbb{T}$-equivariant R module. This means M is an irreducible representation of S_n (with trivial $\mathbb{T}$ action) tensored over $\mathbb{C}$ by a trivial R module $\mathbb{C}$ (with some one-dimensional $\mathbb{T}$ action). $\square$

Further properties of the formal Frobenius series are given by the next proposition. The last one is especially important, as it provides a geometric interpretation of plethystic substiution.

PROPOSITION 5.3. *The formal Frobenius series $\mathcal{F}_M$ has the following properties:*

(1) *If the S_n character χ_λ has multiplicity zero in $M/\mathfrak{m}M$, then $\langle s_\lambda, \mathcal{F}_M \rangle = 0$.*

(2) *Let V be a finite-dimensional doubly graded S_n module. Then (tensoring over $\mathbb{C}$) we have $\mathcal{F}_{V \otimes M} = \mathcal{F}_V * \mathcal{F}_M$, where $*$ is the internal product of symmetric functions.*

(3) *Suppose M is an S_n-equivariant S-module, where S is a finite R-algebra with an S_n action. Suppose $x_1, \ldots, x_n \in S$ is an M-regular sequence, that $\mathbb{T}$ acts on the elements x_i by $(t, q) \cdot x_i = tx_i$, and that S_n acts on these elements by permuting them. Then $\mathcal{F}_{M/(x)M}(X; q, t) = \mathcal{F}_M[X(1-t); q, t]$.*

PROOF. (1) Let Θ_λ be the Reynolds operator, the central idempotent in the group algebra of S_n that acts in each representation as the projection on the S_n isotypic component with character χ^λ. If M is a finitely generated R-module with $S_n \times \mathbb{T}$ action then it has a canonical isotypic decomposition

$$M = \bigoplus_{|\lambda|=n} \Theta_\lambda M,$$

and each $\Theta_\lambda M$ is also a finitely generated R module with $S_n \times \mathbb{T}$ action. The decomposition means that the identity functor is naturally isomorphic to the direct sum of the functors $M \mapsto \Theta_\lambda M$. In particular the functors $M \mapsto \Theta_\lambda M$ are exact, and commute with Tor, since the S_n action does.

Now, comparing the definitions of the formal Hilbert and Frobenius series, and using the fact that $\Phi \chi^\lambda = s_\lambda$, we see that

$$\langle s_\lambda, \mathcal{F}_M \rangle = \frac{1}{\chi_\lambda(1)} \mathcal{H}_{\Theta_\lambda M}.$$

By Nakayama's Lemma, if $\Theta_\lambda(M/\mathfrak{m}M) = \Theta_\lambda M/\mathfrak{m}\Theta_\lambda M = 0$, then $\Theta_\lambda M = 0$.

(2) Recall that the internal product $*$ is defined by $p_\lambda * p_\mu = \langle p_\lambda, p_\mu \rangle p_\lambda$, and satisfies the identity

$$\Phi(\chi \otimes \mu) = \Phi(\chi) * \Phi(\mu)$$

for all characters χ, μ. From this it is clear that (2) holds in the case where M has finite length, by Proposition 5.2. For the general case we have $\mathrm{Tor}_i(V \otimes M, \mathbb{C}) = V \otimes \mathrm{Tor}_i(M, \mathbb{C})$, which reduces the identity to the corresponding one with the finite-length modules $\mathrm{Tor}_i(M, \mathbb{C})$ in place of M.

(3) Let V denote the space spanned by elements of the regular sequence x; as an $S_n \times \mathbb{T}$ module it affords the permutation representation of S_n tensored by the one-dimensional representation of $\mathbb{T}$ with character t. Since x is a regular sequence we have the following exact sequence, the Koszul resolution of $M/(x)M$:

$$0 \to M \otimes \textstyle\bigwedge^n V \to \cdots \to M \otimes \textstyle\bigwedge^1 V \to M \otimes \textstyle\bigwedge^0 V \to M/(x)M \to 0, \qquad (5.2)$$

where the tensor products are over $\mathbb{C}$. The maps in the Koszul complex are $S_n \times \mathbb{T}$ equivariant if we take the $\mathbb{T}$ action on $\bigwedge^k V$ to be multiplication by

t^k. From part (1) of Proposition 5.2, together with (2) above, we deduce that
$\mathcal{F}_{M/(x)M} = \mathcal{F}_M * g(X;t)$, where $g(X;t) = \sum_k (-1)^k t^k \Phi \operatorname{ch} \bigwedge^k V$.

Now I claim that $g(X;t) = h_n[X(1-t)]$. To prove this it suffices to show that
for each power-sum p_λ, $|\lambda| = n$, we have

$$\langle g(X;t), p_\lambda \rangle = \langle h_n[X(1-t)], p_\lambda \rangle = p_\lambda[1-t].$$

The second equality here comes from (2.6). For any character χ, it follows
from the definition of the Frobenius map that $\langle \Phi\chi, p_\lambda \rangle = \chi(w)$, where w is a
permutation whose cycle lengths are the parts of λ. The symmetric group acts
on $\bigwedge^k V$ by signed permutations of the basis of monomials $x_{i_1} \wedge \cdots \wedge x_{i_k}$, and
such a monomial is stabilized by w if and only if, for each cycle of w, either all
or none of the corresponding variables appear in the monomial. In that case
$w(x_{i_1} \wedge \cdots \wedge x_{i_k}) = \pm x_{i_1} \wedge \cdots \wedge x_{i_k}$, the sign being $(-1)^{k+r}$ if the variables for r
of the cycles appear. Hence the trace of w on $\bigwedge^k V$ is the sum of $(-1)^{k+r}$ over
all collections of the parts of λ that add up to k, where r is the number of parts
included. It follows that

$$\sum_k (-1)^k t^k (\operatorname{ch} \bigwedge^k V)(w) = \prod_i (1-t^{\lambda_i}) = p_\lambda[1-t].$$

Finally, we have

$$h_n[X(1-t)] * p_\lambda = \langle h_n[X(1-t)], p_\lambda \rangle p_\lambda = p_\lambda[1-t]p_\lambda = p_\lambda[X(1-t)],$$

where the last equality holds because of the identity $p_k[AB] = p_k[A]p_k[B]$. Since
both sides are linear in f it follows that

$$h_n[X(1-t)] * f = f[X(1-t)]$$

for all symmetric functions f. $\square$

The remainder of this section is devoted to the proof of the following theorem.

THEOREM 5.4. *If the $n!$ conjecture holds for μ, then the Frobenius series of R_μ
is given by*

$$\mathcal{F}_{R_\mu} = \tilde{H}_\mu,$$

so the Macdonald positivity conjecture holds for $K_{\lambda\mu}(q,t)$, for all λ.

For the rest of the discussion we *assume the $n!$ conjecture holds for μ*, so that
X_n is locally Cohen–Macaulay at the point Q_μ.

We take R to be the local ring of $\operatorname{Hilb}^n(\mathbb{C}^2)$ at I_μ. Note that $\operatorname{Hilb}^n(\mathbb{C}^2)$ is
nonsingular, and I_μ is a $\mathbb{T}$ fixed point, since this is equivalent to $I_\mu \subseteq \mathbb{C}[x,y]$ being doubly homogeneous, and thus spanned by monomials. As remarked earlier,
the ideals I_μ are the only such ideals, so I_μ is an isolated fixed point.

The local ring S of X_n at Q_μ is a finite R-algebra on which $\mathbb{T}$ acts equivariantly. The symmetric group S_n also acts on S by R-algebra automorphisms,
commuting with the $\mathbb{T}$ action.

Via the map $X_n \to (\mathbb{C}^2)^n$, the coordinates $x_1, y_1, \ldots, x_n, y_n$ on $(\mathbb{C}^2)^n$ define global regular functions on X_n and thus elements of the local ring S.

LEMMA 5.5. *If X_n is Cohen–Macaulay at Q_μ then $y_1, \ldots, y_n$ is a regular sequence.*

PROOF. We are to show that $y = y_1, \ldots, y_n$ cut out a complete intersection in X_n. Since $\dim(X_n) = 2n$ we must show that $\dim V(y) = n$. Now $V(y)$ consists of those points $(I, P_1, \ldots, P_n) \in X_n$ for which all the points P_i lie on the x-axis, which is the same as saying that $V(I)$ lies on the x-axis.

Ellingsrud and Strömme [Ellingsrud and Strømme 1987], studying the cohomology of $\mathrm{Hilb}^n(\mathbb{P}^2)$, constructed a cell decomposition that contains within it a cell decomposition of the subset $H_X \subseteq \mathrm{Hilb}^n(\mathbb{C}^2)$ consisting of points I for which $V(I)$ lies on the x-axis. Every cell in their decomposition of H_X has dimension n.

Since the locus $V(y) \subseteq X_n$ is finite over H_X its dimension is n as well.   □

The local ring S of X_n at Q_μ is not only finite over R, it is free, and $S/\mathfrak{m}S \cong R_\mu$, by the proof of Theorem 4.5, where $\mathfrak{m}$ is the maximal ideal of R. By Nakayama's lemma, S is freely generated as an R module by any subspace D complementary to the ideal $\mathfrak{m}S$. We may choose D to be $S_n \times \mathbb{T}$ stable (in fact, we can choose D to be the space of derivatives D_μ), and then D will have the same Frobenius series as R_μ. This shows that

$$\mathcal{F}_S(X; q, t) = \mathcal{F}_{R_\mu}(X; q, t)\mathcal{H}_R(q, t). \tag{5.3}$$

Incidentally, the quantity $\mathcal{H}_R(q, t)$ has a tantalizing explicit value. In [Haiman 1998] we constructed an explicit system of doubly homogeneous regular local parameters for R. From their degrees one obtains

$$\mathcal{H}_R(q, t) = \frac{1}{\prod_{s \in D(\mu)}(1 - q^{-a(s)}t^{1+l(s)}) \prod_{s \in D(\mu)}(1 - q^{1+a(s)}t^{-l(s)})},$$

where the *arms* $a(s)$ and *legs* $l(s)$ are as in (2.16). After the replacement $q \mapsto q^{-1}$ the first factor in the denominator is exactly the normalizing factor for the definition of Macdonald's integral forms. This coincidence is a typical example of the links between the numerology associated with Macdonald polynomials and geometrically significant quantities attached to the Hilbert scheme.

Now consider the ring $S/(y)$. Just as S is generated as an R module by any subspace representing $S/\mathfrak{m}S$, $S/(y)$ is generated (no longer freely, however) by representatives of $S/((y)+\mathfrak{m}S) = R_\mu/(y)$. This last space can be identified with the component of R_μ, or of D_μ, homogeneous of degree zero in y. As mentioned in the proof of Proposition 3.7, this space is well-understood, and its Frobenius series is

$$t^{n(\mu)}Q_\mu[X/(1-t^{-1}); t^{-1}] = \sum_\lambda t^{n(\mu)}K_{\lambda\mu}(t^{-1})s_\lambda,$$

where Q_μ is a Hall–Littlewood polynomial and $K_{\lambda\mu}(t)$ denotes the classical one-variable Kostka coefficient. In particular, since $K_{\lambda\mu}(t) = 0$ unless $\lambda \geq \mu$, the space $R_\mu/(\boldsymbol{y})$ contains only those S_n representations χ^λ with $\lambda \geq \mu$. It follows, by Proposition 5.3, part (1), that

$$\mathcal{F}_{S/(\boldsymbol{y})} \in \mathbb{Q}(q,t)\{s_\lambda : \lambda \geq \mu\}.$$

Now using Proposition 5.3, part (3), with q in place of t, this implies

$$\mathcal{F}_S[X(1-q)] \in \mathbb{Q}(q,t)\{s_\lambda : \lambda \geq \mu\},$$

and hence, by (5.3),

$$\mathcal{F}_{R_\mu}[X(1-q)] \in \mathbb{Q}(q,t)\{s_\lambda : \lambda \geq \mu\}.$$

Everything we have done applies symmetrically, with $\boldsymbol{x}$ in place of $\boldsymbol{y}$ and t in place of q, to show that also

$$\mathcal{F}_{R_\mu}[X(1-t)] \in \mathbb{Q}(q,t)\{s_\lambda : \lambda \geq \mu'\}.$$

Finally, since R_μ affords the regular representation (this follows from the $n!$ conjecture by Proposition 4.4), its only S_n invariants are the constants, so

$$\langle \mathcal{F}_{R_\mu}, s_{(n)} \rangle = 1.$$

By Proposition 2.6 these three conditions imply that $\mathcal{F}_{R_\mu} = \tilde{H}_\mu$, and the proof of Theorem 5.4 is complete.

6. The Ideals J and J^m

Let $R = \mathbb{C}[\boldsymbol{x}, \boldsymbol{y}] = \mathbb{C}[x_1, y_1, \ldots, x_n, y_n]$, and let J denote the ideal in R generated by all S_n-alternating polynomials, that is, by the polynomials $\Delta_D(\boldsymbol{x}, \boldsymbol{y})$ of (3.1). Since any alternating polynomial must vanish whenever two of the points (x_i, y_i) and (x_j, y_j) coincide, it follows that

$$J \subseteq \bigcap_{i<j} (x_i - x_j, \, y_i - y_j), \tag{6.1}$$

and more generally

$$J^m \subseteq \bigcap_{i<j} (x_i - x_j, \, y_i - y_j)^m. \tag{6.2}$$

We shall denote the ideal on the right hand side of (6.2) by $J^{(m)}$; it is the m-th *symbolic power* of $J^{(1)}$.

CONJECTURE 6.1. *We have $J^m = J^{(m)}$ for all m, i.e., we have equality in (6.2).*

Here is a result that at least gives the impression of reducing the conjecture to something simpler.

PROPOSITION 6.2. *Suppose that for all $n \geq 3$, $(x_1 - x_2, \, x_2 - x_3)$ is a regular sequence for the R module J^m. Then $J^m = J^{(m)}$.*

PROOF. The proof is by induction on n. Note that for $n = 1$ and $n = 2$ we trivially have $J^m = J^{(m)}$, and we have the remaining cases through $n - 1$ by induction.

First consider the situation locally at a point $P \in \mathbb{C}[\boldsymbol{x}, \boldsymbol{y}]$ where the (x_i, y_i) are not all equal. Without loss of generality we can assume that none of $(x_1, y_1), \ldots, (x_r, y_r)$ is equal to any of $(x_{r+1}, y_{r+1}), \ldots, (x_n, y_n)$. In the local ring R_P, the differences $x_i - x_j$ and $y_i - y_j$ are invertible whenever i is in the first group and j in the second. Hence $J^{(m)}$ reduces locally to the product of the ideals $J^{(m)}$ in the first r indices and the last $n - r$ separately.

Less obvious, but still true, is that J, and hence J^m, decomposes similarly. To see this, let g be a generator of $J' = J(x_1, y_1, \ldots, x_r, y_r) J(x_{r+1}, y_{r+1}, \ldots, x_n, y_n)$, alternating in the first r and last $n - r$ indices, i.e., the subgroup $S_r \times S_{n-r} \subseteq S_n$ acts on g by the sign character. Let h be any polynomial that belongs to the localization J_Q at every point $Q \neq P$ in the S_n orbit of P, but does not vanish at P. Now $f = \text{Alt}\, gh$ belongs to J. The terms in the alternation corresponding to elements $w \in S_n$ that do not stabilize P belong to J_P, by construction of h. Since g is already alternating with respect to the stabilizer of P, the remaining terms sum to $g \sum_{wP=P} wh$, and the sum here is invertible in R_P. This shows $g \in J_P$, and so $J'_P \subseteq J_P$. The reverse inclusion $J \subseteq J'$ is clear.

Using the induction hypothesis, we conclude that $J^m = J^{(m)}$ locally outside the locus V where all n points coincide. Now since $x_1 - x_2$ and $x_2 - x_3$ belong the ideal of V, our hypothesis implies $\text{depth}_V\, J^m \geq 2$, and the local cohomology exact sequence for the sheaf of ideals $\tilde{J}^m$ associated to J^m gives

$$0 = H^0_V(\tilde{J}^m) \to H^0(\mathbb{C}^2, \tilde{J}^m) = J^m \to H^0(U, \tilde{J}^m) \to H^1_V(\tilde{J}^m) = 0, \qquad (6.3)$$

where $U = \mathbb{C}^2 \setminus V$. Thus $J^m = H^0(U, \tilde{J}^m) = H^0(U, \widetilde{J^{(m)}})$. The latter is the ideal of all polynomials whose restrictions to U belong locally to $J^{(m)}$, so we have shown $J^m \supseteq J^{(m)}$. As we had $J^m \subseteq J^{(m)}$ to begin with, we must have $J^m = J^{(m)}$. $\qquad\qquad\qquad\qquad\qquad\qquad\qquad\qquad\qquad\qquad\qquad\qquad\qquad\quad\square$

There is of course nothing special about the choice of $x_1 - x_2$ and $x_2 - x_3$ in the above Proposition; it's just an explicit way to guarantee that $\text{depth}_V\, J^m \geq 2$. This would also follow if $\text{depth}_V(R/J^m) \geq 1$, which means the ideal of V contains an element that is a non-zero-divisor modulo J^m. Note, by the way, that the proof of Proposition 6.2 works equally well with more than two sets of variables.

Some explorations we have done for small values of n using the computer algebra system Macaulay [Bayer and Stillman 1989] suggest the following conjecture.

CONJECTURE 6.3. *If J denotes the ideal generated by the S_n alternants in $\mathbb{C}[\boldsymbol{x}, \boldsymbol{y}, \ldots, \boldsymbol{z}]$, for any number of sets of n variables, and V is the locus where all the points $(x_i, y_i, \ldots, z_i)$, $(x_j, y_j, \ldots, z_j)$ coincide, then $x_1 - x_2$, $x_2 - x_3$, $\ldots$, $x_{n-1} - x_n$ is a maximal J^m-regular sequence in the ideal of V, for all m. In particular, $\text{depth}_V\, J^m = n - 1$.*

We remark that if the sequence x in question is regular then it is maximal: modulo $(x)J$, the Vandermonde determinant $v(x)$ is annihiliated by the ideal of V, so $\text{depth}_V J/(x)J = 0$.

The relevance of all this to X_n and the $n!$ conjecture is given by the following proposition.

PROPOSITION 6.4. *The isospectral Hilbert scheme X_n is the blowup $\operatorname{Proj} R[tJ]$ of $(\mathbb{C}^2)^n$ at the ideal J.*

PROOF. We only outline the proof, as the analogous result for the ordinary Hilbert scheme $\operatorname{Hilb}^n(\mathbb{C}^2)$ was given in [Haiman 1998], and the proof carries over to X_n with only superficial modifications.

First, one shows that the pullback of the ideal J to X_n becomes locally principal. In fact, it can be identified with $\bigwedge^n B$, where B is the tautological bundle (of $\operatorname{Hilb}^n(\mathbb{C}^2)$, lifted to X_n). By the universal property of the blowup, this gives a morphism from X_n to $\operatorname{Proj} R[tJ]$. This map is projective and generically an isomorphism, so it's surjective. To show it's also a closed embedding, we have to show that all regular functions on X_n are pulled back from $\operatorname{Proj} R[tJ]$.

The regular functions on X_n are the coodinates x_i, y_i, which come from R, and the lifts of regular functions on the Hilbert scheme. But the proof of the result for $\operatorname{Hilb}^n(\mathbb{C}^2)$ shows the latter are generated by fractions of the form Δ_D/Δ_μ, that are (local) regular functions on $\operatorname{Proj} R[tJ]$. $\square$

In d sets of variables, the above proposition applies as well to the isospectral Hilbert scheme of points in $\mathbb{C}^d$, with an important qualification. For general d these Hilbert schemes are not irreducible and even have components of dimension greater than dn [Iarrobino 1972]. What the blowup construction gives is the component that is the closure of the locus corresponding to reduced subschemes of n distinct points in $\mathbb{C}^d$. We suspect that this *generic component* may have good geometric properties, indeed may be Cohen–Macaulay and even Gorenstein for all d. Note that this would not imply the $n!$ conjecture in more sets of variables, since the generic component of $\operatorname{Hilb}^n(\mathbb{C}^d)$ may be singular, and thus the isospectral Hilbert scheme need not be flat over it.

In the remainder of this section we prove the following reduction of the $n!$ conjecture to Conjecture 6.3.

THEOREM 6.5. *If Conjecture 6.3 holds in two sets of variables for all m and n, then X_n is Cohen–Macaulay and normal for all n.*

The proof is by induction on n, using geometric properties of the *nested Hilbert scheme* $H^{n-1,n}$ to be defined shortly. We employ a local cohomology argument in the same spirit as the proof of Proposition 6.2. For this we need a large open set where we may assume the result by induction, which the next lemma provides.

LEMMA 6.6. *Let $P = (I, P_1, \ldots, P_n)$ be a point of X_n. Let the distinct points among $P_1, \ldots, P_n$ be $Q_1, \ldots, Q_k$, with multiplicities $r_1, \ldots, r_k$. Then in X_n there is a neigborhood of P isomorphic to an open set in the product $X_{r_1} \times \cdots \times X_{r_k}$.*

PROOF. Without loss of generality we can assume $Q_1 = P_1 = \cdots = P_{r_1}$, $Q_2 = P_{r_1+1} = \cdots = P_{r_1+r_2}$, and so on. For our neighborhood of P we can take the preimage in X_n of the open set $U \subseteq (\mathbb{C}^2)^n$ of points where the only coincidences $P_i = P_j$ that occur have i, j within one of these k consecutive blocks. Then the result is clear from Proposition 6.4, together with the product decomposition, valid on U, of the ideal J as $J_{1,\ldots,r_1} J_{r_1+1,\ldots,r_1+r_2} \cdots$ from the proof of Proposition 6.2. □

For some dimension arguments below we will need the following results.

LEMMA 6.7. *There is a decomposition of $\mathrm{Hilb}^n(\mathbb{C}^2)$ into locally closed affine cells C_μ, such that every point of C_μ contains I_μ in the closure of its $\mathbb{T}$ orbit, and $\dim C_\mu = n + l(\mu)$, where $l(\mu)$ is the number of parts of μ.*

LEMMA 6.8. *There is a decomposition of the zero-fiber $H_0^n = \tau^{-1}(0) \subseteq \mathrm{Hilb}^n(\mathbb{C}^2)$ into locally closed affine cells C_μ', such that every point of C_μ' contains I_μ in the closure of its $\mathbb{T}$ orbit, and $\dim C_\mu' = l(\mu) - 1$.*

PROOF. See [Ellingsrud and Strømme 1987]. □

LEMMA 6.9. *Let G_r be the (closed) locus of ideals $I \in \mathrm{Hilb}^n(\mathbb{C}^2)$ for which some point of $V(I)$ has multiplicity at least r. Then G_r has codimension $r - 1$, and has only one irreducible component of maximal dimension.*

PROOF. It is known [Briançon 1977] that the zero-fiber $H_0^n = \tau^{-1}(0)$ is irreducible of dimension $n - 1$. The locus where all the points coincide is just the product of $\mathbb{C}^2$ (for the choice of origin) by H_0^n, so it is irreducible of dimension $n + 1$.

By Lemma 6.6, it follows that the (locally closed) locus where the multiplicities are $r_1, \ldots, r_k$ has dimension $\sum_i (r_i + 1) = n + k$ and codimension $n - k$. If one multiplicity is at least r, this codimension is at least $r - 1$, with equality only for multiplicities $r, 1, \ldots, 1$. Again by Lemma 6.6, the locus in X_n where $P_1 = \cdots = P_r$, and the other points are distinct from P_1 and each other, is irreducible. It surjects on the locus in $\mathrm{Hilb}^n(\mathbb{C}^2)$ where the multiplicities are $r, 1, \ldots, 1$, so the latter is irreducible as well. □

As a step toward the Cohen–Macaulay property we need normality results for X_n and U_n.

DEFINITION. An ideal $I \in \mathrm{Hilb}^n(\mathbb{C}^2)$ is *curvilinear* if the local rings $\mathcal{O}_{S,P}$ have embedding dimension 1, i.e., their maximal ideals are principal.

This is equivalent to $S = V(I)$ being a subscheme of a smooth curve in $\mathbb{C}^2$, whence the name.

LEMMA 6.10. *The locus W of curvilinear ideals $I \in \operatorname{Hilb}^n(\mathbb{C}^2)$ is open and equal to $\bigcup_z W_z$, where $z = ax + by$ is a linear form, and W_z is the open set of ideals I such that $\{1, z, \ldots, z^{n-1}\}$ is a basis of $\mathbb{C}[x, y]/I$.*

PROOF. If I is curvilinear, then for a generically chosen linear form z, the values $z(P)$ at distinct points $P \in V(I)$ will be distinct, and for each P, $z - z(P)$ will be a local parameter generating the maximal ideal $\mathfrak{m}_P \subseteq \mathcal{O}_{S,P}$. This implies that, as a $\mathbb{C}[z]$ module and hence as a ring,

$$\mathbb{C}[x, y]/I \cong \mathbb{C}[z] \Big/ \prod_P (z - z(P))^{r_P}, \tag{6.4}$$

where r_P is the multiplicity of P, and therefore $\{1, z, \ldots, z^{n-1}\}$ is a basis of $\mathbb{C}[x, y]/I$. Conversely, if $\{1, z, \ldots, z^{n-1}\}$ is a basis, then z generates $\mathbb{C}[x, y]/I$, and I contains a monic polynomial of degree n in z, so (6.4) holds and I is curvilinear. $\qquad\square$

LEMMA 6.11. *The universal scheme U over $\operatorname{Hilb}^n(\mathbb{C}^2)$ is Cohen–Macaulay and normal.*

PROOF. U is Cohen–Macaulay because it is flat over $\operatorname{Hilb}^n(\mathbb{C}^2)$. Hence it is normal if its singular locus has codimension at least 2.

Now I claim that over the curvilinear locus W, U is nonsingular. After a linear transformation of $\mathbb{C}^2$, we can restrict to W_x. For $I \in W_x$, we have y and x^n congruent mod I to unique polynomials of degree at most $n - 1$ in x, so I contains elements

$$\big(x^n - e_1 x^{n-1} + e_2 x^{n-2} - \cdots + (-1)^n e_n, \ y - (a_{n-1} x^{n-1} + \cdots + a_1 x + a_0)\big), \tag{6.5}$$

where the parameters e_i and a_i are regular functions of I on W_x. On the other hand these two equations clearly generate a complete intersection ideal $I \subseteq \mathbb{C}[x, y]$ modulo which $1, x, \ldots, x^{n-1}$ are a basis, so they determine I. This exhibits W_x explicitly as an affine cell with coordinates e_i, a_i. (As a matter of fact, W_x is the cell $C_{(1^n)}$ in Lemma 6.7.) Moreover, regarded as equations on $W_x \times \mathbb{C}^2 = \operatorname{Spec}\mathbb{C}[e, a, x, y]$, equations (6.5) define the universal family U.

Viewing the first equation as eliminating e_n and the second as eliminating y we conclude that the open subset of U lying over W_x is

$$\operatorname{Spec}\mathbb{C}[x, e_1, \ldots, e_{n-1}, a_0, \ldots, a_{n-1}],$$

and in particular is nonsingular. It follows that U is nonsingular over the whole curvilinear locus.

Finally, if I is not curvilinear, then some point of $V(I)$ has to have multiplicity at least 3. By Lemma 6.9 this occurs only on a locus of codimension 2. Since U is finite over $\operatorname{Hilb}^n(\mathbb{C}^2)$, its singular locus also has codimension at least 2. $\qquad\square$

LEMMA 6.12. *If Conjecture 6.3 holds for all m and any given n, then X_n is normal.*

PROOF. Recall that an ideal J in a normal domain R is said to be *integrally closed* if every element $x \in R$ satisfying

$$x^n \in Jx^{n-1} + J^2 x^{n-1} + \cdots + J^n \qquad (6.6)$$

already belongs to J. The above condition for J^m is equivalent to saying that $t^m x$ belongs to the integral closure of $R[tJ]$ in $R[t]$, so all the ideals J^m are integrally closed if and only if $R[tJ]$ is normal.

For our R and J, Conjecture 6.3 implies $J^m = J^{(m)}$, by Proposition 6.2. It is well-known that the powers of an ideal generated by a regular sequence are integrally closed, and it is obvious that an intersection of integrally closed ideals is integrally closed, so $J^{(m)}$ is integrally closed.

This shows that $R[tJ]$ is normal, so $X_n = \operatorname{Proj} R[tJ]$ is, by definition, *arithmetically normal* in the given projective embedding. In particular it is normal. $\square$

Now we come to the geometric construction that supplies the inductive machinery.

DEFINITION. The *nested Hilbert scheme* $H^{n-1,n}$ is the subvariety of pairs

$$H^{n-1,n} = \{(I_{n-1}, I_n) : I_{n-1} \supseteq I_n\} \subseteq \operatorname{Hilb}^{n-1}(\mathbb{C}^2) \times \operatorname{Hilb}^n(\mathbb{C}^2).$$

PROPOSITION 6.13 [Cheah 1998; Tikhomirov 1992]. *The nested Hilbert scheme $H^{n-1,n}$ is irreducible of dimension $2n$ and nonsingular.*

If (I_{n-1}, I_n) is a point of $H_{n-1,n}$ then the corresponding subscheme $V(I_{n-1}) \subseteq \mathbb{C}^2$ is a subscheme of $V(I_n)$, so the multiset $\tau(I_n)$ contains $\tau(I_{n-1})$ along with one additional point, or else with the multiplicity of one of the original points increased by 1. So if the spectrum of I_{n-1} is $(x_1, y_1), \ldots, (x_{n-1}, y_{n-1})$ then that of I_n is $(x_1, y_1), \ldots, (x_{n-1}, y_{n-1}), (x_n, y_n)$ for a *distinguished point* (x_n, y_n). Now both the S_{n-1} invariants $p_{h,k}(x_1, y_1, \ldots, x_{n-1}, y_{n-1})$ and the S_n invariants $p_{h,k}(x_1, y_1, \ldots, x_n, y_n)$ are regular functions on $H^{n-1,n}$, hence so are $x_n = p_1(x_1, \ldots, x_n) - p_1(x_1, \ldots, x_{n-1})$ and similarly y_n. This means we have a morphism

$$H^{n-1,n} \to \mathbb{C}^2 = \operatorname{Spec} \mathbb{C}[x_n, y_n],$$

mapping a pair to its distinguished point. Of course $(x_n, y_n) \in V(I_n)$, and by suitable choice of I_{n-1}, given I_n, the distinguished point can be any point of $V(I_n)$. Hence the combined map $H^{n-1,n} \to \mathbb{C}^2 \times \operatorname{Hilb}^n(\mathbb{C}^2)$ factors $H^{n-1,n} \to \operatorname{Hilb}^n(\mathbb{C}^2)$ through a surjective morphism

$$\alpha: H^{n-1,n} \to U \qquad (6.7)$$

to the universal scheme U over $\operatorname{Hilb}^n(\mathbb{C}^2)$. Where the n points are distinct, this map is locally an isomorphism, so it is birational.

The above map and the map $H^{n-1,n} \to \operatorname{Hilb}^n(\mathbb{C}^2)$ are projective. In fact, given I_n, I_{n-1} is determined by its single generator mod I_n, so $H^{n-1,n}$ is a subvariety of the projective space bundle $\mathbb{P}(B)$, where B is the tautological bundle over $\text{Hilb}^n(\mathbb{C}^2)$. More precisely, given I_n and $P = (x_n, y_n)$, the possible ideals I_{n-1} correspond one-to-one with length-1 ideals in the local ring $\mathcal{O}_{S,P}$ of $V(I_n)$ at P. Such ideals are simply the 1-dimensional subspaces of the *socle*, $\text{soc}\,\mathcal{O}_{S,P} = (0 : \mathfrak{m}_P)$. Thus each fiber of the map (6.7) is a projective space $\mathbb{P}(\text{soc}\,\mathcal{O}_{S,P})$, of dimension $\dim \text{soc}\,\mathcal{O}_{S,P} - 1$.

LEMMA 6.14. *If the dimension of the fiber of* $\alpha\colon H^{n-1,n} \to U$ *over a point* $(I, P) \in U$ *is* d, *then the multiplicity of* P *is at least* $\binom{d+2}{2}$.

PROOF. We are to show that if $T = \mathbb{C}[x,y]/I$ is a local ring of finite length, with $d+1 = \dim \text{soc}\,T$, then the length of T is at least $\binom{d+2}{2}$. This is equivalent to showing that if there exists a fiber of the map $H^{n-1,n} \to U$ with dimension at least d, then $n \geq \binom{d+2}{2}$. Now by the upper-semicontinuity of fiber dimension, and the fact (Lemma 6.7) that every I has one of the ideals I_μ in the closure of its $\mathbb{T}$ orbit, the maximal fiber dimension must occur at some I_μ. There we see immediately that the dimension of the socle is the number of corners of μ, so the result reduces to the fact that if the diagram of a partition of n has k corners then $n \geq \binom{k+1}{2}$. $\square$

LEMMA 6.15. *The map* $\alpha\colon H^{n-1,n} \to U$ *restricts to an isomorphism outside a locus of codimension 2 in* $H^{n-1,n}$.

PROOF. First note that the 2-dimensional and higher fibers of α form a locus of codimension at least 3, by Lemmas 6.9 and 6.14, since for $d \geq 2$ we have $\binom{d+2}{2} - 1 - d \geq 3$.

For I_n curvilinear, $\text{soc}\,\mathcal{O}_{S,P}$ is always 1-dimensional, so α restricts to an bijective morphism on the curvilinear locus, which is then an isomorphism by Zariski's theorem and Lemma 6.11.

As noted in the proof of Lemma 6.11, the noncurvilinear locus in $\text{Hilb}^n(\mathbb{C}^2)$ is contained in G_3, so its codimension is at least 2. If it had codimension exactly 2 then it would contain the whole codimension 2 component of G_3, and in particular, every point where the multiplicities are $3, 1, \ldots, 1$. But there are clearly curvilinear subschemes with these multiplicities, so the co-dimension of the noncurvilinear locus is at least 3. If its preimage in $H^{n-1,n}$ had a component of codimension 1, then every fiber in that component would have dimension at least 2, contradicting the observation made at the outset. Hence the locus where I_n is noncurvilinear has codimension at least 2 in $H^{n-1,n}$, and α restricts to an isomorphism outside it. $\square$

LEMMA 6.16. *The canonical sheaf* ω *of regular* $2n$-*forms on* $\text{Hilb}^n(\mathbb{C}^2)$ *is trivial, i.e., isomorphic to the structure sheaf* $\mathcal{O}$.

PROOF. We again use the description in the proof of Lemma 6.11 of the open set W_x of ideals I modulo which $1, x, \ldots, x^{n-1}$ is a basis: W_x is an affine $2n$-cell with coordinates $e_1, \ldots, e_n, a_0, \ldots, a_{n-1}$, in terms of which I is generated at each point by equations (6.5). On the locus where I is the ideal of a reduced subscheme $S = \{(x_1, y_1), \ldots, (x_n, y_n)\}$, the first equation

$$x^n - e_1 x^{n-1} + e_2 x^{n-2} - \cdots + (-1)^n e_n$$

must be the polynomial $\prod_i (x - x_i)$, so the parameters e_i are the elementary symmetric functions $e_i(x)$. From the second equation,

$$y - (a_{n-1} x^{n-1} + \cdots + a_1 x + a_0),$$

the a_i are the coefficients of the interpolating polynomial $\phi_a(x)$ that satisfies $y_i = \phi_a(x_i)$ when the x_i's are all distinct.

It is well-known that the elementary symmetric functions satisfy $de_1 \wedge \cdots \wedge de_n = v(x) dx_1 \wedge \cdots \wedge dx_n$, where $v(x)$ is the Vandermonde determinant. In particular, since the $e_i(x)$ and $e_i(y)$ are global regular functions on $\mathrm{Hilb}^n(\mathbb{C}^2)$, and $v(x)v(y)$ is S_n invariant, $dx\, dy = dx_1 \wedge \cdots \wedge dx_n \wedge dy_1 \wedge \cdots \wedge dy_n = v(x)^{-1} v(y)^{-1} de_1(x) \wedge \cdots de_n(x) de_1(y) \wedge \cdots de_n(y)$ makes sense as a rational $2n$-form on $\mathrm{Hilb}^n(\mathbb{C}^2)$.

Moreover, the equations $y_i = \phi_a(x_i)$ say that the vector $(y_1, \ldots, y_n)$ is the product of $(a_{n-1}, \ldots, a_0)$ by the Vandermonde matrix, so $da_0 \wedge \cdots \wedge da_{n-1} = dy_1 \wedge \cdots \wedge dy_n / v(x)$, and hence

$$de_1 \wedge \cdots \wedge de_n \wedge da_{n-1} \wedge \cdots \wedge da_0 = dx\, dy.$$

In particular, the rational $2n$-form $dx\, dy$ is regular and has no zeroes on W_x. But $dx\, dy$ is invariant under the action of SL_2 on $\mathbb{C}^2$, so it follows that $dx\, dy$ is regular and nowhere vanishing on every W_z. Since we have already seen that the complement of $\bigcup_z W_z$ has codimension greater than 1, it follows that $dx\, dy$ is regular everywhere and vanishes nowhere, which shows that $\omega = 0$. $\square$

LEMMA 6.17. *The canonical sheaf ω of regular $2n$-forms on $H^{n-1,n}$ is isomorphic to L^{-1}, where L is the line bundle defined by the exact sequence*

$$0 \to L \to B_n \to B_{n-1} \to 0 \tag{6.8}$$

induced on the tautological bundles over $H^{n-1,n}$ by the containment $I_{n-1} \supseteq I_n$.

PROOF. We are to show that the line bundle $L\omega$ is trivial, and it suffices to do this on an open set whose complement has codimension ≥ 2. By Lemma 6.15, we can use the open set where I_n is curvilinear and the map $H^{n-1,n} \to U$ restricts to an isomorphism, which means we can verify it on the curvilinear locus in U.

By Lemma 6.16 and duality for the finite, flat morphism $\pi: U \to \mathrm{Hilb}^n(\mathbb{C}^2)$, we have $\pi_*(L\omega_U) \cong (\pi_* L^{-1})^* \omega_H = (\pi_* L^{-1})^*$, where $\omega_H = 0$ is the canonical sheaf on the Hilbert scheme. So we have to show that $\pi_* L^{-1} \cong B^*$ as a B-module, since $\pi_* \mathcal{O}_U = B$.

Now let's examine L as a subbundle of $\pi^* B$ on the curvilinear locus in U. To avoid confusion, since we already have regular functions x, y on U, we write x', y' for the variables of B, so the fiber of $\pi^* B$ at (I, P) is $\mathbb{C}[x', y']/I(x', y')$. Then the fiber of L at (I, P) is the socle of the summand $\mathcal{O}_{S,P}$ in $\mathbb{C}[x', y']/I(x', y')$, since the generator of I_{n-1} mod I belongs to this socle, which is 1-dimensional. Equivalently, the fiber of L is the ideal $(0 : \mathfrak{m}_P(x', y')) = (0 : (x' - x, y' - y))$ in $\pi^* B(I)$. Dualizing this, we see that $L^{-1} = \pi^* B^*/(x' - x, y' - y)\pi^* B^*$, or $\pi^* B^* \otimes_{\mathbb{C}[x', y']} \mathcal{O}_U$, where $\mathbb{C}[x', y']$ acts on $\mathcal{O}_U$ through the homomorphism $x' \mapsto x, y' \mapsto y$.

Now $\pi_* \pi^* B^* = B \otimes B^*$, so $\pi_* L^{-1} = (B \otimes B^*)/(x' - x, y' - y)(B \otimes B^*)$, where x, y act through B and x', y' act through B^*. But this is just another way of writing $\pi_* L^{-1} = B \otimes_B B^* = B^*$. $\qquad\square$

LEMMA 6.18. *If X_n is Cohen–Macaulay, then it is Gorenstein, and its canonical sheaf ω is the line bundle $\mathcal{O}(-1)$, where $\mathcal{O}(1) = \bigwedge^n B$.*

PROOF. By Proposition 4.2, $X_n = \operatorname{Spec} P$ as a scheme affine over $\operatorname{Hilb}^n(\mathbb{C}^2)$, where $P = \sigma_* \mathcal{O}_{X_n}$ is the image of the sheaf homomorphism $\phi: B^{\otimes n} \to (B^{\otimes n})^* \otimes \mathcal{O}(1)$ in (4.7). If X_n is Cohen–Macaulay this is a vector-bundle homomorphism. By construction, this description of P means the bilinear pairing of vector bundles $P \otimes P \to \mathcal{O}(1)$, given by multiplication followed by alternation, is nondegenerate, so $P^* \cong P \otimes \mathcal{O}(-1)$. By duality for the flat, finite morphism $\sigma: X_n \to \operatorname{Hilb}^n(\mathbb{C}^2)$, the canonical sheaf ω_{X_n} is the sheaf associated to the P module sheaf $P^* \otimes \omega_H$, which is the same as $P \otimes \mathcal{O}(-1)$, by Lemma 6.16. This shows $\omega_{X_n} = \mathcal{O}(-1)$, and since this is a line bundle, X_n is Gorenstein (by definition). $\qquad\square$

We now have all the technical ingredients we need to prove Theorem 6.5. *From this point on we assume Conjecture 6.3 holds,* and we assume X_{n-1} is Cohen–Macaulay by induction. We shall also assume $n \geq 4$. There is no harm in this since it is trivial to verify the $n!$ conjecture for $n \leq 3$.

To carry the induction forward we introduce the fiber product Y_n indicated by the diagram:

$$
\begin{array}{ccc}
Y_n & \longrightarrow & H^{n-1,n} \\
\downarrow & & \downarrow \\
X_{n-1} & \longrightarrow & \operatorname{Hilb}^{n-1}(\mathbb{C}^2).
\end{array}
\tag{6.9}
$$

Since the bottom arrow is flat (by induction), so is the top one. Since Y_n is flat over $H^{n-1,n}$ and generically reduced, it is reduced. A point of Y_n is a pair of ideals $(I_{n-1}, I_n) \in H^{n-1,n}$, together with the spectrum $(x_1, x_n), \ldots, (x_{n-1}, x_{n-1})$ of I_{n-1} in some order. The coordinates of the remaining point (x_n, y_n) in the spectrum of I_n are regular functions on $H^{n-1,n}$ and hence on Y_n, so we obtain a morphism

$$
f: Y_n \to X_n
\tag{6.10}
$$

sending $(I_{n-1}, I_n, \boldsymbol{x}, \boldsymbol{y})$ to $(I_n, \boldsymbol{x}, \boldsymbol{y})$. Note that f is projective, since the map $H^{n-1,n} \to \operatorname{Hilb}^n(\mathbb{C}^2)$ is. Over the locus where the points (x_i, y_i) are all distinct, $X_{n-1} \to (\mathbb{C}^2)^{n-1}$ and $H^{n-1,n} \to \mathbb{C}^2 \times \operatorname{Hilb}^{n-1}(\mathbb{C}^2)$ restrict to isomorphisms, hence so does $Y_n \to X_n \to (\mathbb{C}^2)^n$. The locus where some two points coincide is a proper closed subvariety of the irreducible variety $H^{n-1,n}$. Since Y_n is flat over $H^{n-1,n}$ it cannot have a component contained in the coincidence locus, so the noncoincidence locus is dense in Y_n. This shows Y_n is irreducible and the morphism f is birational.

On Y_n we have, by pullback from $H^{n-1,n}$, the two tautological bundles B_{n-1} and B_n. Set $\mathcal{O}(k,l) = (\bigwedge^{n-1} B_{n-1})^k (\bigwedge^n B_n)^l$. By Lemma 6.17 the canonical sheaf on $H^{n-1,n}$ is $\mathcal{O}(1,-1)$ in this notation. By Lemmas 6.16 and 6.18 the relative canonical sheaf of Y_n over $H^{n-1,n}$, which is pulled back from that of X_{n-1} over $\mathrm{Hilb}^{n-1}(\mathbb{C}^2)$, is $\mathcal{O}(-1,0)$. Hence the canonical sheaf ω_{Y_n} is $\mathcal{O}(0,-1)$. In particular, it is a pullback from X_n.

Now we are going to prove that for the derived functor of the pushforward we have $Rf_*\mathcal{O}_{Y_n} = \mathcal{O}_{X_n}$. Since $\omega_{Y_n} = f^*\mathcal{O}_{X_n}(-1)$ this also proves $Rf_*\omega_{Y_n} = \mathcal{O}_{X_n}(-1)$. By duality for the projective morphism f_* we conclude that the sheaf $\mathcal{O}_{X_n}(-1)$ is the dualizing complex on X_n, so X_n is Cohen–Macaulay. (This also shows X_n is Gorenstein with $\omega_{X_n} = \mathcal{O}(-1)$, so we could have made Lemma 6.18 part of the induction.)

Since f is proper and birational, and X_n is normal by Lemma 6.12, we have $f_*\mathcal{O}_{Y_n} = \mathcal{O}_{X_n}$. We have to prove that $R^i f_*\mathcal{O}_{Y_n} = 0$ for all $i > 0$. Now the fibers of f are also fibers of the map $H^{n-1,n} \to U$, and thus the fiber dimensions d are bounded by $\binom{d+2}{2} \leq n$, by Lemma 6.14. In particular, since we are assuming $n \geq 4$, this implies $d < n-2$ (exercise for the reader). It follows that $R^i f_*\mathcal{O} = 0$ for $i \geq n-2$.

For $i < n-2$ we use the following lemma.

LEMMA 6.19. *Let $f: Y \to X$ be a morphism and let $x_1, \ldots, x_k$ be global regular functions on X (and so also on Y). Suppose that $\boldsymbol{x}$ is an $\mathcal{O}$-regular sequence at every point of $V(\boldsymbol{x})$, both in X and in Y. Let $U = X \setminus V(\boldsymbol{x})$, $W = f^{-1}(U)$, and $f' = f|_W$. Then $Rf'_*\mathcal{O}_Y = \mathcal{O}_X$ implies $R^i f_*\mathcal{O}_Y = 0$ for $0 < i < k-1$.*

PROOF. The hypothesis and conclusion are both local with respect to X, so we can assume X is affine. Then we are to show $H^i(Y, \mathcal{O}) = 0$ for $0 < i < k-1$. Let $V = V(\boldsymbol{x})$ (in both X and Y, by abuse of notation). The regular sequence condition implies $H^i_V(\mathcal{O}) = 0$ for $i < k$, on both X and Y. The hypothesis $Rf'_*\mathcal{O}_Y = \mathcal{O}_X$ implies that $H^i(W, \mathcal{O}_Y) \cong H^i(U, \mathcal{O}_X)$. Then from the local cohomology exact sequences

$$\cdots \to H^i_V(\mathcal{O}_Y) \to H^i(Y, \mathcal{O}) \to H^i(W, \mathcal{O}_Y) \to H^{i+1}_V(\mathcal{O}_Y) \to \cdots$$

and

$$\cdots \to H^i_V(\mathcal{O}_X) \to H^i(X, \mathcal{O}) \to H^i(U, \mathcal{O}_X) \to H^{i+1}_V(\mathcal{O}_X) \to \cdots$$

we obtain $H^i(Y, \mathcal{O}) \cong H^i(W, \mathcal{O}_Y) \cong H^i(U, \mathcal{O}_X) \cong H^i(X, \mathcal{O}) = 0$, for $0 < i < k-1$. $\square$

Conjecture 6.1 implies that $(x_1 - x_2, \ldots, x_{n-1} - x_n)$ is a regular sequence on $R[tJ]$, and hence, by Proposition 6.4, on X_n. To apply the Lemma using this sequence on Y_n and X_n, it remains to prove that the sequence is regular on Y_n, and that $Rf_*\mathcal{O}_Y = \mathcal{O}_X$ outside the locus where all the x_i's coincide.

Note that Y_n can be described directly in terms of X_n as the subscheme of $\mathrm{Hilb}^{n-1}(\mathbb{C}^2) \times X_n$ whose fiber over a point $(I, P_1, \ldots, P_n)$ of X_n consists of all the ideals $I_{n-1} \subseteq I$ for which I_{n-1}/I is a length-1 ideal of the local ring $\mathcal{O}_{S,P_n}$. About a point where the P_i are not all equal, it follows from Lemma 6.6 that there is a neighborhood on which Y_n is locally isomorphic to $X_{r_1} \times \cdots X_{r_{k-1}} \times Y_{r_k}$, where of the distinct points $Q_1, \ldots, Q_k$, Q_k is the one equal to P_n. It also follows that on such a neighborhood the map $f : Y_n \to X_n$ is locally given by the identity on the factors X_{r_i}, times the map $f : Y_{r_k} \to X_{r_k}$. Hence we have $Rf_*\mathcal{O}_Y = \mathcal{O}_X$ by induction on the locus where the points P_i are not all equal, and, a fortiori, on the locus where the coordinates x_i are not all equal.

To conclude, I claim that $x_1, \ldots, x_n$ defines a complete intersection in Y_n, which is Cohen–Macaulay by induction, and hence $\boldsymbol{x}$ is a regular sequence at each point of $V(\boldsymbol{x})$. By shifting the origin of coordinates in $\mathbb{C}^2$, this implies that $(x_1 - x_2, \ldots, x_{n-1} - x_n)$ is a regular sequence at every point where the x_i's are all equal. Thus we have to show that $V(\boldsymbol{x})$ has dimension n. Recall that we already have the analogous result for X_n, Lemma 5.5. By the local product structure described in the preceding paragraph we can assume the result by induction on the open set where the y_i's are not all equal. This reduces us to showing that the dimension of $H_0^{n-1,n} = V(\boldsymbol{x}, \boldsymbol{y})$ is $n - 1$, since the locus where all the y_i's are equal and all the x_i's are zero is $\mathbb{C}^1 \times H_0^{n-1,n}$.

Now we apply Lemma 6.8. By upper-semicontinuity, the maximal fiber dimension of $H_0^{n-1,n} \to H_0^n$ over the cell C'_μ occurs at I_μ. There, by the remarks preceding Lemma 6.14, the fiber dimension is one less than the number of corners of the diagram of μ. Thus to show that the preimage of C'_μ has dimension at most $n - 1$, we just have to check that for any partition of n, we have (number of parts) + (number of corners) $\leq n + 1$. But this is clear, since the number of cells in the first column of the diagram is the number of parts, and at most one of them can be a corner.

7. Diagonal Harmonics

A polynomial function f on a vector space V is said to be *harmonic* with respect to a group G of linear endomorphisms of V if f is annihilated by all G-invariant partial differential operators without constant term. For $V = (\mathbb{Q}^2)^n = \mathbb{Q}^n \oplus \mathbb{Q}^n$, and G the symmetric group S_n acting "diagonally" by simultaneous coordinate permutations in each summand, we refer to the space D_n of harmonic polynomials as the *diagonal harmonics*.

By Weyl's theorem on the ring of invariants $\mathbb{Q}[\boldsymbol{x}, \boldsymbol{y}]^{S_n}$, we may equivalently define D_n as the solution space of the system of differential equations

$$p_{h,k}(\partial \boldsymbol{x}, \partial \boldsymbol{y})f = \sum_i \partial x_i^h \partial y_i^k f = 0, \quad \text{for } 1 \leq h + k \leq n.$$

In particular the diagonal harmonics are solutions of the Laplace equation

$$\sum_i (\partial x_i^2 + \partial y_i^2)f = 0,$$

so they are harmonic polynomials in the classical sense. It is easy to see that the polynomials Δ_μ of Section 3 are diagonal harmonics, and hence the spaces D_μ are subspaces of D_n.

Let $I_n \subseteq \mathbb{Q}[\boldsymbol{x}, \boldsymbol{y}]$ be the ideal generated by the polarized power sums $p_{h,k}$ for $h + k > 0$. We may describe D_n in derivative-free terms as follows.

PROPOSITION 7.1 [Haiman 1994]. *The quotient ring $\mathbb{Q}[\boldsymbol{x}, \boldsymbol{y}]/I_n$ is isomorphic as a doubly graded S_n module to D_n.*

For geometric purposes we will work instead with the ring

$$R_n = \mathbb{C}[\boldsymbol{x}, \boldsymbol{y}]/I_n,$$

I_n being again generated by the polarized power sums, which of course has the same Frobenius series as $\mathbb{Q}[\boldsymbol{x}, \boldsymbol{y}]/I_n$ or D_n.

Computations have suggested a series of surprising combinatorial conjectures concerning the Hilbert and Frobenius series of the rings R_n. As these are treated at length in [Haiman 1994], we here mention only three that are simple to state.

CONJECTURE 7.2. *The dimension of R_n as a vector space, or $\mathcal{H}_{R_n}(1,1)$, is equal to $(n+1)^{n-1}$. Moreover $q^{\binom{n}{2}}\mathcal{H}_{R_n}(q, q^{-1}) = (1 + q + q^2 + \cdots + q^n)^{n-1}$.*

CONJECTURE 7.3. *The specialization $\mathcal{H}_{R_n}(q, 1)$ enumerates spanning trees T on the vertex set $\{0, 1, \ldots, n\}$, each counted with weight $q^{i(T)}$, where $i(T)$ is the number of inversions in T. An inversion is a pair $i < j$ for which vertex j lies on the unique path in T from vertex 0 to vertex i.*

CONJECTURE 7.4. *As an S_n module, R_n is isomorphic to the sign character tensored with the permutation representation of S_n on the finite Abelian group $(\mathbb{Z}/(n+1)\mathbb{Z})^n/H$, where S_n acts by permuting the factors, and $H = (\mathbb{Z}/(n+1)\mathbb{Z}) \cdot (1, 1, \ldots, 1)$ is the subgroup of S_n-invariant elements.*

The above conjectures are corollaries to a pair of more general conjectures giving the Frobenius series specializations $\mathcal{F}_{R_n}(X; q, 1)$ and $q^{\binom{n}{2}}\mathcal{F}_{R_n}(X; q, 1/q)$. In [Garsia and Haiman 1996a] we showed that these specializations are in turn corollaries to the following master formula.

CONJECTURE 7.5. *The Frobenius series of R_n is given by*

$$\mathcal{F}_{R_n}(X; q, t) = \sum_{|\mu|=n} \frac{(1-q)(1-t)B_\mu(q,t)\Pi_\mu(q,t)\tilde{H}_\mu(X; q,t)}{\prod_{s \in D(\mu)}(1-q^{-a(s)}t^{1+l(s)})(1-q^{1+a(s)}t^{-l(s)})}, \qquad (7.1)$$

where μ ranges over partitions of n, the arms and legs $a(s)$, $l(s)$ are as in (2.16), B_μ is given by (2.12), and

$$\Pi_\mu = \Omega[1 - B_\mu] = \prod_{\substack{(h,k)\in D(\mu) \\ (h,k)\neq(0,0)}} (1-q^k t^h).$$

In the remainder of this section we show how formula (7.1) comes about, and prove that it holds if the $n!$ conjecture and a suitable cohomology vanishing hypothesis on X_n are true. The development parallels that in [Haiman 1998], to which we refer for some geometric results. There we studied the specialization of (7.1) to the Hilbert series for the S_n-alternating component, which can be expressed without recourse to Macdonald polynomials as

$$C_n(q,t) = \sum_{|\mu|=n} \frac{(1-q)(1-t)B_\mu(q,t)\Pi_\mu(q,t)t^{n(\mu)}q^{n(\mu')}}{\prod_{s\in D(\mu)}(1-q^{-a(s)}t^{1+l(s)})(1-q^{1+a(s)}t^{-l(s)})}.$$

This turns out to be a two-parameter analog of the *Catalan number* $C_n = \frac{1}{n+1}\binom{2n}{n}$. We proved that $C_n(q,t)$ is a polynomial in q and t, and that under certain cohomology vanishing hypotheses it is the Hilbert series of the S_n-alternating diagonal harmonics. Because we examined only the alternating component we could work on $\mathrm{Hilb}^n(\mathbb{C}^2)$ without introducing the isospectral variety X_n. In essence, what we will now do is to lift these results to X_n.

PROPOSITION 7.6. *Spec R_n is the scheme theoretic fiber $\rho^{-1}(0)$ over the origin, under the canonical map*

$$\rho \colon (\mathbb{C}^2)^n \to S^n\mathbb{C}^2.$$

PROOF. The coordinate ring of $S^n\mathbb{C}^2$ is $\mathbb{C}[\boldsymbol{x},\boldsymbol{y}]^{S_n}$ and the ideal of the origin is the homogeneous maximal ideal $\mathfrak{m} = (p_{h,k} : h + k > 0)$. By definition, the ideal of $\rho^{-1}(0)$ is generated by the image of $\mathfrak{m}$ in $\mathbb{C}[\boldsymbol{x},\boldsymbol{y}]$, or I_n. $\square$

In what follows, we *assume the $n!$ conjecture holds for all μ*, so X_n is flat over $\mathrm{Hilb}^n(\mathbb{C}^2)$.

Consider the fiber square

$$
\begin{array}{ccc}
X_n & \xrightarrow{\ \psi\ } & (\mathbb{C}^2)^n \\
\sigma \downarrow & & \downarrow \rho \\
\mathrm{Hilb}^n(\mathbb{C}^2) & \xrightarrow{\ \tau\ } & S^n\mathbb{C}^2.
\end{array}
$$

Define $X_n^0 = (\rho\psi)^{-1}(0) = (\tau\sigma)^{-1}(0)$ to be the scheme-theoretic fiber of X_n over $0 \in S^n\mathbb{C}^2$. This is a *nonreduced* subscheme of X_n. By Proposition 7.6, ψ induces a morphism

$$X_n^0 \xrightarrow{\psi} \rho^{-1}(0) = \mathrm{Spec}\, R_n,$$

corresponding to a ring homomorphism

$$\psi^\sharp : R_n \to H^0(X_n^0, \mathcal{O}).$$

As a scheme finite over $\mathrm{Hilb}^n(\mathbb{C}^2)$, we have $X_n = \mathrm{Spec}\,\sigma_* \mathcal{O}_{X_n}$, and since we are assuming the $n!$ conjecture, $\sigma_* \mathcal{O}_{X_n}$ is locally free of rank $n!$, that is, it is the sheaf of sections of a vector bundle P, the image of the homomorphism ϕ in (4.7). Then $X_n^0 = \sigma^{-1}(H_0^n) = \mathrm{Spec}\,P|_{H_0^n}$ as a scheme over $H_0^n = \tau^{-1}(0)$. Hence we can identify the global sections $H^0(X_n^0, \mathcal{O})$ with $H^0(H_0^n, P)$.

PROPOSITION 7.7 [Haiman 1998]. *The scheme theoretic zero fiber $H_0^n = \tau^{-1}(0)$ in the Hilbert scheme is reduced, Cohen–Macaulay, and has a $\mathbb{T}$-equivariant resolution by locally free sheaves on $\mathrm{Hilb}^n(\mathbb{C}^2)$*

$$0 \to B \otimes \textstyle\bigwedge^{n+1} V \to \cdots \to B \otimes \textstyle\bigwedge^1 V \to B \to \mathcal{O}_{H_0^n} \to 0, \qquad (7.2)$$

where $V = B' \oplus \mathcal{O}_t \oplus \mathcal{O}_q$, B' is a summand of the tautological bundle $B = B' \oplus \mathcal{O}$, and $\mathcal{O}_t$, $\mathcal{O}_q$ denote the trivial bundle $\mathcal{O}$ tensored by the 1-dimensional representation of $\mathbb{T}$ with character t or q, respectively.

To proceed further we will need to assume the validity of the following conjecture.

CONJECTURE 7.8. *For all $i > 0$ and $k \geq 0$ we have $H^i(X_n, B^{\otimes k}) = 0$, and for $i = 0$, the canonical map*

$$\mathbb{C}[x_1', y_1', \ldots, x_k', y_k', x, y] \to H^0(X_n, B^{\otimes k}) \qquad (7.3)$$

is surjective.

To clarify, recall that $B = \mathbb{C}[x', y']/I$, where $\mathbb{C}[x', y']$ really means the trivial bundle $\mathcal{O} \otimes_{\mathbb{C}} \mathbb{C}[x', y']$, and we use primes to avoid confusion with the variables x, y. The map in (7.3) is induced by the maps $\mathbb{C}[x', y'] \to B$, with the identifications $\mathbb{C}[x', y']^{\otimes k} = \mathbb{C}[x_1', y_1', \ldots, x_k', y_k']$ and $H^0(X_n, \mathbb{C}[x', y']^{\otimes k}) = \mathbb{C}[x', y'] \otimes H^0(X_n, \mathcal{O}) = \mathbb{C}[x', y', x, y]$. Note that $H^0(X_n, \mathcal{O}) = \mathbb{C}[x, y]$ because the map $\psi\colon X_n \to (\mathbb{C}^2)^n$ is proper and birational, and $(\mathbb{C}^2)^n$ is obviously normal. Note also that the exterior power $\bigwedge^k B$ is a summand of $B^{\otimes k}$, so the conjecture extends to tensors of exterior powers as well. We do not use the full strength of Conjecture 7.8 below, only the vanishing property for bundles $B \otimes \bigwedge^k B$ and the surjectivity property for B and $B \otimes B$.

PROPOSITION 7.9. *Assume that Conjecture 7.8 holds. Then the canonical homomorphism*

$$\psi^\sharp \colon R_n \to H^0(X_n^0, \mathcal{O}) = H^0(H_0^n, P)$$

is an isomorphism.

PROOF. Since we are assuming X_n is flat over $\mathrm{Hilb}^n(\mathbb{C}^2)$, the pullback functor σ^* on sheaves is exact. Applying σ^* to the resolution (7.2) we get a resolution on X_n

$$0 \to B \otimes \textstyle\bigwedge^{n+1} V \to \cdots \to B \otimes \textstyle\bigwedge^1 V \to B \to \mathcal{O}_{X_n^0} \to 0. \qquad (7.4)$$

By our vanishing hypothesis, (7.4) is an acyclic resolution of $\mathcal{O}_{X_n^0}$, and therefore, applying H^0, we get an exact sequence

$$0 \to H^0(X_n, B \otimes \textstyle\bigwedge^{n+1} V) \to \cdots \to H^0(X_n, B \otimes \textstyle\bigwedge^1 V) \to$$
$$\to H^0(X_n, B) \to H^0(X_n^0, \mathcal{O}) \to 0. \quad (7.5)$$

There is a *trace map* of $\mathcal{O}$-module sheaves

$$\mathrm{tr}\colon B \to \mathcal{O}$$

which sends a section f of B to the function whose value at a point Q is the trace of multiplication by f on the fiber $B(Q)$, divided by n. On X_n, the joint spectrum of the multiplication operators X, Y is $(x_1, y_1), \ldots, (x_n, y_n)$, and therefore

$$\mathrm{tr}\, f(x', y') = \frac{1}{n} \sum_{i=1}^{n} f(x_i, y_i).$$

By the construction of (7.2) in [Haiman 1998], the map $B \to \mathcal{O}_{X_n^0}$ factors as

$$B \xrightarrow{\ \mathrm{tr}\ } \mathcal{O} \longrightarrow \mathcal{O}_{X_n^0}.$$

Hence $H^0(X_n, B) \to H^0(X_n^0, \mathcal{O})$ factors through $H^0(X_n, \mathcal{O}) = \mathbb{C}[x, y]$, which shows that $\psi^\sharp$ is surjective.

To prove that ψ is injective, we must show that if $f(x', y', x, y)$ represents a global section in the kernel of $H^0(X_n, B) \to H^0(X_n^0, \mathcal{O})$, then $\mathrm{tr}\, f \in I_n$. We are using the surjectivity property in Conjecture 7.8 to assume that such a representative polynomial f exists. By (7.5), the kernel in question is the sum of the images of three maps

$$H^0(X_n, B \otimes B') \to H^0(X_n, B), \quad (7.6)$$
$$H^0(X_n, B \otimes \mathcal{O}_t) \to H^0(X_n, B),$$
$$H^0(X_n, B \otimes \mathcal{O}_q) \to H^0(X_n, B).$$

By the construction of (7.2), the second and third maps are multiplication by x' and y', respectively. For any $f(x', y', x, y)$, the trace map satisfies the identity

$$\mathrm{tr}\, f - f(0, 0, x, y) \in I_n. \quad (7.7)$$

To prove this it is sufficient to take $f = (x')^h (y')^k$, since these monomials generate $\mathbb{C}[x', y', x, y]$ as a $\mathbb{C}[x, y]$ module, and the operation we are performing on f is $\mathbb{C}[x, y]$-linear. We obtain

$$\mathrm{tr}(x')^h (y')^k - 0^h 0^k = \begin{cases} 0 & \text{if } h + k = 0, \\ p_{h,k}(x, y) & \text{if } h + k > 0. \end{cases}$$

In particular if f belongs to the ideal (x', y') then $f(0, 0, x, y) = 0$ and $\mathrm{tr}\, f \in I_n$.

This leaves us only to consider the first map (7.6). The summand B' is defined to be the kernel of the trace map. Hence if $f(x', y', x'', y'', \boldsymbol{x}, \boldsymbol{y})$ represents a section in $H^0(X_n, B \otimes B')$, then

$$\sum_i f(x', y', x_i, y_i, \boldsymbol{x}, \boldsymbol{y})$$

is the zero section in $H^0(X_n, B)$. Now for each j there is a homomorphism of sheaves of $\mathcal{O}_{X_n}$ algebras $B \to \mathcal{O}_{X_n}$ mapping (x', y') to (x_i, y_i). This is so because $X_n \subseteq U^{\times n}$, and the homomorphism $B \to \mathcal{O}_{X_n}$ corresponds to the projection $X_n \to U$ on the j-th factor. Applying these homomorphisms to the sum above, we see that

$$\sum_i f(x_j, y_j, x_i, y_i, \boldsymbol{x}, \boldsymbol{y}) = 0 \quad \text{in } H^0(X_n, \mathcal{O}) = \mathbb{C}[\boldsymbol{x}, \boldsymbol{y}],$$

for all j. By (7.7) this implies that $f(x_j, y_j, 0, 0, \boldsymbol{x}, \boldsymbol{y}) \in I_n$ for each j. Summing over j and using (7.7) again we find that $f(0, 0, 0, 0, \boldsymbol{x}, \boldsymbol{y}) \in I_n$.

The map $H^0(B \otimes B') \to H^0(X, B)$ is multiplication in B, which sends $f(x', y', x'', y'', \boldsymbol{x}, \boldsymbol{y})$ to $f(x', y', x', y', \boldsymbol{x}, \boldsymbol{y})$. Modulo I_n, the trace map

$$H^0(X_n, B) \to H^0(X_n, \mathcal{O})$$

is the same as evaluation at $(x', y') = (0, 0)$, again by (7.7). Hence the image of $f(x', y', x'', y'', \boldsymbol{x}, \boldsymbol{y})$ in $H^0(X_n, \mathcal{O})$ is given modulo I_n by $f(0, 0, 0, 0, \boldsymbol{x}, \boldsymbol{y})$, and since the latter belongs to I_n the proof is complete. $\qquad \square$

THEOREM 7.10. *Assuming the $n!$ conjecture and Conjecture 7.8 hold, the Frobenius series of R_n is given by the master formula (7.1) in Conjecture 7.5.*

PROOF. In [Haiman 1998] we derived an Atiyah–Bott type Lefschetz formula for $\mathbb{T}$-equivariant vector bundles on H_0^n, using the resolution (7.2) and explicit local parameters for $\mathrm{Hilb}^n(\mathbb{C}^2)$ at the $\mathbb{T}$-fixed points I_μ. This formula takes the form

$$\sum_i (-1)^i \mathcal{F}_{H^i(H_0^n, V)}(X; q, t)$$

$$= \sum_{|\mu|=n} \frac{(1-q)(1-t) B_\mu(q,t) \Pi_\mu(q,t) \mathcal{F}_{V(I_\mu)}(X; q, t)}{\prod_{s \in D(\mu)} (1 - q^{-a(s)} t^{1+l(s)})(1 - q^{1+a(s)} t^{-l(s)})}. \quad (7.8)$$

Actually, this formula was derived for Hilbert series, but when V is a bundle of S_n modules it generalizes immediately to Frobenius series. The q, t-Catalan numbers studied in [Haiman 1998] correspond to the line bundle $V = \mathcal{O}(1)$.

If the $n!$ conjecture and Conjecture (7.8) hold, then by Proposition 7.9, the Frobenius series of R_n is equal to $\mathcal{F}_{H^0(H_0^n, P)}$, where $P = \sigma_* \mathcal{O}_{X_n}$. Moreover, using the resolution (7.4), we see that Conjecture 7.5 implies $H^i(X_n^0, \mathcal{O}) = 0$ for $i > 0$, or equivalently, since σ is finite, $H^i(H_0^n, P) = 0$. Therefore the Euler characteristic on the left-hand side of (7.8) reduces to $\mathcal{F}_{R_n}(X; q, t)$.

By Theorem 5.4, the $n!$ conjecture implies that $\mathcal{F}_{P(I_\mu)}(X; q, t) = \tilde{H}_\mu(X; q, t)$, and the result follows.   □

We conclude with some remarks on the vanishing hypothesis, Conjecture 7.8. Strong vanishing theorems such as this are a relatively rare phenomenon. The conjecture is the analog, for the tautological bundle B on the isospectral Hilbert scheme, of a theorem that does hold for the tautological (quotient) bundle on a Grassmann variety.

In the case of the Hilbert scheme there is some favorable computational evidence. Namely, assuming the $n!$ conjecture — which has been verified for $n \leq 8$ — one can use (7.8) to compute the Frobenius series Euler characteristic of any explicit enough bundle V. If V has nonvanishing higher cohomology, we should expect to see some negative terms. For the bundles referred to in the conjecture, and reasonable values of n and k, we have done a number of these computations and the results invariably have positive coefficients. Note also that if the $n!$ conjecture were to fail, we should not even expect to obtain a polynomial in (7.8). For the specialization to Hilbert series, the formula can be evaluated for values of n much larger than those for which we can check the $n!$ conjecture. A. Garsia and I have done some of these computations for n as large as 20, always obtaining polynomials with positive coefficients. I regard this as strong evidence for both the $n!$ conjecture and Conjecture 7.8.

8. The Commuting Variety

The material in this section is based on my conversations with I. Grojnowski, and represents joint work in progress. At present our results are not definitive, but we have made some observations and conjectures that I will discuss briefly.

DEFINITION. The *commuting variety* C_n is the variety of pairs of $n \times n$ matrices (X, Y) such that $XY = YX$.

Little is known about C_n, except that it is irreducible of dimension $n^2 + n$ [Motzkin and Taussky 1952; Richardson 1979]. It is not even known whether the equations $XY = YX$ generate its ideal, although this is conjectured to be true. There is also a conjecture, generally attributed to Hochster, that C_n is Cohen–Macaulay.

We will be interested in the open set C_n^0 of pairs for which the vectors $X^h Y^k e_1$ span $\mathbb{C}^n$, where e_1 is the first unit coordinate vector. This is an open set, since its complement is defined by the vanishing of the $n \times n$ minors of the matrix whose columns are $X^h Y^k e_1$.

Given an ideal $I \subseteq \mathbb{C}[x, y]$ belonging to $\mathrm{Hilb}^n(\mathbb{C}^2)$, fix a basis $\{1, v_2, \ldots, v_n\}$ of $\mathbb{C}[x, y]/I$. With respect to this basis, the operators X and Y of multiplication by x and y are represented by commuting matrices, and the pair (X, Y) belongs to C_n^0 because we took our first basis vector to be 1. Conversely, given $(X, Y) \in C_n^0$, we have a surjective map $\theta: \mathbb{C}[x, y] \to \mathbb{C}^n$ sending $p(x, y)$ to $p(X, Y)e_1$, whose kernel is an ideal $I \in \mathrm{Hilb}^n(\mathbb{C}^2)$. Then θ induces an isomorphism $\mathbb{C}[x,y]/I \to \mathbb{C}^n$, under which the unit coordinate basis $e_1, \ldots, e_n$ of $\mathbb{C}^n$ corresponds to a basis $1, v_2, \ldots, v_n$ of $\mathbb{C}[x,y]/I$. It is easy to see that these two constructions are mutually inverse and so define a smooth fibration

$$C_n^0 \to \mathrm{Hilb}^n(\mathbb{C}^2)$$

with fiber G, where $G \subseteq GL_n$ is the stabilizer of e_1 (so G parametrizes ordered bases of $\mathbb{C}^n$ whose first vector is given). In particular this shows that C_n^0 is nonsingular.

DEFINITION. The *isospectral commuting variety* IC_n is the variety of tuples $(X, Y, \boldsymbol{a}, \boldsymbol{b}) \in C_n \times \mathbb{C}^{2n}$ such that $(a_1, b_1), \ldots, (a_n, b_n)$ is the joint spectrum of X and Y in some order. In other words, we have the identity

$$\det(I + rX + sY) = \prod_{i=1}^n (1 + ra_i + sb_i), \qquad (8.1)$$

where r, s are indeterminates.

Note that if X and Y commute there is a $g \in GL_n$ such that $g^{-1}Xg$ and $g^{-1}Yg$ are both upper triangular, by Lie's theorem. In particular they have a joint spectrum as defined above, given by the diagonal entries of the triangular form. Note also that there is an action of S_n on IC_n, permuting the pairs (a_i, b_i). Under this action we have $IC_n/S_n = C_n$, since the invariants $p_{h,k}(\boldsymbol{a}, \boldsymbol{b})$ are equal to $\mathrm{tr}\, X^h Y^k$ and so reduce to functions on C_n.

Let IC_n^0 denote the open subset of IC_n lying over C_n^0. From the definition of the isospectral Hilbert scheme X_n it follows immediately that we have (set-theoretically) a fiber square

$$
\begin{array}{ccc}
IC_n^0 & \longrightarrow & X_n \\
\downarrow & & \downarrow \\
C_n^0 & \longrightarrow & \mathrm{Hilb}^n(\mathbb{C}^2).
\end{array}
$$

Since the bottom arrow is a smooth morphism, so is the top arrow in the scheme-theoretic fiber square. Hence the scheme-theoretic fiber product is reduced and therefore equal to the set-theoretic fiber product. This proves

PROPOSITION 8.1. *The open set IC_n^0 in IC_n is Cohen–Macaulay (and hence Gorenstein) if and only if X_n is.*

CONJECTURE 8.2. *The isospectral commuting variety IC_n is Gorenstein.*

Note that this implies the conjecture that the commuting variety $C_n = IC_n/S_n$ is Cohen–Macaulay, as well as the $n!$ conjecture. We should point out here that the ideal of IC_n is certainly *not* generated by the ideal of C_n (conjecturally $XY = YX$) together with equations (8.1). This fails even for $n = 2$.

References

[Batyrev and Dais 1996] V. V. Batyrev and D. I. Dais, "Strong McKay correspondence, string-theoretic Hodge numbers and mirror symmetry", *Topology* **35**:4 (1996), 901–929.

[Bayer and Stillman 1989] D. Bayer and M. Stillman, Macaulay (a computer algebra system for algebraic geometry), version 3.0, 1989. See ftp://math.columbia.edu/pub/bayer/Macaulay or ftp://math.harvard.edu/Macaulay/.

[Bergeron and Garsia 1992] N. Bergeron and A. M. Garsia, "On certain spaces of harmonic polynomials", pp. 51–86 in *Hypergeometric functions on domains of positivity, Jack polynomials, and applications* (Tampa, FL, 1991), edited by D. S. P. Richards, Contemporary Mathematics **138**, Amer. Math. Soc., Providence, RI, 1992.

[Briançon 1977] J. Briançon, "Description de $\mathrm{Hilb}^n C\{x, y\}$", *Invent. Math.* **41**:1 (1977), 45–89.

[Cheah 1998] J. Cheah, "Cellular decompositions for nested Hilbert schemes of points", *Pacific J. Math.* **183**:1 (1998), 39–90.

[De Concini and Procesi 1981] C. De Concini and C. Procesi, "Symmetric functions, conjugacy classes and the flag variety", *Invent. Math.* **64**:2 (1981), 203–219.

[Ellingsrud and Strømme 1987] G. Ellingsrud and S. A. Strømme, "On the homology of the Hilbert scheme of points in the plane", *Invent. Math.* **87**:2 (1987), 343–352.

[Fogarty 1968] J. Fogarty, "Algebraic families on an algebraic surface", *Amer. J. Math* **90** (1968), 511–521.

[Garsia and Haiman 1993] A. M. Garsia and M. Haiman, "A graded representation model for Macdonald's polynomials", *Proc. Nat. Acad. Sci. U.S.A.* **90**:8 (1993), 3607–3610.

[Garsia and Haiman 1996a] A. M. Garsia and M. Haiman, "A remarkable q, t-Catalan sequence and q-Lagrange inversion", *J. Algebraic Combin.* **5**:3 (1996), 191–244.

[Garsia and Haiman 1996b] A. M. Garsia and M. Haiman, "Some natural bigraded S_n-modules and q, t-Kostka coefficients", *Electron. J. Combin.* **3**:2 (1996), RP24.

[Garsia and Haiman 1998] A. M. Garsia and M. Haiman, "A random q, t-hook walk and a sum of Pieri coefficients", *J. Combin. Theory Ser. A* **82**:1 (1998), 74–111.

[Garsia and Procesi 1992] A. M. Garsia and C. Procesi, "On certain graded S_n-modules and the q-Kostka polynomials", *Adv. Math.* **94**:1 (1992), 82–138.

[Garsia and Remmel 1998] A. M. Garsia and J. Remmel, "Plethystic formulas and positivity for q, t-Kostka coefficients", pp. 245–262 in *Mathematical essays in honor of Gian-Carlo Rota* (Cambridge, MA, 1996), edited by B. E. Sagan and R. P. Stanley, Progress in Mathematics **161**, Birkhäuser, Boston, 1998.

[Garsia and Tesler 1996] A. M. Garsia and G. Tesler, "Plethystic formulas for Macdonald q, t-Kostka coefficients", *Adv. Math.* **123**:2 (1996), 144–222.

[Gordan 1900] M. Gordan, "Les invariants des formes binaires", *J. Math. Pures Appl.* **6** (1900), 141–156.

[Gotzmann 1978] G. Gotzmann, "Eine Bedingung für die Flachheit und das Hilbert-polynom eines graduierten Ringes", *Math. Z.* **158**:1 (1978), 61–70.

[Grothendieck 1961]   A. Grothendieck, "Techniques de construction et théorèmes d'existence en géométrie algébrique, IV: Les schémas de Hilbert", pp. 249–276 (exposé no. 221) in *Séminaire Bourbaki* 1960/1961 (exposés 205–222), IHP, Paris, 1961. Reprinted by Benjamin, New York, 1966, and Soc. Math. France, Paris, 1995.

[Haiman 1994]   M. D. Haiman, "Conjectures on the quotient ring by diagonal invariants", *J. Algebraic Combin.* **3**:1 (1994), 17–76.

[Haiman 1998]   M. Haiman, "t, q-Catalan numbers and the Hilbert scheme", *Discrete Math.* **193**:1-3 (1998), 201–224.

[Hotta and Springer 1977]   R. Hotta and T. A. Springer, "A specialization theorem for certain Weyl group representations and an application to the Green polynomials of unitary groups", *Invent. Math.* **41**:2 (1977), 113–127.

[Iarrobino 1972]   A. Iarrobino, "Reducibility of the families of 0-dimensional schemes on a variety", *Invent. Math.* **15** (1972), 72–77.

[Kirillov and Noumi 1998]   A. N. Kirillov and M. Noumi, "Affine Hecke algebras and raising operators for Macdonald polynomials", *Duke Math. J.* **93**:1 (1998), 1–39.

[Knop 1997]   F. Knop, "Integrality of two variable Kostka functions", *J. Reine Angew. Math.* **482** (1997), 177–189.

[Kraft 1981]   H. Kraft, "Conjugacy classes and Weyl group representations", pp. 191–205 in *Young tableaux and Schur functions in algebra and geometry* (Toruń, Poland, 1980), Astérisque **87–88**, Soc. Math. France, Paris, 1981.

[Macdonald 1988]   I. G. Macdonald, "A new class of symmetric functions", pp. 131–171 in *Actes du 20e Séminaire Lotharingien*, I.R.M.A. Publ. 372/S–20, Strasbourg, 1988.

[Macdonald 1995]   I. G. Macdonald, *Symmetric functions and Hall polynomials*, 2nd ed., Oxford Univ. Press, 1995.

[Motzkin and Taussky 1952]   T. S. Motzkin and O. Taussky, "Pairs of matrices with property L", *Trans. Amer. Math. Soc.* **73** (1952), 108–114.

[Nakamura 1996]   I. Nakamura, "Simple singularities, McKay correspondence, and Hilbert schemes of G-orbits", preprint, Hokkaido University, 1996.

[Reid 1997]   M. Reid, "McKay correspondence", preprint, 1997. Available at http://xxx.lanl.gov/abs/math.AG/9702016.

[Richardson 1979]   R. W. Richardson, "Commuting varieties of semisimple Lie algebras and algebraic groups", *Compositio Math.* **38**:3 (1979), 311–327.

[Sahi 1996]   S. Sahi, "Interpolation, integrality, and a generalization of Macdonald's polynomials", *Internat. Math. Res. Notices* **10** (1996), 457–471.

[Springer 1978]   T. A. Springer, "A construction of representations of Weyl groups", *Invent. Math.* **44**:3 (1978), 279–293.

[Tikhomirov 1992]   A. S. Tikhomirov, "On Hilbert schemes and flag varieties of points on algebraic surfaces", preprint, 1992.

[Weyl 1946]   H. Weyl, *The classical groups: their invariants and representations*, 2nd ed., Princeton University Press, Princeton, NJ, 1946.

New Perspectives in Geometric Combinatorics
MSRI Publications
Volume **38**, 1999

Enumeration of Matchings:
Problems and Progress

JAMES PROPP

Dedicated to the memory of David Klarner (1940–1999)

ABSTRACT. This document is built around a list of thirty-two problems in enumeration of matchings, the first twenty of which were presented in a lecture at MSRI in the fall of 1996. I begin with a capsule history of the topic of enumeration of matchings. The twenty original problems, with commentary, comprise the bulk of the article. I give an account of the progress that has been made on these problems as of this writing, and include pointers to both the printed and on-line literature; roughly half of the original twenty problems were solved by participants in the MSRI Workshop on Combinatorics, their students, and others, between 1996 and 1999. The article concludes with a dozen new open problems.

1. Introduction

How many perfect matchings does a given graph G have? That is, in how many ways can one choose a subset of the edges of G so that each vertex of G belongs to one and only one chosen edge? (See Figure 1(a) for an example of a perfect matching of a graph.) For general graphs G, it is computationally hard to obtain the answer [Valiant 1979], and even when we have the answer, it is not so clear that we are any the wiser for knowing this number. However, for many infinite families of special graphs the number of perfect matchings is given by compellingly simple formulas. Over the past ten years a great many families of this kind have been discovered, and while there is no single unified result that encompasses all of them, many of these families resemble one another, both in terms of the form of the results and in terms of the methods that have been useful in proving them.

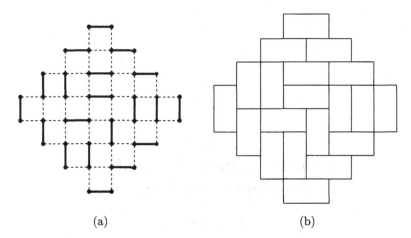

(a)   (b)

Figure 1. The Aztec diamond of order 4.

The deeper significance of these formulas is not clear. Some of them are related to results in representation theory or the theory of symmetric functions, but others seem to be self-contained combinatorial puzzles. Much of the motivation for this branch of research lies in the fact that we are still unable to predict ahead of time which enumerative problems lead to beautiful formulas and which do not; each new positive result seems like an undeserved windfall.

Hereafter, I will use the term "matching" to signify "perfect matching". (See the book of Lovász and Plummer [1986] for general background on the theory of matchings.)

As far as I have been able to determine, problems involving enumeration of matchings were first examined by chemists and physicists in the 1930s, for two different (and unrelated) purposes: the study of aromatic hydrocarbons and the attempt to create a theory of the liquid state.

Shortly after the advent of quantum chemistry, chemists turned their attentions to molecules like benzene composed of carbon rings with attached hydrogen atoms. For these researchers, matchings of a graph corresponded to "Kekulé structures", i.e., ways of assigning single and double bonds in the associated hydrocarbon (with carbon atoms at the vertices and tacit hydrogen atoms attached to carbon atoms with only two neighboring carbon atoms). See for example the article of Gordon and Davison [1952], whose use of nonintersecting lattice paths anticipates certain later work [Gessel and Viennot 1985; Sachs 1990; John and Sachs 1985]. There are strong connections between combinatorics and chemistry for such molecules; for instance, those edges which are present in comparatively few of the matchings of a graph turn out to correspond to the bonds that are least stable, and the more matchings a polyhex graph possesses the more stable is the corresponding benzenoid molecule. Since hexagonal rings are so predominant in the structure of hydrocarbons, chemists gave most of their attention to counting matchings of subgraphs of the infinite honeycomb grid.

At approximately the same time, scientists were trying to understand the behavior of liquids. As an extension of a more basic model for liquids containing only molecules of one type, Fowler and Rushbrooke [1937] devised a lattice-based model for liquids containing two types of molecules, one large and one small. In the case where the large molecule was roughly twice the size of the small molecule, it made sense to model the small molecules as occupying sites of a three-dimensional grid and the large molecules as occupying pairs of adjacent sites. In modern parlance, this is a monomer-dimer model. In later years, the two-dimensional version of the model was found to have applicability to the study of molecules adsorbed on films; if the adsorption sites are assumed to form a lattice, and an adsorbed molecule is assumed to occupy two such sites, then one can imagine fictitious molecules that occupy all the unoccupied sites (one each).

Major progress was made when Temperley and Fisher [1961] and Kasteleyn [1961] independently found ways to count pure dimer configurations on subgraphs of the infinite square grid, with no monomers present. Although the physical significance of this special case was (and remains) unclear, this result, along with Onsager's earlier exact solution of the two-dimensional Ising model [Onsager 1944], paved the way for other advances such as Lieb's exact solution of the six-vertex model [Lieb 1967], culminating in a new field at the intersection of physics and mathematics: exactly solved statistical mechanics models in two-dimensional lattices. (Intriguingly, virtually none of the three- and higher-dimensional analogues of these models have succumbed to researchers' efforts at obtaining exact solutions.) For background on lattice models in statistical mechanics, see the book by Baxter [1982].

An infinite two-dimensional grid has many finite subgraphs; in choosing which ones to study, physicists were guided by the idea that the shape of boundary should be chosen so as to minimize the effect of the boundary — that is, to maximize the number of configurations, at least in the asymptotic sense. For example, Kasteleyn, in his study of the dimer model on the square grid, counted the matchings of the m-by-n rectangle (see the double-product formula at the beginning of Section 5) and of the m-by-n rectangular torus, and showed that the two numbers grow at the same rate as m, n go to infinity, namely C^{mn} for a known constant C. (Analytically, C is $e^{G/\pi}$, where G is Catalan's constant $1 - \frac{1}{9} + \frac{1}{25} - \frac{1}{49} + \frac{1}{81} - \cdots$; numerically, C is approximately 1.34.)

Kasteleyn [1961] wrote: "The effect of boundary conditions is, however, not entirely trivial and will be discussed in more detail in a subsequent paper." (See the article of Cohn, Kenyon and Propp [Cohn et al. 1998a] for a rigorous mathematical treatment of boundary conditions.) Kasteleyn never wrote such a followup paper, but other physicists did give some attention to the issue of boundary shape, most notably Grensing, Carlsen and Zapp [Grensing et al. 1980]. These authors considered a one-parameter family of graphs of the kind shown in Figure 1(a), and they asserted that every graph in this family has $2^{N/4}$

matchings, where N is the number of vertices. They did not give a proof, nor
did they indicate whether they had one. The result was rediscovered in the late
1980s by Elkies, Kuperberg, Larsen, and Propp [Elkies et al. 1992], who gave four
proofs of the formula. This article led to a great deal of work among enumerative
combinatorialists, who refer to graphs like the one shown in Figure 1 as "Aztec
diamond graphs", or sometimes just Aztec diamonds for short. (It should be
noted that Elkies et al. [1992] used the term "Aztec diamond" to denote regions
like the one shown in Figure 1(b). The two sorts of Aztec diamonds are dual to
one another; matchings of Aztec diamond graphs correspond to domino tilings
of Aztec diamond regions.)

At about the same time, it became clear that there had been earlier work
within the combinatorial community that was pertinent to the study of matchings, though its relevance had not hitherto been recognized. For instance, Mills,
Robbins and Rumsey [Mills et al. 1983], in their work on alternating sign matrices, had counted pairs of "compatible" ASMs of consecutive size; these can
be put into one-to-one correspondence with matchings of an associated Aztec
diamond graph [Elkies et al. 1992].

Looking into earlier mathematical literature, one can even see intimations of
enumerative matching theory in the work of MacMahon [1915–16], who nearly
a century ago found a formula for the number of plane partitions whose solid
Young diagram fits inside an a-by-b-by-c box, as will be discussed in Section 2.
(See the book by Andrews [1976] and the article by Stanley [1971] for background
on plane partitions.) Such a Young diagram is nothing more than an assemblage
of cubes, and it has long been known in the extra-mathematical world that such
assemblages, viewed from a distant point, looks like tilings (consider Islamic art,
for instance). Thus it was natural for mathematicians to interpret MacMahon's
theorem on plane partitions as a result about tilings of a hexagon by rhombuses.
This insight may have occurred to a number of people independently; the earliest
chain of oral communication that I have followed leads back to Klarner (who did
not publish his observation but relayed it to Stanley in the 1970s), and the earliest
published statement I have found is in a paper by David and Tomei [1989].

In any case, each of the Young diagrams enumerated by MacMahon corresponds to a tiling of a hexagon by rhombuses, where the hexagon is semiregular
(its opposite sides are parallel and of equal length, with all internal angles equal
to 120 degrees) and has side-lengths a, b, c, a, b, c, and where the rhombuses have
all side-lengths equal to 1. These tilings in turn correspond to matchings of
the "honeycomb" graph that is dual to the dissection of the hexagon into unit
equilateral triangles; see Figure 2, which shows a matching of the honeycomb
graph and the associated tiling of a hexagon. Kuperberg [1994] was the first to
exploit the connection between plane partitions and the dimer model. (Interestingly, some of the same graphs that Kuperberg studied had been investigated
independently by chemists in their study of benzenoids hydrocarbons; Cyvin and
Gutman [1988] give a survey of this work.)

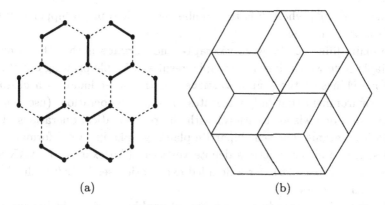

(a)   (b)

Figure 2. A matching and its associated tiling.

Similarly, variants of MacMahon's problem in which the plane partition is subjected to various symmetry constraints (considered by Macdonald, Stanley, and others [Stanley 1986a; 1986b]) correspond to the problem of enumerating matchings possessing corresponding kinds of symmetry. Kuperberg [1994] used this correspondence in solving one of Stanley's open problems, and this created further interest in matchings among combinatorialists.

One of Kuperberg's chief tools was an old result of Kasteleyn, which showed that for any planar graph G, the number of matchings of G is equal to the Pfaffian of a certain matrix of zeros and ones associated with G. A special case of this result, enunciated by Percus [1969], can be used when G is bipartite; in this case, one can use a determinant instead of a Pfaffian. Percus' determinant is a modified version of the bipartite adjacency matrix of the graph, in which rows correspond to "white" vertices and columns correspond to "black" vertices (under a coloring scheme whereby white vertices have only black neighbors and vice versa); the (i, j)-th entry is ± 1 if the i-th white vertex and j-th black vertex are adjacent, and 0 otherwise. For more details on how the signs of the entries are chosen, see the expositions of Kasteleyn [1967] and Percus [1969].

Percus' theorem, incorporated into computer software, makes it easy to count the matchings of many planar graphs and look for patterns in the numbers that arise. Two such programs are vaxmaple, written by Greg Kuperberg, David Wilson and myself, and vaxmacs, written by David Wilson. Most of the patterns described below were discovered with the aid of this software, which is available from http://math.wisc.edu/~propp/software.html. Both programs treat subgraphs of the infinite square grid; this might seem restrictive, but it turns out that counting the matchings of an arbitrary bipartite planar graph can be fit into this framework, with a bit of tweaking. The mathematically interesting part of each program is the routine for choosing the signs of the nonzero entries. There are many choices that would work, but Wilson's sign-rule is far and away the simplest: If an edge is horizontal, we give it weight +1, and if an edge is vertical, joining a vertex in one row to a vertex in the row below it, we give the edge weight $(-1)^k$, where k is the number of vertices in the upper row to the left of the vertical edge.

The main difference between vaxmaple and vaxmacs is that the former creates Maple code which, if sent to Maple, results in Maple printing out the number of matchings of the graph; vaxmacs, on the other hand, is a customized EMACS environment that fully integrates text-editing operations (used for defining the graph one wishes to study) with the mathematical operations of interest. Both programs represent bipartite planar graphs in "VAX-format", where V's, A's, X's, and other letters denote vertices. (An example of VAX-format can be found on page 261; for a detailed explanation see http://math.wisc.edu/~propp/vaxmaple.doc.)

Quite recently, the study of matchings of nonbipartite graphs has been expedited by the programs graph and planemaple, created by Matt Blum and Ben Wieland, respectively. These programs make it easy to define a planar graph by pointing and clicking, after which one can count its matchings using an efficient implementation of Kasteleyn's Pfaffian method. This makes it easy to try out new ideas and look for patterns, outside of the better-explored bipartite case.

Interested readers with access to the World Wide Web can obtain copies of all of these programs via http://math.wisc.edu/~propp/software.html.

Most of the formulas that have been discovered express the number of matchings of a graph as a product of many comparatively small factors. Even before one has conjectured (let alone proved) such a formula, one can frequently infer its existence from the fact that the number of matchings has only small prime factors. Numbers that are large compared to their largest prime factor are sometimes called "smooth" or "round"; the latter term will be used here. The definition of roundness is not precise, since it is not intended for use as a technical term. Its vagueness is intended to capture the uncertainties and the suspense of formula-hunting, and the debatable issue of whether the occurrence of a single larger-than-expected prime factor rules out the existence of a product formula. (For an example of a number whose roundness lies in this gray area, see the table of numbers given in Problem 8.) It is worth noting that Kuperberg [1998, Section VII-A] has shown that rigorous proofs of roundness need not always yield explicit product formulas.

Christian Krattenthaler has written a Mathematica program called RATE that greatly expedites the process of guessing patterns in experimental data on enumeration of matchings; see http://radon.mat.univie.ac.at/People/kratt/rate/rate.html.

A great source of the appeal of research on enumeration of matchings is the ease with which undergraduate research assistants can participate in the hunt for formulas and proofs; many members of the M.I.T. Tilings Research Group (composed mostly of undergraduates like Blum and Wieland) played a role in the developments that led to the writing of this article. Enumeration of matchings has turned out to be a rich avenue of combinatorial inquiry, and many more beautiful patterns undoubtedly await discovery.

Updates on the status of these problems can be found on the Web at http://math.wisc.edu/~propp/update.ps.gz.

2. Lozenges

We begin with problems related to lozenge tilings of hexagons. A *lozenge* is a rhombus of side-length 1 whose internal angles measure 60 and 120 degrees; all the hexagons we will consider will tacitly have integer side-lengths and internal angles of 120 degrees. Every such hexagon H can be dissected into unit equilateral triangles in a unique way, and one can use this dissection to define a graph G whose vertices correspond to the triangles and whose edges correspond to pairs of triangles that share an edge; this is the "finite honeycomb graph" dual to the dissection. It is easy to see that the tilings of H by lozenges are in one-to-one correspondence with the matchings of G.

The a, b, c semiregular hexagon is the hexagon whose side lengths are, in cyclical order, a, b, c, a, b, c. Lozenge tilings of this region are in correspondence with plane partitions with at most a rows, at most b columns, and no part exceeding c. Such hexagons are represented in VAX-format by diagrams like

```
       AVAVAVAVA
      AVAVAVAVAVA
     AVAVAVAVAVAVA
    AVAVAVAVAVAVAVA
    VAVAVAVAVAVAVAV
     VAVAVAVAVAVAV
      VAVAVAVAVAV
       VAVAVAVAV
```

where A's and V's represent upward-pointing and downward-pointing triangles, respectively. In this article we will use triangles instead:

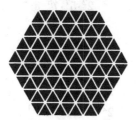

MacMahon [1915–16] showed that the number of such plane partitions is

$$\prod_{i=0}^{a-1}\prod_{j=0}^{b-1}\prod_{k=0}^{c-1}\frac{i+j+k+2}{i+j+k+1}.$$

(This form of MacMahon's formula is due to Macdonald; a short, self-contained proof is given by Cohn et al. [1998b, Section 2].)

PROBLEM 1. Show that in the $2n-1$, $2n$, $2n-1$ semiregular hexagon, the central location (consisting of the two innermost triangles) is covered by a lozenge in exactly one-third of the tilings.

(Equivalently: Show that if one chooses a random matching of the dual graph, the probability that the central edge is contained in the matching is exactly $\frac{1}{3}$.)

PROGRESS. Two independent and very different solutions of this problem have been found; one by Mihai Ciucu and Christian Krattenthaler and the other by Harald Helfgott and Ira Gessel. Ciucu and Krattenthaler [1999] compute more generally the number of rhombus tilings of a hexagon with sides a, a, b, a, a, b that contain the central unit rhombus, where a and b must have opposite parity (the special case $a = 2n-1$, $b = 2n$ solves Problem 1). The same generalization was obtained (in a different but equivalent form) by Helfgott and Gessel [1999], using a completely different method. One might still try to look for a proof whose simplicity is comparable to that of the answer "one-third". Also worthy of note is the paper of Fulmek and Krattenthaler [1998a], which generalizes the result of Ciucu and Krattenthaler [1999].

The hexagon of side-lengths n, $n+1$, n, $n+1$, n, $n+1$ cannot be tiled by lozenges at all, for in the dissection into unit triangles, the number of upward-pointing triangles differs from the number of downward-pointing triangles. However, if one removes the central triangle, one gets a region that can be tiled, and the sort of numbers one gets for small values of n are striking. Here they are, in factored form:

$$2$$
$$2 \cdot 3^3$$
$$2^5 \cdot 3^3 \cdot 5$$
$$2^5 \cdot 5^7$$
$$2^2 \cdot 5^7 \cdot 7^5$$
$$2^8 \cdot 3^3 \cdot 5 \cdot 7^{11}$$
$$2^{13} \cdot 3^9 \cdot 7^{11} \cdot 11$$
$$2^{13} \cdot 3^{18} \cdot 7^5 \cdot 11^7$$
$$2^8 \cdot 3^{18} \cdot 11^{13} \cdot 13^5$$
$$2^2 \cdot 3^9 \cdot 11^{19} \cdot 13^{11}$$
$$2^{10} \cdot 3^3 \cdot 11^{19} \cdot 13^{17} \cdot 17$$
$$2^{16} \cdot 11^{13} \cdot 13^{23} \cdot 17^7$$

These are similar to the numbers one gets from counting lozenge tilings of an n, n, n, n, n, n hexagon, in that the largest prime factor seems to be bounded by a linear function of n.

PROBLEM 2. Enumerate the lozenge tilings of the region obtained from the n, $n+1$, n, $n+1$, n, $n+1$ hexagon by removing the central triangle.

PROGRESS. Mihai Ciucu has solved the more general problem of counting the rhombus tilings of an $(a, b+1, b, a+1, b, b+1)$-hexagon with the central triangle removed [Ciucu 1998]. Ira Gessel proved this result independently using the non-intersecting lattice-paths method [Helfgott and Gessel 1999]. Soichi Okada and Christian Krattenthaler have solved the even more general problem of counting the rhombus tilings of an $(a, b+1, c, a+1, b, c+1)$-hexagon with the central triangle removed [Okada and Krattenthaler 1998].

One can also take a $2n$, $2n+3$, $2n$, $2n+3$, $2n$, $2n+3$ hexagon and make it lozenge-tilable by removing a triangle from the middle of each of its three long sides, as shown:

Here one obtains an equally tantalizing sequence of factorizations:

$$1$$
$$2^7 \cdot 7^2$$
$$2^2 \cdot 7^4 \cdot 11^4 \cdot 13^2$$
$$2^{10} \cdot 3^3 \cdot 5^8 \cdot 13^2 \cdot 17^4 \cdot 19^2$$
$$2^2 \cdot 5^2 \cdot 7^2 \cdot 11^3 \cdot 13^4 \cdot 17^4 \cdot 19^8 \cdot 23^4$$

PROBLEM 3. Enumerate the lozenge tilings of the region obtained from the $2n$, $2n+3$, $2n$, $2n+3$, $2n$, $2n+3$ hexagon by removing a triangle from the middle of each of its long sides.

PROGRESS. Theresia Eisenkölbl solved this problem. What she does in fact is to compute the number of all rhombus tilings of a hexagon with sides a, $b+3$, c, $a+3$, b, $c+3$, where an arbitrary triangle is removed from each of the "long" sides of the hexagon (not necessarily the triangle in the middle). For the proof of her formula [Eisenkölbl 1997] she uses nonintersecting lattice paths, determinants, and the Jacobi determinant formula [Turnbull 1960]. However, I still know of no conceptual explanation for why these numbers are so close (in the multiplicative sense) to being perfect squares.

We now return to ordinary a, b, c semiregular hexagons. When $a = b = c$, there are not two but six central triangles. There are two geometrically distinct ways in which we can choose to remove an upward-pointing triangle and downward-pointing triangle from these six, according to whether the triangles are opposite or adjacent:

Such regions may be called "holey hexagons" of two different kinds. Matt Blum tabulated the number of lozenge tilings of these regions, for small values of $a = b = c$. In the first ("opposite") case, the number of tilings of the holey hexagon is a nice round number (its greatest prime factor appears to be bounded by a linear function of the size of the region). In the second ("adjacent") case, the number of tilings is not round. Note, however, that in the second case, the number of tilings of the holey hexagon divided by the number of tilings of the unaltered hexagon (given to us by MacMahon's formula) is equal to the probability that a random lozenge tiling of the hexagon contains a lozenge that covers these two triangles; this probability tends to $\frac{1}{3}$ for large a, at least on average [Cohn et al. 1998b]. Following this clue, we examine the difference between the aforementioned probability (with its messy, un-round numerator) and the number $\frac{1}{3}$. The result is a fraction in which the numerator is now a nice round number. So, in both cases, we have reason to think that there is an exact product formula.

PROBLEM 4. Determine the number of lozenge tilings of a regular hexagon from which two of its innermost unit triangles (one upward-pointing and one downward-pointing) have been removed.

PROGRESS. Theresia Eisenkölbl solved the first case of Problem 4 and Markus Fulmek and Christian Krattenthaler solved the second case. Eisenkölbl [1998] solves a generalization of the problem by applying Mihai Ciucu's matchings factorization theorem, nonintersecting lattice paths, and a nontrivial determinant evaluation. Fulmek and Krattenthaler [1998b] compute the number of rhombus tilings of a hexagon with sides a, b, a, b, a (with a and b having the same parity) that contain the rhombus that touches the center of the hexagon and lies symmetric with respect to the symmetry axis that runs parallel to the sides of length b. For the proof of their formula they compute Hankel determinants featuring Bernoulli numbers, which they do by using facts about continued fractions, orthogonal polynomials, and, in particular, continuous Hahn polynomials. The special case $a = b$ solves the second part of Problem 4.

I mentioned earlier that Kasteleyn's method, as interpreted by Percus, allows one to write the number of matchings of a bipartite planar graph as the determinant of a signed version of the bipartite adjacency matrix. In the case of lozenge tilings of hexagons and the associated matchings, it turns out that there is no need to modify signs of entries; the ordinary bipartite adjacency matrix will do. Greg Kuperberg [1998] has noticed that when row-reduction and column-reduction are systematically applied to the Kasteleyn–Percus matrix of an a, b, c semiregular hexagon, one can obtain the b-by-b Carlitz matrix [Carlitz and Stanley 1975] whose (i, j)-th entry is $\binom{a+c}{a+i-j}$. (This matrix can also be recognized as the Gessel–Viennot matrix that arises from interpreting each tiling as a family of nonintersecting lattice paths [Gessel and Viennot 1985].) Such reductions do not affect the determinant, so we have a pleasing way of understanding the relationship between the Kasteleyn–Percus matrix method and the Gessel–Viennot lattice-path method. In fact, such reductions do not affect the *cokernel* of the matrix (an abelian group whose order is the determinant). On the other hand, the cokernel of the Kasteleyn–Percus matrix for the a, b, c hexagon is clearly invariant under permuting a, b, and c. This gives rise to three different Carlitz matrices that nontrivially have the same cokernel. For example, if $c = 1$, then one gets an a-by-a matrix and a b-by-b matrix that both have the same cokernel, whose structure can be determined "by inspection" if one notices that the third Carlitz matrix of the trio is just a 1-by-1 matrix whose sole entry is (plus or minus) a binomial coefficient. In this special case, the cokernel is just a cyclic group.

Greg Kuperberg poses this challenge:

PROBLEM 5. Determine the cokernel of the Carlitz matrix, or equivalently of the Kasteleyn–Percus matrix of the a, b, c hexagon, and if possible find a way to interpret the cokernel in terms of the tilings.

This combines Questions 1 and 2 of Kuperberg [1998]. As he points out in that article, in the case $a = b = c = 2$, one gets the noncyclic group $\mathbb{Z}/2\mathbb{Z} \times \mathbb{Z}/10\mathbb{Z}$ as the cokernel.

As was remarked above, one nice thing about the Kasteleyn–Percus matrices of honeycomb graphs is that it is not necessary to make any of the entries negative. For general graphs, however, there is no canonical way of defining K, in the sense that there may be many ways of modifying the signs of certain entries of the bipartite adjacency matrix of a graph so that all nonzero contributions to the determinant have the same sign. Thus, one should not expect the eigenvalues of K to possess combinatorial significance. However, the spectrum of K times its adjoint K^* is independent of which Kasteleyn–Percus matrix K one chooses (as was independently shown by David Wilson and Horst Sachs). Thus, digressing somewhat from the topic of lozenge tilings, we find it natural to ask:

PROBLEM 6. What is the significance of the spectrum of KK^*, where K is any Kasteleyn–Percus matrix associated with a bipartite planar graph?

PROGRESS. Nicolau Saldanha [1997] has proposed a combinatorial interpretation of the spectrum of KK^*. Horst Sachs says (personal communication) that KK^* may have some significance in the chemistry of polycyclic hydrocarbons (so-called benzenoids) and related compounds as a useful approximate measure of the "degree of aromaticity".

Returning now to lozenge tilings, or equivalently, matchings of finite subgraphs of the infinite honeycomb, consider the hexagon graph with $a = b = c = 2$:

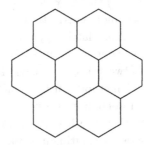

This is the graph whose 20 matchings correspond to the 20 tilings of the regular hexagon of side 2 by rhombuses of side 1. If we look at the probability of each individual vertical edge belonging to a matching chosen uniformly at random ("edge-probabilities"), we get

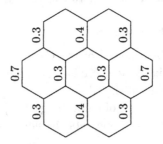

Now look at this table of numbers as if it described a distribution of mass. If we assign the three rows y-coordinates -1 through 1, we find that the weighted sum of the squares of the y-coordinates is equal to

$$(0.3+0.4+0.3)(-1)^2 + (0.7+0.3+0.7+0.3)(0)^2 + (0.3+0.4+0.3)(1)^2 = 2.$$

If we assign to the seven columns x-coordinates -3 through 3, we find that the weighted sum of the squares of the x-coordinates is equal to $(0.7)(-3)^2 + (0.6)(-2)^2 + (0.3)(-1)^2 + (0.8)(0)^2 + (0.3)(1)^2 + (0.6)(2)^2 + (0.7)(3)^2 = 20$. You can do a similar (but even easier) calculation yourself for the case $a = b = c = 1$, to see that the "moments of inertia" of the vertical edge-probabilities around the horizontal and vertical axes are 0 and 1, respectively. Using vaxmaple to study the case $a = b = c = n$ for larger values of n, I find that the moment of inertia about the horizontal axis goes like

$$0, 2, 12, 40, 100, \ldots$$

and the moment of inertia about the vertical axis goes like

$$1, 20, 93, 296, 725, \ldots.$$

It is easy to show that the former moments of inertia are given in general by the polynomial $(n^4 - n^2)/6$ (in fact, the number of vertical lozenges that have any particular y-coordinate does not depend on the tiling chosen). The latter moments of inertia are subtler; they are not given by a polynomial of degree 4, though it is noteworthy that the n-th term is an integer divisible by n, at least for the first few values of n.

PROBLEM 7. Find the "moments of inertia" for the mass on edges arising from edge-probabilities for random matchings of the a, b, c honeycomb graph.

3. Dominoes

Now let us turn from lozenge-tiling problems to domino-tiling problems. A *domino* is a 1-by-2 or 2-by-1 rectangle. Although lozenge tilings (in the guise of constrained plane partitions) were studied first, it was really the study of domino tilings in Aztec diamonds that gave current work on enumeration of matchings its current impetus. Here is the Aztec diamond of order 5:

A tiling of such a region by dominos is equivalent to a matching of a certain (dual) subgraph of the infinite square graph. This grid is bipartite, and it is convenient to color its vertices alternately black and white; equivalently, it is convenient to color the 1-by-1 squares alternately black and white, so that every domino contains one 1-by-1 square of each color. Elkies, Kuperberg, Larsen, and Propp showed in [Elkies et al. 1992] that the number of domino tilings of such a region is $2^{n(n+1)/2}$ (where $2n$ is the number of rows), and Gessel, Ionescu, and Propp proved in [Gessel et al. $\geq$ 1999] an exact formula (originally conjectured by Jockusch) for the number of tilings of regions like

in which two innermost squares of opposite color have been removed. (For some values of n, the number of tilings is exactly $\frac{1}{4}$ times $2^{n(n+1)/2}$; in the other cases, there is an exact product formula for the difference between the number of tilings and $\left(\frac{1}{4}\right)2^{n(n+1)/2}$. It is this latter fact that motivated the idea of trying something similar in the case of lozenge tilings, as described in the paragraph preceding the statement of Problem 4.)

Now suppose one removes two squares from the middle of an Aztec diamond of order n in the following way:

(The two squares removed are a knight's-move apart, and subject to that constraint, they are as close to being in the middle as they can be. Up to symmetries of the square, there is only one way of doing this.) The numbers of tilings one gets are as follows (for $n = 2$ through 10):

$$2$$
$$2^3$$
$$2^5 \cdot 5$$
$$2^9 \cdot 3^2$$
$$2^{17} \cdot 3$$
$$2^{22} \cdot 3^2$$
$$2^{24} \cdot 3^2 \cdot 73$$
$$2^{31} \cdot 3^2 \cdot 5^2 \cdot 11$$
$$2^{47} \cdot 3^2 \cdot 5$$

Only the presence of the large prime factor 73 makes one doubt that there is a general product formula; the other prime factors are reassuringly small.

PROBLEM 8. Count the domino tilings of an Aztec diamond from which two close-to-central squares, related by a knight's move, have been deleted.

PROGRESS. Harald Helfgott has solved this problem; it follows from the main result in his thesis [1998]. The formula is somewhat complicated, as the prime factor 73 might have led us to expect. (One of the factors in Helfgott's product formula is a single-indexed sum; 73 arises as $128 - 60 + 5$.)

One can also look at "Aztec rectangles" from which squares have been removed so as to restore the balance between black and white squares (a necessary condition for tileability). For instance, one can remove the central square from an a-by-b Aztec rectangle in which a and b differ by 1, with the larger of a, b odd:

PROBLEM 9. Find a formula for the number of domino tilings of a $2n$-by-$(2n+1)$ Aztec rectangle with its central square removed.

PROGRESS. This had already been solved when I posed the problem; it is a special case of a result of Ciucu [1997, Theorem 4.1]. Eric Kuo solved the problem independently.

What about $(2n-1)$-by-$2n$ rectangles? For these regions, removing the central square does not make the region tilable. However, if one removes any one of the four squares adjacent to the middle square, one obtains a region that is tilable, and moreover, for this region the number of tilings appears to be a nice round number.

PROBLEM 10. Find a formula for the number of domino tilings of a $(2n-1)$-by-$2n$ Aztec rectangle with a square adjoining the central square removed.

PROGRESS. This problem was solved independently three times: by Harald Helfgott and Ira Gessel [1999], by Christian Krattenthaler [1997], and by Eric Kuo (private communication). Gessel and Helfgott solve a more general problem than Problem 10. Krattenthaler's preprint gives several results concerning the enumeration of matchings of Aztec rectangles where (a suitable number of) collinear vertices are removed, of which Problem 10 is just a special case. There is some overlap between the results of Helfgott and Gessel and the results of Krattenthaler.

At this point, some readers may be wondering why m-by-n rectangles have not played a bigger part in the story. Indeed, one of the surprising facts of life in the study of enumeration of matchings is that Aztec diamonds and their kin have been much more fertile ground for exact combinatorics that the seemingly more natural rectangles. There are, however, a few cases I know of in which something rather nice turns up. One is the problem of Ira Gessel that appears as Problem 20 in this document. Another is the work done by Jockusch [1994] and, later, Ciucu [1997] on why the number of domino tilings of the square is always either a perfect square or twice a perfect square. In the spirit of the work of Jockusch and Ciucu, I offer here a problem based on Lior Pachter's observation [Pachter and Kim 1998] that the region on the left below, obtained by removing 8 dominos from a 16-by-16 square, has exactly one tiling. What if we make the intrusion half as long, as in the region on the right?

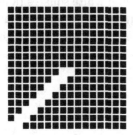 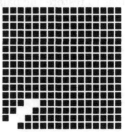

That is, we take a $2n$-by-$2n$ square (with n even) and remove $n/2$ dominos from it, in a partial zig-zag pattern that starts from the corner. Here are the numbers we get, in factored form, for $n = 2, 4, 6, 8, 10$:

$$2 \cdot 3^2$$
$$2^2 \cdot 3^6 \cdot 13^2$$
$$2^3 \cdot 3^2 \cdot 5^4 \cdot 7^2 \cdot 3187^2$$
$$2^4 \cdot 11771899^2 \cdot 27487^2$$
$$2^5 \cdot 2534588575976069659^2$$

The factors are ugly, but the exponents are nice: we get $2^{n/2}$ times an odd square.

Perhaps this is a special case of a two-parameter fact that says that you can take an intrusion of length m in a $2n$-by-$2n$ square and the number of tilings of the resulting region will always be a square or twice a square.

PROBLEM 11. What is going on with "intruded Aztec diamonds"? In particular, why is the number of tilings so square-ish?

It should also be noted that the square root of the odd parts of these numbers $(3, 3^3 \cdot 13,$ etc.) alternate between 1 and 3 mod 4. Perhaps these quantities are continuous functions of n in the 2-adic sense, as is the case for intact $2n$-by-$2n$ squares [Cohn 1999]; however, the presence of large prime factors means that no simple product formula is available, and that the analysis will require new techniques.

We now return to the Kasteleyn–Percus matrices discussed earlier. Work of Rick Kenyon and David Wilson [Kenyon 1997] has shown that the *inverses* of these matrices are loaded with combinatorial information, so it would be nice to get our hands on them. Unfortunately, there are many nonzero entries in the inverse-matrices. (Recall that the Kasteleyn–Percus matrices themselves, being nothing more than adjacency matrices in which some of the 1's have been strategically replaced by -1's, are sparse; their inverses, however, tend to have most if not all of their entries nonzero.) Nonetheless, some exploratory "numerology" leaves room for hope that this is do-able.

Consider the Kasteleyn–Percus matrix K_n for the Aztec diamond of order n, in which every vertical domino with its white square on top (relative to some fixed checkerboard coloring) has its sign inverted — that is, the corresponding 1 in the bipartite adjacency matrix is replaced by -1.

PROBLEM 12. Show that the sum of the entries of the matrix inverse of K_n is $\frac{1}{2}(n-1)(n+3) - 2^{n-1} + 2$.

(This formula works for $n = 1$ through $n = 8$.)

PROGRESS. Harald Helfgott has solved a similar problem using the main result of his thesis [1998], and it is likely that the result asserted in Problem 12 can be proved similarly. (A slight technical hurdle arises from the fact that Helfgott's thesis uses a different sign-convention for the Kasteleyn–Percus matrix, which results in different signs, and a different sum, for the inverse matrix; however, Helfgott's methods are quite general, so there is no conceptual obstacle to applying them to Problem 12.)

I should mention that my original reason for examining the sum of the entries of the inverse Kasteleyn–Percus matrix was to see whether there might be formulas governing the individual entries themselves. Helfgott's work provides such formulas.

Also, in this connection, Greg Kuperberg and Douglas Zare have some high-tech ruminations on the inverses of Kasteleyn–Percus matrices, and there is a chance that representation-theory methods will give a different way of proving the result.

Now we turn to a class of regions I call "pillows". Here are a "0 mod 4" pillow of "order 5" and a "2 mod 4" pillow of "order 7":

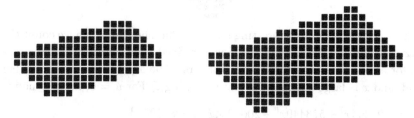

It turns out (empirically) that the number of tilings of the 0-mod-4 pillow of order n is a perfect square times the coefficient of x^n in the Taylor expansion of $(5+3x+x^2-x^3)/(1-2x-2x^2-2x^3+x^4)$. This fact came to light in several steps. First it was noticed that the number of tilings has a comparatively small square-free part. Then it was noticed that in the derived sequence of square-free parts, many terms were roughly three times the preceding term. Then it was noticed that, by judiciously including some of the square factors, one could obtain a sequence in which each term was roughly three times the preceding term. Finally it was noticed that this approximately geometric sequence satisfied a fourth-order linear recurrence relation.

Similarly, it appears that the number of tilings of the 2-mod-4 pillow of order n is a perfect square times the coefficient of x^n in the Taylor expansion of $(5+6x+3x^2-2x^3)/(1-2x-2x^2-2x^3+x^4)$. (If you are wondering about "odd pillows", I should mention that there is a nice formula for the number of tilings, but this is not an interesting result, because an odd pillow splits up into many small noncommunicating sub-regions such that a tiling of the whole region corresponds to a choice of tiling on each of the sub-regions.)

PROBLEM 13. Find a general formula for the number of domino tilings of even pillows.

Jockusch looked at the Aztec diamond of order n with a 2-by-2 hole in the center, for small values of n; he came up with a conjecture for the number of domino tilings, subsequently proved by Gessel, Ionescu, and Propp [Gessel et al. $\geq$ 1999]. One way to generalize this is to make the hole larger, as was suggested by Douglas Zare and investigated by David Wilson. Here is an abridged and adapted version of the report David Wilson sent me on October 15, 1996:

Define the Aztec window with outer order y and inner order x to be the Aztec diamond of order y with an Aztec diamond of order x deleted from its center. For example, this is the Aztec window with orders 8 and 2:

There are a number of interesting patterns that show up when we count tilings of Aztec windows. For one thing, if w is a fixed even number, and $y = x+w$, then for any w the number of tilings appears to be a polynomial in x. (When w is odd, and x is large enough, there are no tilings.) For $w = 6$, the polynomial is

$8192x^8 + 98304x^7 + 573440x^6 + 2064384x^5 + 4988928x^4$
$$+8257536x^3 + 9175040x^2 + 6291456x + 2097152.$$

This can be written as

$$2^{17}\left(\tfrac{1}{2}\left(x+\tfrac{3}{2}\right)^2 + \tfrac{7}{8}\right)^4$$

or as

$$2^{17}x^4 \; \circ \; \tfrac{1}{2}x+\tfrac{7}{8} \; \circ \; \left(x+\tfrac{3}{2}\right)^2,$$

where it is understood that these three polynomials get composed.

More generally, all the polynomials in x that arise in this fashion appear to "factor" in the sense of functional composition. Here are the factored forms of the polynomials for $n = 2, 4, 6, 8, 10$:

$$2^3 x^4 \quad \circ \qquad 1 \qquad \circ \quad \left(x + \tfrac{1}{2}\right)^2$$

$$2^8 x^2 \quad \circ \qquad x + 1 \qquad \circ \quad \left(x + 1\right)^2$$

$$2^{17} x^4 \quad \circ \qquad \tfrac{1}{2}x + \tfrac{7}{8} \qquad \circ \quad \left(x + \tfrac{3}{2}\right)^2$$

$$2^{28} x^2 \quad \circ \quad \tfrac{1}{144}x^4 + \tfrac{7}{72}x^3 + \tfrac{41}{144}x^2 + \tfrac{11}{18}x + 1 \quad \circ \quad \left(x + 2\right)^2$$

$$2^{43} x^4 \quad \circ \quad \tfrac{1}{144}x^3 + \tfrac{61}{576}x^2 + \tfrac{451}{2304}x + \tfrac{967}{1024} \quad \circ \quad \left(x + \tfrac{5}{2}\right)^2$$

In general the rightmost polynomial is $(x + w/4)^2$, and the leftmost polynomial is either a perfect square, twice a fourth power, or half a fourth power, depending on $w \bmod 8$. A pattern for the middle polynomial however is elusive.

PROBLEM 14. Find a general formula for the number of domino tilings of Aztec windows.

PROGRESS. Constantin Chiscanu found a polynomial bound on the number of domino tilings of the Aztec window of inner order x and outer order $x + w$ [Chiscanu 1997]. Douglas Zare used the transfer-matrix method to show that the number of tilings is not just bounded by a polynomial, but given by a polynomial, for each fixed w [Zare 1997–98].

4. Miscellaneous

Now we come to some problems involving tiling that fit neither the domino-tiling nor the lozenge-tiling framework. Here the more general picture is that we have some periodic dissection of the plane by polygons, such that an even number of polygons meet at each vertex, allowing us to color the polygons alternately black or white. We then make a suitable choice of a finite region R composed of equal numbers of black and white polygons, and we look at the number of "diform" tilings of the region, where a *diform* is the union of two polygonal cells that share an edge. In the case of domino tilings, the underlying dissection of the infinite plane is the tiling by squares, 4 around each vertex; in the case of lozenge tilings, the underlying dissection of the infinite plane is the tiling by equilateral triangles, 6 around each vertex.

Other sorts of periodic dissections have already played a role in the theory of enumeration of matchings. For instance, there is a tiling of the plane by isosceles right triangles associated with a discrete reflection group in the plane; in this case, the right choice of R (see Figure 3) gives us a region that can be tiled in $5^{n^2/4}$ ways when n is even and in $5^{(n^2-1)/2}$ or $2 \cdot 5^{(n^2-1)/2}$ ways when n is odd [Yang 1991].

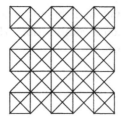

Figure 3. A fortress of order 5, with 2×5^6 diform tilings.

Similarly, in the tiling of the plane by triangles that comes from a 30 degree, 60 degree, 90 degree right triangle by repeatedly reflecting it in its edges, a certain region called the "Aztec dungeon" (see Figure 4) gives rise to a tiling problem in which powers of 13 occur (as was proved in not-yet-published work of Mihai Ciucu).

A key feature of these regions R is revealed by looking at the colors of those polygons in the dissection that share an edge with the border of R. One sees that the border splits up into four long stretches such that along each stretch, all the polygons that touch the border have the same color. It is not clear why regions with this sort of property should be the ones that give rise to the nicest enumerations, but this appears to happen in practice.

One interesting case arises from a rather symmetric dissection of the plane into equilateral triangles, squares, and regular hexagons, with 4 polygons meeting at each vertex and with no two squares sharing an edge. A typical diform tiling of this region (called a "dragon") is shown in Figure 5. Empirically, one finds that the number of diform tilings is $2^{n(n+1)}$.

PROBLEM 15. Prove that the number of diform tilings of the dragon of order n is $2^{n(n+1)}$.

PROGRESS. Ben Wieland solved this problem (private communication).

Incidentally, the tiling shown in Figure 5 was generated using an algorithm that generates each of the possible diform tilings of the region with equal probability. It is no fluke that the tiling looks so orderly in the left and right corners of the region; this appears to be typical behavior in situations of this kind. This phenomenon has been analyzed rigorously for two tiling-models: lozenge tilings

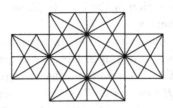

Figure 4. An Aztec dungeon of order 2, with 13^3 diform tilings.

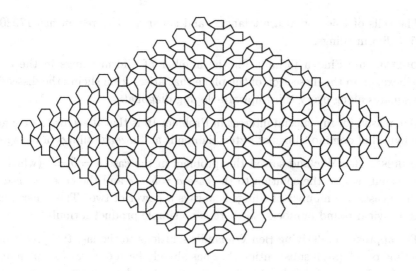

Figure 5. A dragon of order 10 (tiled).

of hexagons [Cohn et al. 1998b] and domino tilings of Aztec diamonds [Cohn et al. 1996].

One way to get a new dissection of the plane from an old one is to refine it. For instance, starting from the dissection of the plane into squares, one can draw in every k-th southwest-to-northeast diagonal. When k is 1, this is just a distortion of the dissection of the plane into equilateral triangles. When k is 2, this is a dissection that leads to finite regions for which the number of diform tilings is a known power of 2, thanks to a theorem of Chris Douglas [1996]. But what about $k = 3$ and higher?

For instance, we have the roughly hexagonal region shown in Figure 6; certain boundary vertices have been marked with a dot so as to bring out the large-scale $2, 3, 2, 2, 3, 2$ hexagonal structure more clearly.

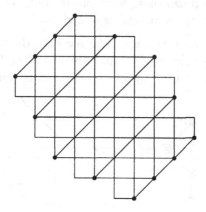

Figure 6. A region for Problem 16.

The cells of this region are triangles and squares. The region has $17920 = 2^9 \cdot 5 \cdot 7$ diform tilings.

PROBLEM 16. Find a formula for the number of diform tilings in the a, b, c quasihexagon in the dissection of the plane that arises from slicing the dissection into squares along every third upward-sloping diagonal.

One reason for my special interest in Problem 16 is that it seems to be a genuine hybrid of domino tilings of Aztec diamonds and lozenge tilings of hexagons.

PROGRESS. Ben Wieland solved this problem in the case $a = b = c$ (which, as it turns out, is also the solution to the case $a = b < c$ and the case $a = c < b$). In these cases the number of tilings is always a power of two. The general case does not yield round numbers, so there is no simple product formula.

The approach underlying Ben Wieland's solutions to the last two problems is a method of subgraph substitution that has already been of great use in enumeration of matchings of graphs. I will not go into great detail here on this method [Propp 1996; ≥ 1999], but here is an overview: One studies graphs with weights assigned to their edges, and one does weighted enumeration of matchings, where the weight of a matching is the product of the weights of the constituent edges. One then looks at local substitutions of subgraphs within a graph that preserve the sum of the weights of the matchings, or more generally, multiply the sum of the weights of the matchings by some predictable factor. Then the problem of weight-enumerating matchings of one graph reduces to the problem of weight-enumerating matchings of another graph. Iterating this procedure, one can often eventually reduce the graph to something easier to understand.

Problems 15 and 16 are just two instances of a broad class of problems arising from periodic graphs in the plane. A unified understanding of this class of problems has begun to emerge, by way of subgraph substitution. The most important open problem connected with this class of results is the following:

PROBLEM 17. Characterize those local substitutions that have a predictable effect on the weighted sum of matchings of a graph.

The most useful local substitution so far has been the one shown in Figure 7, where unmarked edges have weight 1 and where A, B, C, D are respectively obtained from a, b, c, d by dividing by $ad + bc$; if G and G' denote the graph before

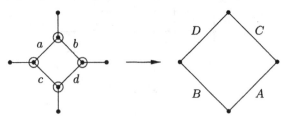

Figure 7. The "urban renewal" substitution.

and after the substitution, one can check that the sum of the weights of the matchings of G' equals the sum of the weights of the matchings of G divided by $ad+bc$.

It is required that the four innermost vertices have no neighbors other than the four vertices shown; this constraint is indicated by circling them. Noncircled vertices may have any number of neighbors.

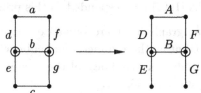

Figure 8. Rick Kenyon's substitution.

The substitution shown in Figure 8 (a straightforward generalization of a clever substitution due to Rick Kenyon) has also been of use. Here the new weights are not entirely determined by the old, but have a single degree of freedom; the relevant formulas can be written as

$$A = \frac{abc+aeg+cdf}{bc+eg}, \quad B = b, \quad D = \frac{dg}{bc+eg}E, \quad F = ef\frac{1}{E}, \quad G = (bc+eg)\frac{1}{E},$$

with E free. As before, the circled vertices must not have any neighbors other than the ones shown. In this case, the sum of the weights in the before-graph G is exactly equal to the sum of the weights in the after-graph G'; there is no need for a correction factor like the $1/(ad+bc)$ that arises in urban renewal.

The extremely powerful "wye-delta" substitution of Colbourn, Provan, and Vertigan [Colbourn et al. 1995] should also be mentioned.

Up till now we have been dealing exclusively with bipartite planar graphs. We now turn to the less well-explored nonbipartite case.

For instance, one can look at the triangle graph of order n, shown in Figure 9 in the case $n = 4$. (Here n is the number of vertices in the longest row.)

Let $M(n)$ denote the number of matchings of the triangle graph of order n. When n is 1 or 2 mod 4, the graph has an odd number of vertices and $M(n)$ is 0; hence let us only consider the cases in which n is 0 or 3 mod 4. Here are the first few values of $M(n)$, expressed in factored form: 2, $2 \cdot 3$, $2 \cdot 2 \cdot 3 \cdot 3 \cdot 61$, $2 \cdot 2 \cdot 11 \cdot 29 \cdot 29$, $2^3 \cdot 3^3 \cdot 5^2 \cdot 7^2 \cdot 19 \cdot 461$, $2^3 \cdot 5^2 \cdot 37^2 \cdot 41 \cdot 139^2$, $2^4 \cdot 73 \cdot 149 \cdot 757 \cdot 33721 \cdot 523657$,

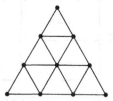

Figure 9. The triangle graph.

$2^4 \cdot 3^8 \cdot 17 \cdot 37^2 \cdot 703459^2, \ldots$. It is interesting that $M(n)$ seems to be divisible by $2^{\lfloor(n+1)/4\rfloor}$ but no higher power of 2; it is also interesting that when we divide by this power of 2, in the case where n is a multiple of 4, the quotient we get, in addition to being odd, is a perfect square times a small number $(3, 11, 41, 17, \ldots)$.

PROBLEM 18. How many matchings does the triangle graph of order n have?

PROGRESS. Horst Sachs [1997] has responded to this problem.

One can also look at graphs that are bipartite but not planar. A natural example is the n-cube (that is, the n-dimensional cube with 2^n vertices). It has been shown that the number of matchings of the n-cube goes like 1, 2, $9 = 3^2$, $272 = 16 \cdot 17$, $589185 = 3^2 \cdot 5 \cdot 13093, \ldots$.

PROBLEM 19. Find a formula for the number of matchings of the n-cube.

(This may be intractable; after all, the graph has exponentially many vertices.)

PROGRESS. László Lovász gave a simple proof of my (oral) conjecture that the number of matchings of the n-cube has the same parity as n itself. Consider the orbit of a particular matching of the n-cube under the group generated by the n standard reflections of the n-cube. If all the edges are parallel (which can happen in exactly n ways), the orbit has size 1; otherwise the size of the orbit is of the form 2^k (with $k \geq 1$) — an even number. The claim follows, and similar albeit more complex reasoning should allow one to compute the enumerating sequence modulo any power of 2. Meanwhile, L. H. Clark, J. C. George, and T. D. Porter have shown [Clark et al. 1997] that if one lets $f(n)$ denote the number of 1-factors in the n-cube, then

$$f(n)^{2^{1-n}} \sim n/e$$

as $n \to \infty$. It was subsequently pointed out by Bruce Sagan that the main result of Clark et al. [1997] is a special case of the theorem cited by Lovász and Plummer [1986, top of page 312].

Finally, we turn to a problem involving domino tilings of rectangles, submitted by Ira Gessel (what follows are his words):

We consider dimer coverings of an $m \times n$ rectangle, with m and n even. We assign a vertical domino from row i to row $i+1$ the weight $\sqrt{y_i}$ and a horizontal domino from column j to column $j+1$ the weight $\sqrt{x_j}$. For example, the covering

for $m = 2$ and $n = 10$ has weight $y_1^2 x_2 x_5 x_7$. (The weight will always be a product of integral powers of the x_i and y_j.)

Now I'll define what I call "dimer tableaux." Take an $m/2$ by $n/2$ rectangle and split it into two parts by a path from the lower left corner to the upper right corner. For example (with $m = 6$ and $n = 10$)

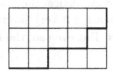

Then fill in the upper left part with entries from $1, 2, \ldots, n-1$ so that for adjacent entries $\boxed{i\,j}$ we have $i < j-1$ and for adjacent entries $\boxed{\genfrac{}{}{0pt}{}{i}{j}}$ we have $i \le j+1$, and fill in the lower-right partition with entries from $1, 2 \ldots, m-1$ with the reverse inequalities ($\boxed{i\,j}$ implies $i \le j+1$ and $\boxed{\genfrac{}{}{0pt}{}{i}{j}}$ implies $i < j-1$). We weight an i in the upper-left part by x_i and a j in the lower-right part by y_j.

THEOREM 1. *The sum of the weights of the $m \times n$ dimer coverings is equal to the sum of the weights of the $m/2 \times n/2$ dimer tableaux.*

My proof is not very enlightening; it essentially involves showing that both of these are counted by the same formula.

PROBLEM 20. Is there an "explanation" for this equality? In particular, is there a reasonable bijective proof? Notes:

(1) The case $m = 2$ is easy: the 2×10 dimer covering above corresponds to the 1×5 dimer tableau

x_2	x_5	x_7	y_1	y_1

(there's only one possibility!).

(2) If we set $x_i = y_i = 0$ when i is even (so that every two-by-two square of the dimer covering may be chosen independently), then the equality is equivalent to the identity

$$\prod_{i,j}(x_i + y_j) = \sum_{\lambda} s_\lambda(x) s_{\tilde{\lambda}'}(y);$$

compare [Macdonald 1995, p. 37]. This identity can be proved by a variant of Schensted's correspondence, so a bijective proof of the general equality would be essentially a generalization of Schensted. Several people have looked at the problem of a Schensted generalization corresponding to the case in which $y_i = 0$ when i is even.

(3) The analogous results in which m or n is odd are included in the case in which m and n are both even. For example, if we take $m = 4$ and set $y_3 = 0$, then the fourth row of a dimer covering must consist of $n/2$ horizontal dominoes, which contribute $\sqrt{x_1 x_3 \cdots x_{n-1}}$ to the weight, so we are essentially looking at dimer coverings with three rows.

PROGRESS. A special case of the Robinson–Schensted algorithm given by Sundquist et al. [1997] can be used to get a bijection for a special case of the problem,
in which one sets $y_i = 0$ for all i even, so that we are looking at dimer coverings
(or domino tilings) in which every vertical domino goes from row $2i+1$ to row
$2i+2$ for some i. These tilings are not very interesting because they break up
into tilings of 2-by-n rectangles. But even so, the Robinson–Schensted bijection
is nontrivial.

5. New Problems

Let $N(a,b)$ denote the number of matchings of the a-by-b rectangular grid.
Kasteleyn showed that $N(a,b)$ is equal to the square root of the absolute value
of

$$\prod_{j=1}^{a} \prod_{k=1}^{b} \left(2\cos\frac{\pi j}{a+1} + 2i\cos\frac{\pi k}{b+1} \right).$$

Some number-theoretic properties of $N(a,b)$ follow from this representation (see,
e.g., [Cohn 1999]) but lack a combinatorial explanation. The next two problems
describe two such facts.

PROBLEM 21.   Give a combinatorial proof of the fact that $N(a,b)$ divides
$N(A,B)$ whenever $a+1$ divides $A+1$ and $b+1$ divides $B+1$.

PROGRESS. Bruce Sagan has given an answer in the "Fibonacci case" $a = 2$. A
matching of a 2-by-$(kn-1)$ grid either splits up as a matching of a 2-by-$(n-1)$
grid on the left and a 2-by-$(kn-n)$ grid on the right or it splits up as a matching
of a 2-by-$(n-2)$ grid on the left, a horizontal matching of a 2-by-2 grid in the
middle, and a matching of a 2-by-$(kn-n-1)$ grid on the right. Hence

$$N(2, kn-1) = N(2, n-1)N(2, kn-n) + N(2, n-2)N(2, (k-1)n-1).$$

From this formula one can prove that $N(2, n-1)$ divides $N(2, kn-1)$ by induction on k. Volker Strehl has approached the problem in a different way; his ideas
make it seem likely that a better combinatorial understanding of resultants, in
combination with known interpretations of Chebyshev polynomials, would be
helpful in approaching this problem.

PROBLEM 22.   Give a combinatorial proof of the fact that $N(a, 2a)$ is always
congruent to 1 mod 4.

(Pachter [1997] has demonstrated the sort of combinatorial methods one can
use in such problems.)

Even without Kasteleyn's formula, it is easy to show (e.g., via the transfer-
matrix method) that for any fixed a, the sequence of numbers $N(a,b)$ (with b
varying) satisfies a linear recurrence relation with constant coefficients. Indeed,
consider all 2^a different ways of removing some subset of the a rightmost vertices

in the a-by-b grid; this gives us 2^a "mutilated" versions of the graph. We can set up recurrences that link matchings of mutilated graphs of width b with matchings of mutilated graphs of width b and $b-1$, and standard algebraic methods allow us to turn this system of joint mutual recurrences of low degree into a single recurrence of high degree governing the particular sequence of interest, which enumerates matchings of unmutilated rectangles. The recurrence obtained in this way is not, however, best possible, as one can see even in the simple case $a = 2$.

PROBLEM 23 (STANLEY). Prove or disprove that the minimum degree of a linear recurrence governing the sequence $N(a,1), N(a,2), N(a,3), \ldots$ is $2^{\lfloor (a+1)/2 \rfloor}$.

PROGRESS. Observations made by Stanley [1985, p. 87] imply that the conjecture is true when $a+1$ is an odd prime.

The idea of mutilating a graph by removing some vertices along its boundary leads us to the next problem. It has been observed for small values of n that if one removes equal numbers of black and white vertices from the boundary of a $2n$-by-$2n$ square grid, the number of matchings of the mutilated graph is less than the number of matchings of the original graph. In fact, it appears to be true that one can delete *any* subset of the vertices of the square grid and obtain an induced graph with strictly fewer matchings than the original.

It is worth pointing out that not every graph shares this property with the square grid. For instance, if G is the Aztec diamond graph of order 5 and G' is the graph obtained from G by deleting the middle vertices along the northwest and northeast borders, then G has 32768 matchings while G' has 59493.

PROBLEM 24. Prove or disprove that every subgraph of the $2n$-by-$2n$ grid graph has strictly fewer matchings.

Next we come to a variant on the Aztec dungeon region shown in Figure 4. Figure 10 shows an "hexagonal dungeon" with sides $2, 4, 4, 2, 4, 4$. Matt Blum's investigation of these shapes has led him to discover many patterns; the most striking of these patterns forms the basis of the next problem.

PROBLEM 25. Show that the hexagonal dungeon with sides $a, 2a, b, a, 2a, b$ has exactly

$$13^{2a^2} 14^{\lfloor a^2/2 \rfloor}$$

diform tilings, for all $b \geq 2a$.

Unmatchable bipartite graphs can sometimes give rise to interesting quasi-matching problems, either by way of KK^* (see Problem 6) or by systematic addition or deletion of vertices or edges. The former sort of problem simply asks for the determinant of KK^* (where we may assume that K has more columns than rows). When the underlying graph has equal numbers of black and white

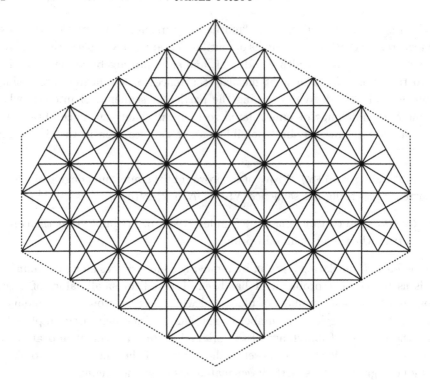

Figure 10. An hexagonal dungeon.

vertices, this is just the square of the number of matchings, but when K is a rectangular matrix, KK^* will in general have a nonzero determinant, even though
the graph has no matchings.

PROBLEM 26. Calculate the determinant of KK^* where K is the Kasteleyn–
Percus matrix of the a, b, c, d, e, f honeycomb graph.

(Note that in this case we can simply take K to be the bipartite adjacency
matrix of the graph.)

Cases of special interest are $a, b+1, c, a+1, b, c+1$ and a, b, a, b, a, b hexagons.
These two cases overlap in the one-parameter family of $a, a+1, a, a+1, a, a+1$
hexagons. For instance, in the case of the $3, 4, 3, 4, 3, 4$ hexagon, $\det(KK^*)$ is
$2^8 \cdot 3^3 \cdot 7^6$.

PROBLEM 27. Calculate the determinant of KK^* where K is the Kasteleyn–
Percus matrix of an m-by-n Aztec rectangle, or where K is the Kasteleyn–Percus
matrix of the "fool's diamond" of order n. (The fool's diamond of order 3 is the
following region:

Fool's diamonds of higher orders are defined in a similar way.)

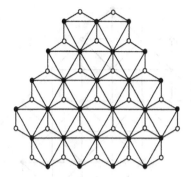

Figure 11. A hexagon with extra edges.

PROGRESS. In the case of Aztec rectangles, Matt Blum has found general formulas for $\det(KK^*)$ when m is 1, 2, or 3. For fool's diamonds, we get

$$1$$
$$2$$
$$3 \cdot 5$$
$$2^7 \cdot 3$$
$$3^2 \cdot 5^3 \cdot 29$$
$$2^9 \cdot 3 \cdot 5 \cdot 7 \cdot 13^2$$
$$7^3 \cdot 13^4 \cdot 29^2$$
$$2^{25} \cdot 3 \cdot 7^2 \cdot 17^3$$

(One might also look at "fool's rectangles".)

Another thing one can do with an unmatchable graph is add extra edges. Even when this ruins the bipartiteness of the graph, there can still be interesting combinatorics. For instance, consider the $2, 4, 2, 4, 2, 4$ hexagon-graph; it has an even number of vertices, but it has a surplus of black vertices over white vertices. We therefore introduce edges between every black vertex and the six nearest black vertices. (That is, in each hexagon of the honeycomb, we draw a triangle connecting the three black vertices, as in Figure 11.) Then the graph has $5187 = 3 \cdot 7 \cdot 13 \cdot 19$ matchings.

PROBLEM 28. Count the matchings of the a, b, c, d, e, f hexagon-graph in which extra edges have been drawn connecting vertices of the majority color.

What works for honeycomb graphs works (or seems to work) for square-grid graphs as well. If one adds edges joining each vertex of majority color to the four nearest like-colored vertices in the n by $n+2$ Aztec rectangle graph as in Figure 12, one gets a graph for which the number of matchings grows like $2^2 \cdot 3$, $2^3 \cdot 3 \cdot 7$, $2^7 \cdot 3 \cdot 11$, $2^{17} \cdot 5 \cdot 31$, etc. If one does the same for the holey $2n-1$ by $2n$ Aztec rectangle from which the central vertex has been removed, as in Figure 13, one gets the numbers $2^6 \cdot 7$, $2^9 \cdot 3^2 \cdot 13 \cdot 17$, $2^{23} \cdot 5^3 \cdot 31$, etc.

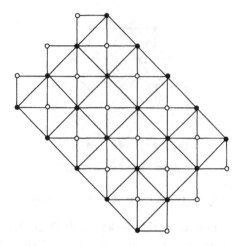

Figure 12. An Aztec rectangle with extra edges.

PROBLEM 29. Count the matchings of the a by b Aztec rectangle (with $a+b$ even) in which extra edges have been drawn connecting vertices of the majority color. Do the same for the $2n-1$ by $2n$ holey Aztec rectangle.

Other examples of nonbipartite graphs for which the number of matchings has only small prime factors arise when one takes the quotient of a symmetrical bipartite graph modulo a symmetry-group at least one element of which interchanges the two colors; Kuperberg [1994] gives some examples of this. In general, there seem to be fewer product-formula enumerations of matchings for nonbipartite graphs than for bipartite graphs. Nevertheless, even in cases where no product formula has been found, there can be patterns in need of explanation.

Consider the one-parameter family of graphs illustrated in Figure 14 for the case $n = 7$ (based on the same nonbipartite infinite graph as Figures 12 and 13). Such a graph has an even number of vertices whenever n is congruent to 0 or 3

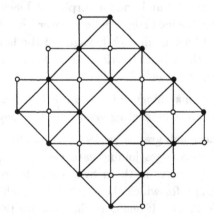

Figure 13. A holey Aztec rectangle with extra edges.

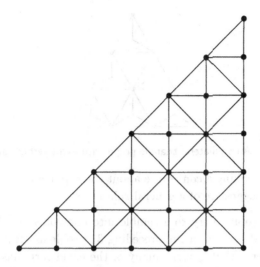

Figure 14. An isosceles right triangle graph with extra edges.

modulo 4. Here are the data for the first few cases, courtesy of Matt Blum:

n	number of matchings	factorization
3	3	3
4	6	$2 \cdot 3$
7	1065	$3 \cdot 5 \cdot 71$
8	6276	$2^2 \cdot 3 \cdot 523$
11	45949563	$3^2 \cdot 11 \cdot 464137$
12	807343128	$2^3 \cdot 3^2 \cdot 1109 \cdot 10111$
15	221797080594801	$3^2 \cdot 24644120066089$
16	11812299253803024	$2^4 \cdot 3 \cdot 246089567787563$
19	117066491250943949567763	$3 \cdot 89 \cdot 28289 \cdot 15499002371714201$
20	191008032503971486607852640	$2^5 \cdot 3^2 \cdot 5 \cdot 41 \cdot 367 \cdot 881534305952328473$

The following problem describes some of Blum's conjectures:

PROBLEM 30. Show that for the isosceles right triangle graph with extra edges, the number of matchings is always a multiple of 3. Furthermore, show that the exact power of 2 dividing the number of matchings is $2^{n/4}$ when n is 0 modulo 4, and $2^0 (= 1)$ when n is 3 modulo 4.

This property of divisibility by 3 pops up in another problem of a similar flavor. Consider the graph shown in Figure 15, which is just like the one shown in Figure 9, except that half of the triangular cells have an extra vertex in them, connected to the three nearest vertices. (Note also the resemblance to Figure 11.)

PROBLEM 31. Show that for the equilateral triangle graph with extra vertices and edges, the number of matchings is always a multiple of 3.

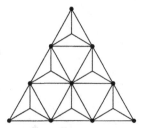

Figure 15. An equilateral triangle graph with extra vertices and edges.

(I refrain from making a conjecture about the exponent of 2, though the data contain patterns suggestive of a general rule.)

It may be too soon to try to assemble into one coherent picture all the diverse phenomena discussed in the preceding 31 problems. But I have noticed a gratuitous symmetry that governs many of the exact formulas, and I will close by pointing it out. Consider, for example, the MacMahon–Macdonald product

$$M_n = \prod_{i=0}^{n-1} \prod_{j=0}^{n-1} \prod_{k=0}^{n-1} \frac{i+j+k+2}{i+j+k+1}$$

that counts matchings of the n, n, n semiregular honeycomb graph. We find that the "second quotient" $M_{n-1}M_{n+1}/M_n^2$ is the rational function

$$\frac{27}{64} \frac{(3n-2)(3n-1)^2(3n+1)^2(3n+2)}{(2n-1)^3(2n+1)^3}$$

which is an even function of n.

The right hand side in Bo-Yin Yang's theorem (giving the number of diabolo tilings of a fortress of order n) has a power of 5 whose exponent is $n^2/4$ when n is even and $(n^2-1)/4$ when n is odd; this too is an even function of n.

Domino tilings of Aztec diamonds are enumerated by the formula $2^{n(n+1)}$. Here the symmetry is a bit different: replacing n by $-1-n$ leaves the answer unaffected.

The right hand side of Mihai Ciucu's theorem (giving the number of diform tilings of an Aztec dungeon of order n) has a power of 13 whose exponent is $(n+1)^2/3$ or $n(n+2)/3$ (according to whether or not n is 2 mod 3). so that the symmetry corresponds to replacing n by $-2-n$.

There are other instances of this kind that arise, in which some base is raised to the power of some quadratic function of n; in each case, the quadratic function admits a symmetry that preserves the integrality of n (unlike, say, the quadratic function $n(3n+1)/2$, which as a function from integers to integers does not possess such a symmetry).

PROBLEM 32. For many of our formulas, the "algebraic" (right hand) side is invariant under substitutions that make the "combinatorial" (left hand) side meaningless, insofar as one cannot speak of graphs with negative numbers of vertices or edges. Might this invariance nonetheless have some deeper significance?

Cohn [1999] has found another example of gratuitous symmetry related to tilings.

Acknowledgements

This research was conducted with the support of the National Science Foundation, the National Security Agency, and the M.I.T. Class of 1922 Career Development fund. I am deeply indebted to the past and present members of the Tilings Research Group for their many forms of assistance: Pramod Achar, Karen Acquista, Josie Ammer, Federico Ardila, Rob Blau, Matt Blum, Carl Bosley, Ruth Britto-Pacumio, Constantin Chiscanu, Henry Cohn, Chris Douglas, Edward Early, Nicholas Eriksson, David Farris, Lukasz Fidkowski, Marisa Gioioso, David Gupta, Harald Helfgott, Sharon Hollander, Dan Ionescu, Sameera Iyengar, Julia Khodor, Neelakantan Krishnaswami, Eric Kuo, Yvonne Lai, Ching Law, Andrew Menard, Alyce Moy, Anne-Marie Oreskovich, Ben Raphael, Vis Taraz, Jordan Weitz, Ben Wieland, Lauren Williams, David Wilson, Jessica Wong, Jason Woolever, and Laurence Yogman. I also acknowledge the helpful comments on this manuscript given by Henry Cohn and Richard Stanley, and the information provided by Jerry Dias, Michael Fisher and Horst Sachs concerning the connections between matching theory and the physical sciences.

References

Many of the papers in the bibliography are included as e-prints in the xxx mathematics archive; this is indicated by the inclusion of a `math.CO` identifier. Such e-prints can currrently be accessed across the World Wide Web by appending the identifier to the URL http://front.math.ucdavis.edu/ or http://xxx.lanl.gov/abs/.

[Andrews 1976] G. E. Andrews, *The theory of partitions*, vol. 2, Encyclopedia of Mathematics and its Applications, Addison-Wesley, Reading, MA, 1976. Reprinted by Cambridge University Press, New York, 1998.

[Baxter 1982] R. J. Baxter, *Exactly solved models in statistical mechanics*, Academic Press, London, 1982. Reprinted 1989.

[Carlitz and Stanley 1975] L. Carlitz and R. P. Stanley, "Branchings and partitions", *Proc. Amer. Math. Soc.* **53**:1 (1975), 246–249.

[Chiscanu 1997] C. Chiscanu, "Comments on Propp's problem 14", 1997. Available at http://math.wisc.edu/~propp/chiscanu.ps.gz.

[Ciucu 1997] M. Ciucu, "Enumeration of perfect matchings in graphs with reflective symmetry", *J. Combin. Theory Ser. A* **77**:1 (1997), 67–97.

[Ciucu 1998] M. Ciucu, "Enumeration of lozenge tilings of punctured hexagons", *J. Combin. Theory Ser. A* **83**:2 (1998), 268–272.

[Ciucu and Krattenthaler 1999]   M. Ciucu and C. Krattenthaler, "The number of centered lozenge tilings of a symmetric hexagon", *J. Combin. Theory Ser. A* **86** (1999), 103–126.

[Clark et al. 1997]   L. H. Clark, J. C. George, and T. D. Porter, "On the number of 1-factors in the *n*-cube", *Congr. Numer.* **127** (1997), 67–69.

[Cohn 1999]   H. Cohn, "2-adic behavior of numbers of domino tilings", *Electron. J. Combin.* **6** (1999), R14.

[Cohn et al. 1996]   H. Cohn, N. Elkies, and J. Propp, "Local statistics for random domino tilings of the Aztec diamond", *Duke Math. J.* **85**:1 (1996), 117–166.

[Cohn et al. 1998a]   H. Cohn, R. Kenyon, and J. Propp, "A variational principle for domino tilings", 1998. Available at http://www.math.wisc.edu/~propp/variational.ps Submitted to *J. Amer. Math. Soc.*

[Cohn et al. 1998b]   H. Cohn, M. Larsen, and J. Propp, "The shape of a typical boxed plane partition", *New York J. Math.* **4** (1998), 137–165.

[Colbourn et al. 1995]   C. J. Colbourn, J. S. Provan, and D. Vertigan, "A new approach to solving three combinatorial enumeration problems on planar graphs", *Discrete Appl. Math.* **60** (1995), 119–129.

[Cyvin and Gutman 1988]   S. J. Cyvin and I. Gutman, *Kekulé structures in benzenoid hydrocarbons*, Lecture Notes in Chemistry **46**, Springer, Berlin, 1988.

[David and Tomei 1989]   G. David and C. Tomei, "The problem of the calissons", *Amer. Math. Monthly* **96** (1989), 429–431.

[Douglas 1996]   C. Douglas, "An illustrative study of the enumeration of tilings: conjecture discovery and proof techniques", 1996. Available at http://www.math.wisc.edu/~propp/Enumeration.ps.

[Eisenkölbl 1997]   T. Eisenkölbl, "Rhombus tilings of a hexagon with three fixed border tiles", 1997. Available at math.CO/9712261. To appear in *J. Combin. Theory Ser. A*.

[Eisenkölbl 1998]   T. Eisenkölbl, "Rhombus tilings of a hexagon with two triangles missing on the symmetry axis", 1998. Available at math.CO/9810019. To appear in *J. Combin. Theory Ser. A*.

[Elkies et al. 1992]   N. Elkies, G. Kuperberg, M. Larsen, and J. Propp, "Alternating sign matrices and domino tilings", *J. Algebraic Combin.* **1** (1992), 111–132 and 219–234.

[Fowler and Rushbrooke 1937]   R. H. Fowler and G. S. Rushbrooke, "An attempt to expand the statistical theory of perfect solutions", *Trans. Faraday Soc.* **33** (1937), 1272–1294.

[Fulmek and Krattenthaler 1998a]   M. Fulmek and C. Krattenthaler, "The number of rhombus tilings of a symmetric hexagon which contain a fixed rhombus on the symmetry axis, I", *Ann. Combin.* **2** (1998), 19–40. Available at http://radon.mat.univie.ac.at/People/kratt/artikel/fixrhom2.html or math.CO/9712244.

[Fulmek and Krattenthaler 1998b]   M. Fulmek and C. Krattenthaler, "The number of rhombus tilings of a symmetric hexagon which contain a fixed rhombus on the symmetry axis, II", 1998. Available at http://radon.mat.univie.ac.at/People/kratt/artikel/fixrhom3.html.

[Gessel and Viennot 1985] I. Gessel and G. Viennot, "Binomial determinants, paths, and hook length formulae", *Adv. Math.* **58** (1985), 300–321.

[Gessel et al. ≥ 1999] I. Gessel, A. Ionescu, and J. Propp, In preparation.

[Gordon and Davison 1952] M. Gordon and W. H. T. Davison, "Theory of resonance topology of fully aromatic hydrocarbons, I", *J. of Chem. Phys.* **20** (1952), 428–435.

[Grensing et al. 1980] D. Grensing, I. Carlsen, and H.-C. Zapp, "Some exact results for the dimer problem on plane lattices with non-standard boundaries", *Phil. Mag. A* **41** (1980), 777–781.

[Helfgott 1998] H. Helfgott, *Edge effects on local statistics in lattice dimers: a study of the Aztec diamond finite case*, Bachelor's thesis, Brandeis University, 1998. Available at http://www.math.wisc.edu/~propp/helfgott.ps.

[Helfgott and Gessel 1999] H. Helfgott and I. Gessel, "Tilings of diamonds and hexagons with defects", *Electron. J. Combin.* **6** (1999), R16.

[Jockusch 1994] W. Jockusch, "Perfect matchings and perfect squares", *J. Combin. Theory Ser. A* **67** (1994), 100–115.

[John and Sachs 1985] P. John and H. Sachs, "Wegesysteme und Linearfaktoren in hexagonalen und quadratischen Systemen", pp. 85–101 in *Graphen in Forschung und Unterricht*, Barbara Franzbecker Verlag, Bad Salzdetfurth, 1985.

[Kasteleyn 1961] P. W. Kasteleyn, "The statistics of dimers on a lattice I: The number of dimer arrangements on a quadratic lattice", *Physica* **27** (1961), 1209–1225.

[Kasteleyn 1967] P. W. Kasteleyn, "Graph theory and crystal physics", in *Graph theory and theoretical physics*, edited by F. Harary, Academic Press, 1967.

[Kenyon 1997] R. Kenyon, "Local statistics of lattice dimers", *Ann. Inst. H. Poincaré Probab. Statist.* **33** (1997), 591–618.

[Krattenthaler 1997] C. Krattenthaler, "Schur function identities and the number of perfect matchings of holey Aztec rectangles", 1997. Available at http://radon.mat.univie.ac.at/People/kratt/artikel/holeyazt.html or math.CO/9712204.

[Kuperberg 1994] G. Kuperberg, "Symmetries of plane partitions and the permanent-determinant method", *J. Combin. Theory Ser. A* **68**:1 (1994), 115–151.

[Kuperberg 1998] G. Kuperberg, "An exploration of the permanent-determinant method", *Electron. J. Combin.* **5** (1998), R46.

[Lieb 1967] E. H. Lieb, "Residual entropy of square ice", *Phys. Rev.* **162** (1967), 162–172.

[Lovász and Plummer 1986] L. Lovász and M. D. Plummer, *Matching theory*, Elsevier, Amsterdam and New York, 1986.

[Macdonald 1995] I. G. Macdonald, *Symmetric functions and Hall polynomials*, 2nd ed., Oxford Univ. Press, 1995.

[MacMahon 1915–16] P. A. MacMahon, *Combinatory analysis* (2 v.), Cambridge University Press, 1915–16. Reprinted by Chelsea, New York, 1960.

[Mills et al. 1983] W. H. Mills, D. P. Robbins, and H. Rumsey Jr., "Alternating sign matrices and descending plane partitions", *J. Combin. Theory Ser. A* **34** (1983), 340–359.

[Okada and Krattenthaler 1998] S. Okada and C. Krattenthaler, "The number of rhombus tilings of a 'punctured' hexagon and the minor summation formula", *Adv. in Appl. Math.* **21** (1998), 381–404.

[Onsager 1944] L. Onsager, "Crystal statistics, I: a two-dimensional model with an order-disorder transition", *Phys. Rev.* **65** (1944), 117–149.

[Pachter 1997] L. Pachter, "Combinatorial approaches and conjectures for 2-divisibility problems concerning domino tilings of polyominoes", *Electron. J. Combin.* **4** (1997), R29.

[Pachter and Kim 1998] L. Pachter and P. Kim, "Forcing matchings on square grids", *Discrete Math.* **190**:1-3 (1998), 287–294.

[Percus 1969] J. Percus, "One more technique for the dimer problem", *J. Math. Phys.* **10** (1969), 1881–1888.

[Propp 1996] J. Propp, "Counting constrained domino tilings of Aztec diamonds", 1996. Available at http://math.wisc.edu/~propp/fpsac96.ps.gz.

[Propp ≥ 1999] J. Propp, "Memo on urban renewal". Available at http://math.wisc.edu/~propp/renewal.ps.gz.

[Sachs 1990] H. Sachs, "Counting perfect matchings in lattice graphs", pp. 577–584 in *Topics in combinatorics and graph theory*, edited by R. Bodendiek and R. Henn, Physica-Verlag, Heidelberg, 1990.

[Sachs 1997] H. Sachs, "A contribution to problem 18", 1997. Available at http://math.wisc.edu/~propp/kekule.

[Saldanha 1997] N. Saldanha, "Generalized Kasteleyn matrices and their singular values", 1997. Available at http://www.upa.ens-lyon.fr/~nsaldanh/kk.ps.gz or http://www.impa.br/~nicolau/kk.ps.gz.

[Stanley 1971] R. Stanley, "Theory and application of plane partitions", *Stud. Appl. Math.* **50** (1971), 167–188 and 259–279.

[Stanley 1985] R. Stanley, "On dimer coverings of rectangles of fixed width", *Discrete Appl. Math.* **12** (1985), 81–87.

[Stanley 1986a] R. Stanley, "A baker's dozen of conjectures concerning plane partitions", pp. 285–293 in *Combinatoire énumerative*, edited by G. Labelle and P. Leroux, Lecture Notes in Math. **1234**, Springer, Berlin, 1986.

[Stanley 1986b] R. Stanley, "Symmetries of plane partitions", *J. Combin. Theory Ser. A* **43** (1986), 103–113.

[Sundquist et al. 1997] T. S. Sundquist, D. G. Wagner, and J. West, "A Robinson-Schensted algorithm for a class of partial orders", *J. Combin. Theory Ser. A* **79** (1997), 36–52.

[Temperley and Fisher 1961] H. N. V. Temperley and M. E. Fisher, "Dimer problem in statistical mechanics — an exact result", *Phil. Mag.* **6** (1961), 1061–1063.

[Turnbull 1960] H. W. Turnbull, *Theory of determinants, matrices, and invariants*, Dover, New York, 1960.

[Valiant 1979] L. Valiant, "The complexity of computing the permanent", *Theoret. Comput. Sci.* **8** (1979), 189–201.

[Yang 1991] B.-Y. Yang, *Two enumeration problems about the Aztec diamonds*, Ph.D. thesis, Mass. Inst. Tech., 1991.

[Zare 1997–98] D. Zare, email from May 20, 1997 and March 2, 1998. Available at http://math.wisc.edu/~propp/zare.

JAMES PROPP
DEPARTMENT OF MATHEMATICS
UNIVERSITY OF WISCONSIN
MADISON, WI 53706
UNITED STATES
    propp@math.wisc.edu

New Perspectives in Geometric Combinatorics
MSRI Publications
Volume **38**, 1999

The Generalized Baues Problem

VICTOR REINER

ABSTRACT. We survey the generalized Baues problem of Billera and Sturmfels. The problem is one of discrete geometry and topology, and asks about
the topology of the set of subdivisions of a certain kind of a convex polytope. Along with a discussion of most of the known results, we survey the
motivation for the problem and its relation to triangulations, zonotopal
tilings, monotone paths in linear programming, oriented matroid Grassmannians, singularities, and homotopy theory. Included are several open
questions and problems.

1. Introduction

The generalized Baues problem, or GBP for short, is a question arising in the
work of Billera and Sturmfels [1992, p. 545] on *fiber polytopes*; see also [Billera
et al. 1994, § 3]. The question asks whether certain partially ordered sets whose
elements are subdivisions of polytopes, endowed with a certain topology [Björner
1995], have the homotopy type of spheres. Cases are known [Rambau and Ziegler
1996] where this fails to be true, but the general question of when it is true or
false remains an exciting subject of current research.

The goal of this survey is to review the motivation for fiber polytopes and the
GBP, and discuss recent progress on the GBP and the open questions remaining. Some recommended summary sources on this subject are the introductory
chapters in the doctoral theses [Rambau 1996; Richter-Gebert 1992], Lecture 9
in [Ziegler 1995], and the paper [Sturmfels 1991]. The articles [Billera et al.
1990; 1993], though not discussed in the text, are nonetheless also relevant to
the GBP.

Before diving into the general setting of fiber polytopes and the GBP, it is
worthwhile to ponder three motivating classes of examples.

Partially supported by Sloan Foundation and University of Minnesota McKnight Land Grant
Fellowships.

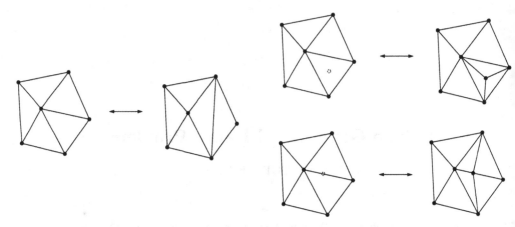

Figure 1. Typical bistellar operations (also known as perestroikas or modifications) in $\mathbb{R}^2$. Left: diagonal flip. Right: vertex insertion/removal.

Triangulations. Let $\mathcal{A}$ denote a finite set of points in $\mathbb{R}^d$. A *triangulation* of $\mathcal{A}$ is, roughly speaking, a polyhedral subdivision of the convex hull of $\mathcal{A}$ into simplices, each having the property that their vertices lie in $\mathcal{A}$. Note that not every point of $\mathcal{A}$ need appear as a vertex of one of the simplices in the triangulation. The set of all triangulations of $\mathcal{A}$ is in general a difficult object to compute, but one that arises in many applications; see [de Loera 1995b]. One approach to the study and computation of triangulations is to consider an extra structure on them, namely the connections between them by certain moves called *bistellar operations* (or *perestroikas* or *modifications*). For triangulations of $\mathcal{A}$ in $\mathbb{R}^2$, typical bistellar operations are shown in Figure 1, where points of $\mathcal{A}$ that are not being used as a vertex in the triangulation are shown dotted. Figure 2, adapted from [de Loera 1995b], depicts the set of all triangulations of a particular configuration of six points in $\mathbb{R}^2$, and the bistellar moves which connect them. We remark that the precise coordinates of the points of $\mathcal{A}$ are important in determining which triangulations and bistellar operations are possible, since we are talking about triangulations using straight geometric simplices. This is different from the point of view in the theory of triangulated *planar maps* (see [Goulden and Jackson 1983, § 2.9], for example) and also different from the bistellar equivalences of triangulations of PL-manifolds as considered by Nabutovsky [1996] or Pachner [1991].

The most well-studied example of triangulations occurs when $\mathcal{A}$ is the vertex set of a convex n-gon in $\mathbb{R}^2$. It is well-known that the number of triangulations is the *Catalan number* $\frac{1}{n-1}\binom{2n-4}{n-2}$ (see, for example, [Stanton and White 1986, § 3.1; Stanley 1999, Exercise 6.19] for this and for bijections between triangulations and other standard objects counted by the Catalan number). This is essentially the only nontrivial example of an infinite family of point configurations whose number of triangulations is known (but see Conjecture 6.5). The only possible bistellar operations in this case are the *diagonal flips* from Figure 1, and it is easy to see that any two triangulations can be connected by a sequence

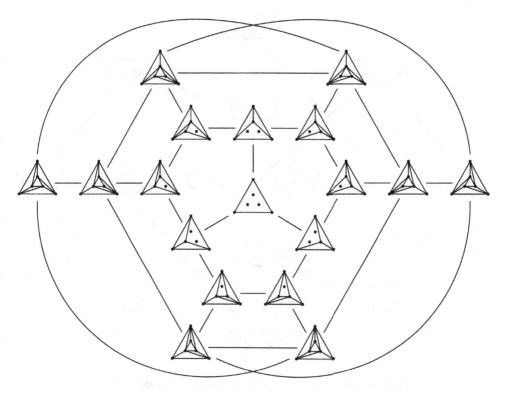

Figure 2. All triangulations and bistellar operations for a set $\mathcal{A}$ of 6 points in $\mathbb{R}^2$. The points of $\mathcal{A}$ form the vertices of two homothetic and concentric equilateral triangles. Adapted from [de Loera 1995b].

of such flips. There is a well-known bijection between triangulations of an n-gon and nonassociative bracketings of a product $a_1 a_2 \cdots a_{n-1}$, and under this identification bistellar operations correspond to "rebracketings". From this point of view, the graph of triangulations of an n-gon and diagonal flips was perhaps first studied in the 1950's by Tamari [1951] and later in collaboration with others [Tamari 1962; Friedman and Tamari 1967; Huang and Tamari 1972; Huguet and Tamari 1978]. These authors distinguished a direction on each rebracketing and defined a poset on the triangulations having these directed edges as its cover relations. They were able to show that this *Tamari poset* is a lattice [Friedman and Tamari 1967; Huang and Tamari 1972]. Its Hasse diagram is depicted in Figure 3 for $n = 6$, for a choice of a particular convex 6-gon whose vertices lie on a semicircle.

These authors seem also to have been aware (without proof) that this graph appears to be the 1-skeleton of a cellular $(n-4)$-sphere, and proved results about how its "facial" structure interacts with the Tamari lattice structure. Meanwhile, similar issues of associativity appeared in the early 1960's in Stasheff's work [1963] on homotopy associativity. Stasheff vindicated this apparent sphericity

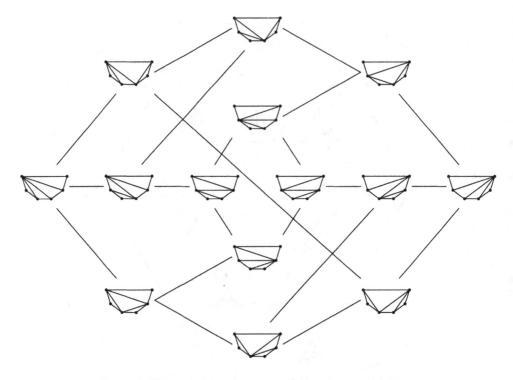

Figure 3. Triangulations of a convex 6-gon: the associahedron.

by showing (essentially) that the set of all *polygonal subdivisions* of an n-gon
indexes the cells in a *regular cell complex* [Björner 1995, (12.3)] homeomorphic
to the $(n-4)$-sphere. Note that in this way of thinking, a diagonal flip bistellar
operation corresponds to a polygonal subdivision whose maximal cells are all
triangles except for one quadrangle (containing the flipping diagonal), and less
refined subdivisions of the n-gon correspond to higher dimensional cells in the
sphere. In an unpublished work (see [Kapranov 1993, p. 120]), Milnor produced
a set of vertex coordinates for the vertices of this $(n-4)$-sphere which embed
it as the boundary complex of an $(n-3)$-dimensional polytope. Unfortunately,
the existence of this polytopal embedding seems to have been unknown in the
combinatorial geometry community, and was rediscovered in the mid 1980's after
Perles posed the problem of whether this complex was polytopal; see [Lee 1989].
Independently, Haiman [1984] and Lee [1989] constructed this polytope, which
Haiman dubbed the *associahedron*. In [Kapranov and Voevodsky 1991; Gel'fand
et al. 1994] it is sometimes called the *Stasheff polytope*. Kapranov and Saito
[1997] document its occurrence in other surprising geometric contexts.

The associahedron also makes its appearance in computer science, where triangulations of an n-gon show up in the equivalent guise of *binary trees*, and
bistellar operations correspond to an operation on binary trees called *rotation*.
Here Sleator, Tarjan and Thurston [Sleator et al. 1988] were able to determine

the diameter of the 1-skeleton of the associahedron (it is at most $2n - 10$ for $n \geq 13$ and is exactly $2n - 10$ for infinitely many values of n). Pallo [1987; 1988; 1990; 1993] studied computational aspects of this 1-skeleton and in particular computed the *Möbius function* [Stanley 1997, § 3.7] of the Tamari lattice.

In one of Lee's constructions of the associahedron, he employs the method of *Gale diagrams* [Lee 1989, § 4]. Around the same time, Gelfand, Kapranov and Zelevinsky were using these methods for studying triangulations as part of their theory of $\mathcal{A}$-*discriminants*, $\mathcal{A}$-*resultants*, and $\mathcal{A}$-*determinants* (see [Gel'fand et al. 1994] and the references therein). Briefly, the principal $\mathcal{A}$-*determinant* is a polynomial $E_{\mathcal{A}}$ in a variable set $\{c_a\}_{a \in \mathcal{A}}$ indexed by $\mathcal{A}$, which vanishes whenever the sparse d-variate polynomial in $x_1, \ldots, x_d$

$$f := \sum_{a \in \mathcal{A}} c_a x^a$$

has a root $(x_1, \ldots, x_n)$ in common with all of the derived polynomials

$$x_1 \frac{\partial f}{\partial x_1}, \ \ldots, \ x_d \frac{\partial f}{\partial x_d}.$$

Their work showed that the *Newton polytope* of $E_{\mathcal{A}}$, that is, the convex hull in $\mathbb{R}^{\mathcal{A}}$ of the set of exponent vectors of the monomials having nonzero coefficients in $E_{\mathcal{A}}$, is an $(n - d - 1)$-dimensional polytope whose vertices correspond to a subset of the triangulations of $\mathcal{A}$ called the *regular* (and later called *coherent*) triangulations. A triangulation T of $\mathcal{A}$ is *coherent* if there exists a choice of *heights* α_a in $\mathbb{R}$ for each $a \in \mathcal{A}$ which induces T in the following fashion: after "lifting" the points a in $\mathbb{R}^d$ to the points $(a, \alpha_a) \in \mathbb{R}^{d+1}$ and taking the convex hull to form a polytope P_α, the "lower" facets of P_α (i.e., those facets whose normal vector has negative $(d + 1)$-coordinate) project to the maximal simplices of T under the projection $\mathbb{R}^{d+1} \to \mathbb{R}^d$. Figure 4, borrowed from [Rambau 1996],

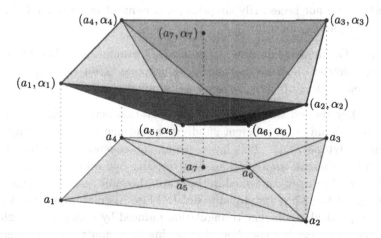

Figure 4. A coherent triangulation induced by a choice of heights. From [Rambau 1996].

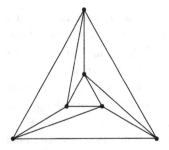 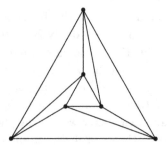

Figure 5. The two incoherent triangulations lurking among those in Figure 2 (on the far left and far right of that figure).

illustrates a coherent triangulation of a set $\mathcal{A}$ in $\mathbb{R}^2$ along with a choice of heights α which induces it.

After seeing the definition, it is perhaps not obvious that one can have *incoherent* triangulations! However, the standard examples, discussed extensively in [Connelly and Henderson 1980; Schönhardt 1928], already occur as two of the triangulations appearing in Figure 2. We have isolated these two triangulations and depicted them separately in Figure 5. It is a nontrivial exercise to check the impossibility of assigning six heights to these points in such a way as to induce either of these triangulations. In general, checking whether a triangulation is coherent involves checking whether there exists a solution to a certain system of linear inequalities in the heights α_a, where the coefficients in the inequalities depend upon the coordinates of the points in $\mathcal{A}$; see [Hastings 1998, Chapter 2; de Loera 1995b, § 1.3].

Gelfand, Kapranov and Zelevinsky called the Newton polytope of $E_{\mathcal{A}}$ the *secondary polytope* $\Sigma(\mathcal{A})$. Knowing that the vertices of $\Sigma(\mathcal{A})$ correspond to the coherent triangulations of $\mathcal{A}$, it is perhaps not surprising that the higher dimensional faces of $\Sigma(\mathcal{A})$ correspond to *coherent subdivisions*, that is, subdivisions into polytopes which are not necessarily simplices, but induced in a similar fashion by a choice of heights α_a for a in $\mathcal{A}$.

THEOREM 1.1 [Gel'fand et al. 1994, Chapter 7, Theorem 2.4]. *The faces of the secondary polytope $\Sigma(\mathcal{A})$ are indexed by the coherent subdivisions of $\mathcal{A}$, and reverse inclusion of faces of $\Sigma(\mathcal{A})$ corresponds to refinement of subdivisions.*

In particular, they showed that every bistellar operation between coherent triangulations corresponds to a coherent subdivision and hence forms an edge in the secondary polytope $\Sigma(\mathcal{A})$. This has a strong consequence: it implies that the subgraph of coherent triangulations and bistellar operations is connected (and even $(n-d-1)$-vertex-connected in the graph-theoretic sense by Balinski's Theorem [Ziegler 1995, 3.5]). Polytopality of $\Sigma(\mathcal{A})$ also has nice implications for computing the particular coherent triangulation induced by a choice of heights α_a, such as the *Delaunay triangulation* of $\mathcal{A}$ arising in computational geometry applications; see [Edelsbrunner and Shah 1992].

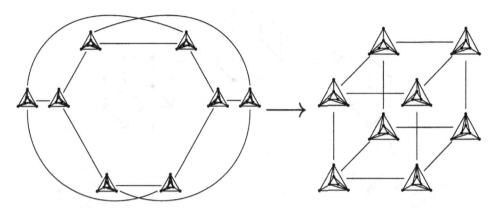

Figure 6. The incoherent triangulations and their neighbors: two equivalent pictures.

We remark that in the case where $\mathcal{A}$ is the set of vertices of a convex n-gon, every subdivision is coherent, and hence the secondary polytope $\Sigma(\mathcal{A})$ is the associahedron encountered earlier.

The fact that the subgraph of coherent triangulations and bistellar operations is highly connected and forms the 1-skeleton of a cellular (even polytopal) sphere raises the following basic question:

QUESTION 1.2. *Is the graph of all triangulations of $\mathcal{A}$ and their bistellar operations connected?*

A glance at Figure 2 illustrates that even in small cases where there are incoherent triangulations, the graph still appears to be connected. We can provide some motivation for the Generalized Baues Problem by performing the following mental exercise while staring at Figure 2. First picture the planar subgraph of coherent triangulations, by ignoring the two vertices corresponding to the incoherent triangulations in Figure 2 (call them T_1 and T_2). When one imagines this planar subgraph as a two-dimensional spherical cell complex, that is the boundary of the three-dimensional secondary polytope $\Sigma(\mathcal{A})$, the union of the neighbors of T_1 and T_2 form the vertices of a hexagonal cell, corresponding to the unbounded region in the planar embedding; see Figure 6. Now "inflate" this hexagonal cell on the 2-sphere into a cubical 3-dimensional cell with the extra two vertices corresponding to T_1, T_2. This gives a 3-dimensional cell complex which is still homotopy equivalent (but not homeomorphic) to a 2-sphere.

Roughly speaking, the Baues question in this context asks whether this behavior is general: Do the *incoherent* triangulations and subdivisions of $\mathcal{A}$ attach themselves to the spherical boundary of $\Sigma(\mathcal{A})$ in such a way as to not change its homotopy type?

Zonotopal tilings. Consider Figure 7, similar to [Billera and Sturmfels 1992, Figure 1], depicting the tilings of a centrally symmetric octagon having unit side

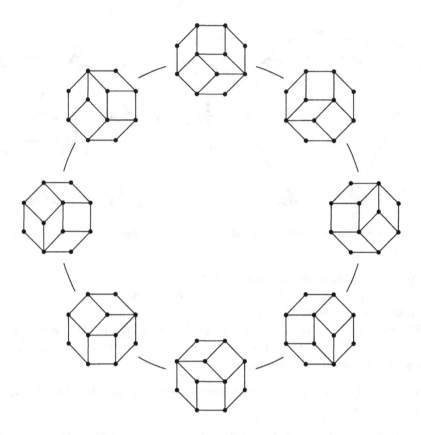

Figure 7. The rhombic tilings of an octagon.

lengths by unit rhombi. As in the case of triangulations of a point set, we have drawn in edges between the tilings corresponding to certain natural operations connecting them, illustrated in Figure 8. Similarly, the graph whose vertices are the tilings of a 10-gon and whose edges are these operations is depicted in Figure 9, which may not look very planar, but is in fact the 1-skeleton of a 3-dimensional polyotope.

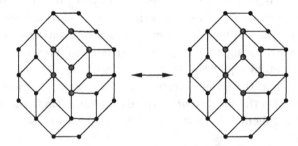

Figure 8. A typical cube flip, also known as a mutation, triangle switch, 1-move, braid relation, Yang–Baxter relation, elementary flip, or localized phason.

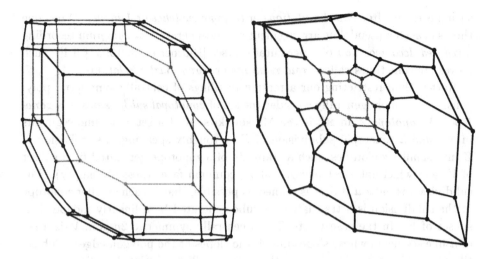

Figure 9. The graph of tilings of a decagon, seen in $\mathbb{R}^3$ from infinity (left) and from a nearby point. Compare [Ziegler 1993, Figure 3].

These operations have been given various names in the literature, depending upon the context in which the tilings arise. In the crystalline physics literature [Destainville et al. 1997; Mosseri and Bailly 1993], where the set of tilings is a model for the possible states of a crystalline solid, these moves are called *elementary flips* or *localized phasons*. Rather than considering tilings of a $2n$-gon, an equivalent (and useful) viewpoint comes from consideration of *arrangements of pseudolines* (see [Björner et al. 1993, Chapter 6] for definition, background and references). An arrangement of n affine pseudolines in the plane labelled $1, 2, \ldots, n$ counterclockwise gives rise to a rhombic tiling of a centrally symmetric $2n$-gon which is "dual" to the line arrangement in the sense of planar maps; see Figure 10.

In the pseudoline picture, the move depicted in Figure 8 corresponds to moving one pseudoline locally across the nearby crossing point of two other pseudolines;

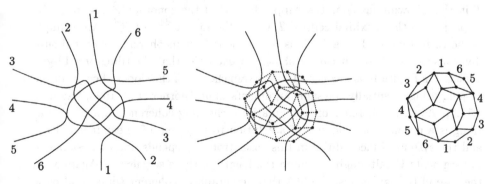

Figure 10. A configuration of affine pseudolines and its associated tiling.

such moves are often called *mutations* or *triangle-switches* or *1-moves*. When one thinks of such a pseudoline arrangement as a degenerate *braid diagram* recording a *reduced decomposition* of a permutation (see [Björner et al. 1993, § 6.4]), such moves are sometimes called *braid relations* or *Yang–Baxter relations*.

Rather than restricting our attention to tilings of centrally symmetric polygons, we can more generally consider the set of *zonotopal subdivisions* of a *zonotope*. A *zonotope* Z in $\mathbb{R}^d$ is the Minkowski sum of a set V of line segments in $\mathbb{R}^d$, and a zonotopal subdivision of Z is, roughly speaking, a subdivision of Z into smaller zonotopes, each a translate of a zonotope generated by a subset of V, and which intersect pairwise along common faces (possibly empty). The subdivision is *cubical* if it is as refined as possible, that is each smaller zonotope in the subdivision is a translate of a cube generated by a linearly independent subset of V. In the case where Z is a centrally symmetric $2n$-gon, V is a set of n line segments whose slopes match the slopes of the polygon edges. Cubical tilings in this case coincide with the rhombic tilings depicted earlier, and the "cube flip" moves which formed the edges in the graphs of Figures 7 and 9 correspond to zonotopal subdivisions of Z in which all of the smaller zonotopes are cubes except for one which is hexagonal.

Note that the graph of tilings in Figure 7 is circular, and the graph of tilings in Figure 9 appears to be planar and possibly even polytopal. This reflects the fact that for centrally symmetric octagons and decagons, all zonotopal subdivisions are *coherent* in a sense which will be described below. A special case of Billera and Sturmfels' fiber polytope construction [Billera and Sturmfels 1992, § 5] states that the subset of coherent zonotopal subdivisions of a d-dimensional zonotope having n generators index the faces an $(n - d)$-dimensional polytope (which happens to be itself a zonotope). Thus the graphs in Figures 7 and 9 are the 1-skeleta of these *fiber zonotopes*.

Coherence of a zonotopal subdivision is defined similarly to coherence of a triangulation. A zonotopal subdivision T of a zonotope Z in $\mathbb{R}^d$ having generating line segments V is *coherent* if there exists a choice of segments $\hat{V}$ in $\mathbb{R}^{d+1}$ which project down to V under the forgetful projection $\mathbb{R}^{d+1} \to \mathbb{R}^d$ and induce T in the following fashion: the "upper facets" of the zonotope $\hat{Z}$ generated by $\hat{V}$ project to the maximal cells of T under the map $\mathbb{R}^{d+1} \to \mathbb{R}^d$. An example is shown in Figure 11. Again, it is not obvious that incoherent zonotopal subdivisions can exist, but it can be shown for example, that the tiling of a 12-gon depicted in Figure 10 is incoherent for certain choices of the slopes of edges in the 12-gon, using essentially the same arguments as in [Björner et al. 1993, Example 1.11.2]. As with coherence of triangulations, checking coherence of a particular tiling is a problem of existence of a solution to a system of linear inequalities, and the system of inequalities in this case strongly depends upon the slopes of the segments V (although not upon the length of these segments). Again as in the case of triangulations, the fact that the graph of coherent tilings and cube flips is the 1-skeleton of a polytope has strong consequences for its connectivity.

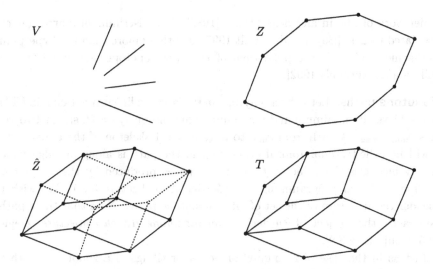

Figure 11. A coherent tiling T, induced by lifting into $\mathbb{R}^3$ the generating segments V of the 2-dimensional zonotope Z, then projecting the upper facets of the resulting 3-dimensional zonotope $\hat{Z}$ back into the plane.

This raises the analogous question to Question 1.2:

QUESTION 1.3. *Is the graph of all cubical tilings of a zonotope and their cube flips connected?*

One can also view coherence of two-dimensional tilings in terms of pseudolines and straight lines. A coherent tiling is one whose pseudoline arrangement is isomorphic to a (straight) line arrangement in which each line has slope perpendicular to the slope of the edge in the polygon to which it corresponds (that is, to the edge of the polygon labelled with the same number in Figure 10). Some of this viewpoint is explained in the instructions for the delightful puzzle Hexa-Grid [MRI n.d.], which supplies foam rubber versions of the rhombic tiles occurring in Figure 10, and asks the consumer to assemble them into a tiling of a zonotopal 12-gon!

In studying tilings and zonotopal subdivisions of higher dimensional zonotopes, the oriented matroid point of view has become indispensable; see [Björner et al. 1993, § 2.2; Ziegler 1996]. The Bohne–Dress Theorem [Bohne 1992; Richter-Gebert and Ziegler 1994] states that zonotopal subdivisions of Z biject with the *single-element liftings* of the *realized* oriented matroid $\mathcal{M}$ associated with the generating segments V, or using oriented matroid duality, to the *single-element extensions* of the dual oriented matroid $\mathcal{M}^*$ (see [Björner et al. 1993, § 7.1]). From this point of view, the subset of coherent zonotopal subdivisions of Z corresponds to the *coherent liftings* of V [Billera and Sturmfels 1992, § 5]. If one views realized oriented matroids and their liftings in terms of sphere and pseudosphere arrangements, then the notion of a coherent lifting was explored in the work of Bayer and Brandt [1997] on *discriminantal arrangements*, generalizing earlier work of Manin and Schechtman [1989]. The discriminantal arrangement associated to Z in [Bayer and Brandt 1997] is nothing more than the hyperplane arrangement which is the polar dual of the fiber zonotope associated to Z in [Billera and Sturmfels 1992].

Monotone paths. Let P be a polytope in $\mathbb{R}^d$, and f a linear functional in $(\mathbb{R}^d)^*$ that achieves its minimum and maximum values uniquely on P, say at two vertices $v_{\min}, v_{\max}$. A path from $v_{\min}$ to $v_{\max}$ in the 1-skeleton of the boundary of P will be called f-*monotone* if every step in the path is along an edge which strictly increases the value of f (as in the paths produced by the *simplex algorithm* for linear programming- see [Ziegler 1995, Lecture 3.2]). We wish to consider the structure of the set of all f-monotone paths. Note that these paths are exactly the subject of Ziegler's *strict monotone Hirsch conjecture* [Ziegler 1995, Conjecture 3.9].

Just as in the case of triangulations of $\mathcal{A}$ or tilings of a zonotope Z, there is a natural set of moves which connect f-monotone paths: if two paths agree in most of their steps and differ only by following opposite paths around some 2-dimensional face of P, we say that the two paths differ by a *polygon move*. Figure 12 illustrates the graph of f-monotone paths and polygon moves where P is the 3-cube $[0,1]^3 \subset \mathbb{R}^3$ and $f(x_1, x_2, x_3) = x_1 + x_2 + x_3$.

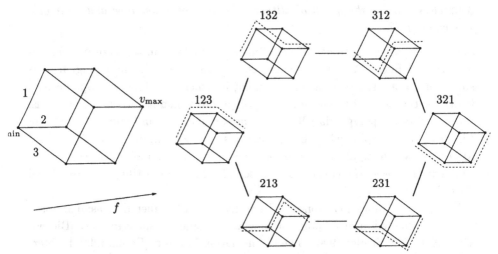

Figure 12. Monotone paths in the 3-cube. Edges pointing in the original three coordinate directions have been labelled 1, 2, 3, and then monotone paths are labelled by the sequence of directions of the steps.

In general if P is the n-cube $[0,1]^n \subset \mathbb{R}^n$ and $f(x) = \sum_{i=1}^n x_i$, the f-monotone paths biject with permutations of $\{1, 2, \ldots, n\}$: one obtains a permutation by recording which coordinate axis is parallel to each step of the path in sequence, as in Figure 12. The polygon moves across square faces of the n-cube then correspond to *adjacent transpositions* of the permutations, and the whole graph

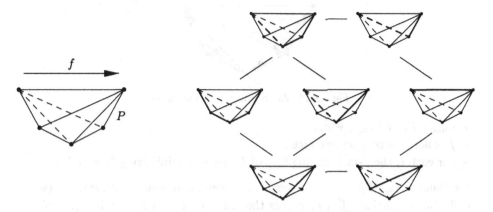

Figure 13. Monotone paths (thick lines) in a cyclic 3-polytope with five vertices.

is isomorphic to the 1-skeleton of the well-known *permutohedron* [Ziegler 1995, Example 0.10]; see also [Milgram 1966].

What happens for other polytopes P and functionals f? Figure 13 shows the graph of f-monotone paths in a (cyclic) 3-polytope with five vertices in $\mathbb{R}^3$ which is the convex hull of the points $\{(t, t^2, t^3) : t = -2, -1, 0, 1, 2\}$, and $f(x_1, x_2, x_3) = x_1$.

Although this graph is connected, it is perhaps disappointing that it is not circular as in the case of Figure 12. Once again, geometry comes to the rescue in singling out a well-behaved subset of f-monotone paths. Say that an f-monotone path γ on P is *coherent* if there exists some linear functional $g \in (\mathbb{R}^d)^*$ which induces γ in the following way: each point of γ (not necessarily a vertex) is the g-maximal point among all those points of P with the same f value, or in other words, γ is the union over all points x in $f(P) \subset \mathbb{R}$ of the g-maximal points in the fibers $f^{-1}(x)$. With this definition, the monotone path in the middle of the graph in Figure 13 is incoherent. To see this, assume there is some functional g inducing this monotone path, and identify g as the dot product with some fixed vector. Then this vector must point roughly toward the front (the visible side) of the polytope P in order to induce the right portion of the path, but also point toward the back (the invisible side) in order to induce the left portion of the path; contradiction. The remaining six paths in Figure 13 are easily seen to be coherent (by imagining appropriate functionals g) and the subgraph on the corresponding six vertices is indeed circular.

In general, it follows as a special case of Billera and Sturmfels' fiber poly-tope construction [Billera and Sturmfels 1992, §7] that the graph of coherent f-monotone paths in a polytope P is the 1-skeleton of a polytope called the *monotone path polytope*. Higher dimensional faces of the monotone path poly-tope correspond to objects called *coherent cellular strings* on P with respect to f. A *cellular string* on P with respect to f is a sequence $(F_1, \ldots, F_r)$ of boundary faces of P with the following properties:

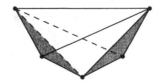

Figure 14. An incoherent cellular string.

- $v_{\min} \in F_1$ and $v_{\max} \in F_r$.
- f is not constant on any face F_i.
- For each i, the f-maximizing face of F_i is the f-minimizing face of F_{i+1}.

A cellular string $(F_1, \ldots, F_r)$ is *coherent* if there exists some functional $g \in (\mathbb{R}^d)^*$ such that the union $\bigcup_{i=1}^r F_i$ equals the union over all points x in $f(P)$ of the g-maximal points in the fibers $f^{-1}(x)$.

As an exercise to get a feeling for how the cellular strings fit into the graph of f-monotone paths, and to further motivate the Baues problem, we invite the reader to try the following labelling exercise. Label each edge in the graph of Figure 13 by a cellular string containing mostly 1-faces along with exactly one triangular 2-face corresponding to the polygon move for that edge. Having done this, there is only one other possible cellular string, consisting of two triangles and pictured in Figure 14. This cellular string should label a square 2-cell attached to the four leftmost vertices and edges in Figure 13 (see also the middle picture in Figure 15). Notice that the resulting 2-dimensional cell complex is homotopy equivalent to the circular subgraph indexed by coherent cellular strings. This raises the following question.

QUESTION 1.4. *For a polytope P and functional f, is the graph of f-monotone paths and their polygon moves connected? Is it part of a complex homotopy equivalent to a $(d-2)$-sphere?*

This question includes the original question asked by Baues [1980] as a special case. Specifically, Baues asked if the poset of cellular strings on the *permutohedron* with respect to a generic linear functional f has the homotopy type of a sphere (after endowing the poset with a certain topology: see Section 2). His question is a natural extension of ideas of Adams [1956] and Milgram [1966] involving edge paths in polytopes as models for loop spaces and iterated loop spaces: see [Rambau 1996, § 1.2] for a nice sketch of the ideas involved.

As we will see in Section 4, this original Baues question was answered positively by Billera, Kapranov and Sturmfels [Billera et al. 1994], who resolved it not only for cellular strings on the permutohedron, but on arbitrary polytopes. Another proof, for the case of arbitrary zonotopes, was given by Björner [1992].

2. Fiber Polytopes and the Baues Problem

The theory of fiber polytopes [Billera and Sturmfels 1992] provides a common framework in which to discuss triangulations, tilings, and monotone paths, and also a common notion of coherence for these objects. The fiber polytope $\Sigma(P \xrightarrow{\pi} Q)$, is a polytope naturally associated to any linear projection of polytopes $\pi : P \to Q$. Let P be a d'-dimensional polytope in $\mathbb{R}^{d'}$, Q a d-dimensional polytope in $\mathbb{R}^d$ and $\pi : \mathbb{R}^{d'} \to \mathbb{R}^d$ a linear map with $\pi(P) = Q$. A *polytopal subdivision* of Q is a polytopal complex which subdivides Q. A polytopal subdivision of Q is π-*induced* if

(i) it is of the form $\{\pi(F) : F \in \mathcal{F}\}$ for some specified collection $\mathcal{F}$ of faces of P having all $\pi(F)$ distinct, and

(ii) $\pi(F) \subseteq \pi(F')$ implies $F = F' \cap \pi^{-1}(\pi(F))$, and in particular $F \subseteq F'$.

It is possible that different collections $\mathcal{F}$ of faces of P project to the same subdivision $\{\pi(F) : F \in \mathcal{F}\}$ of Q, so we distinguish these subdivisions by labelling them with the family $\mathcal{F}$. We partially order the π-induced subdivisions of Q by $\mathcal{F}_1 \leq \mathcal{F}_2$ if and only if $\bigcup \mathcal{F}_1 \subseteq \bigcup \mathcal{F}_2$. The resulting partially ordered set is denoted by $\omega(P \xrightarrow{\pi} Q)$ and called the *Baues poset*. The minimal elements in this poset are the *tight* subdivisions, that is those for which F and $\pi(F)$ have the same dimension for all F in $\mathcal{F}$.

We next explain how π-induced subdivisions of Q generalize triangulations, tilings, and monotone paths. This is perhaps easiest to see for monotone paths and cellular strings. Given a polytope P and functional f, let Q be the 1-dimensional polytope $f(P)$ in $\mathbb{R}^1$. Then a cellular string $(F_1, \ldots, F_r)$ on P with respect to f gives rise to a family $\mathcal{F}$ satisfying the definition for a π-induced subdivision of Q as follows: $\mathcal{F}$ consist of the F_i's along with their f-minimizing and f-maximizing faces. Tight π-induced subdivisions of Q correspond to monotone paths on P.

For triangulations and tilings, there is a concealed projection of polytopes lurking in the background. Given a point set $\mathcal{A}$ in $\mathbb{R}^d$ with cardinality n, let Q denote its convex hull. There is a natural surjection $\pi : \Delta^{n-1} \to Q$ from a simplex Δ^{n-1} having n vertices, which sends each vertex of the simplex to one of the points of $\mathcal{A}$. One can then check that the π-induced subdivisions of Q as defined above correspond to the following notion of a subdivision of $\mathcal{A}$, which replaces the naive definition given in Section 1. A subdivision of $\mathcal{A}$ is a collection of pairs $\{(Q_\alpha, \mathcal{A}_\alpha)\}$, where

- $\mathcal{A}_\alpha$ are subsets of $\mathcal{A}$,
- each Q_α is the convex hull of $\mathcal{A}_\alpha$ and is d-dimensional,
- the union of the Q_α covers Q, and
- for any α, β, the intersection $Q_\alpha \cap Q_\beta$ is a face F (possibly empty) of each, and $\mathcal{A}_\alpha \cap F = \mathcal{A}_\beta \cap F$.

Tight π-induced subdivisions of Q correspond to triangulations of $\mathcal{A}$. Furthermore, the Baues poset corresponds to the natural *refinement* ordering on subdivisions of $\mathcal{A}$: $\{(Q_\alpha, \mathcal{A}_\alpha)\} \leq \{(Q'_\beta, \mathcal{A}'_\beta)\}$ if and only if for every α there exists some β with $\mathcal{A}_\alpha \subseteq \mathcal{A}'_\beta$ (and hence also $Q_\alpha \subseteq Q'_\beta$).

Let Z be a zonotope in $\mathbb{R}^d$ generated by n line segments V. Without loss of generality we may assume that these segments all have one endpoint at the origin, and we can think of them as vectors pointing in a certain direction rather than segments. There is then a natural surjection $\pi : I^n \to Z$ of the n-cube I^n in $\mathbb{R}^n$ onto Z which sends the standard basis vectors in $\mathbb{R}^n$ onto the vectors V. Then the π-induced subdivisions of Q as defined above correspond to a notion of zonotopal subdivision of Z explained carefully in [Richter-Gebert and Ziegler 1994, Definitions 1.3 and 1.4], and which replaces the naive definition we gave in Section 1. Tight π-induced subdivisions then correspond to cubical tilings of Z, and the Baues poset corresponds to a natural *refinement* ordering on zonotopal subdivisions. As was remarked in Section 1, the Bohne–Dress Theorem shows that the zonotopal subdivisions, or equivalently π-induced subdivisions of Z, are the same as single-element liftings of the realized oriented matroid $\mathcal{M}$ corresponding to the vectors V. It is furthermore true that the Baues poset corresponds to the usual *weak map* ordering [Björner et al. 1993, §7.2] on single-element liftings of $\mathcal{M}$, or equivalently on the single-element extensions of the dual $\mathcal{M}^*$.

Returning to our general set-up of a projection $\pi : P \to Q$, we wish to define when a π-induced subdivision is π-*coherent*, generalizing the notion of coherence for triangulations, tilings, and monotone paths. There is more than one way to say this, and we start with one of the descriptions from [Billera and Sturmfels 1992]. Choose a linear functional $g \in (\mathbb{R}^{d'})^*$. For each point q in Q, the fiber $\pi^{-1}(q)$ is a convex polytope which has a unique face $\overline{F}_q$ on which the value of g is minimized. This face lies in the relative interior of a unique face F_q of P and the collection of faces $\mathcal{F} = \{F_q\}_{q \in Q}$ projects under π to a subdivision of Q. Subdivisions of Q which arise from a functional g in this fashion are called π-*coherent*. Note that this definition of π-coherence clearly generalizes our earlier notion of coherence for cellular strings on P.

It is possible to rephrase the definition of π-coherent subdivisions given in [Billera and Sturmfels 1992] as follows (see also [Ziegler 1995, §9.1]). Having chosen the functional $g \in (\mathbb{R}^{d'})^*$ as above, form the graph of the linear map $\hat{\pi} : P \to \mathbb{R}^{d+1}$ given by $p \mapsto (\pi(p), g(p))$. The image of this map is a polytope $\hat{Q}$ in $\mathbb{R}^{d+1}$ which maps onto Q under the forgetful map $\mathbb{R}^{d+1} \to \mathbb{R}^d$. Therefore, the set of *lower faces* of $\hat{Q}$ (those faces whose normal cone contains a vector with negative last coordinate) form a polytopal subdivision of Q. We identify this subdivision of Q with the family of faces $\mathcal{F} = \{F\}$ in P which are the inverse images under $\hat{\pi}$ of the lower faces of $\hat{Q}$. Under this identification, it is not hard to check that the subdivision of Q is exactly the same as the π-coherent subdivision induced by g, described in the previous paragraph. A glance at the definitions shows that this second definition of π-coherence generalizes the ones we gave for coherent subdivisions of a point set $\mathcal{A}$ and for coherent zonotopal subdivisions of a zonotope Z.

Let $\omega_{\mathrm{coh}}(P \overset{\pi}{\to} Q)$ denote the induced subposet of the Baues poset $\omega(P \overset{\pi}{\to} Q)$ on the set of π-coherent subdivisions of Q. The following beautiful result of Billera and Sturmfels which explains all of our pretty polytopal pictures is the following:

THEOREM 2.1 [Billera and Sturmfels 1992, Theorem 3.1]. *Let P be a d'-polytope, Q a d-polytope, and $\pi : P \to Q$ a linear surjection. Then the poset*

$$\omega_{\mathrm{coh}}(P \overset{\pi}{\to} Q)$$

is the face poset of a $(d' - d)$-polytope $\Sigma(P \overset{\pi}{\to} Q)$.

In particular, the tight π-coherent subdivisions of Q correspond to the vertices of $\Sigma(P \overset{\pi}{\to} Q)$.

The $(d' - d)$-polytope $\Sigma(P \overset{\pi}{\to} Q)$ is called the *fiber polytope* of the surjection π. It generalizes the *secondary polytopes* $\Sigma(\mathcal{A})$, *fiber zonotopes*, and *monotone path polytopes* encountered in Section 1. A striking feature of $\Sigma(P \overset{\pi}{\to} Q)$ is that it can also be constructed as the "Minkowski average" over points $q \in Q$ (in a well-defined sense; see [Billera and Sturmfels 1992, §2]) of all of the polytopal fibers $\pi^{-1}(q)$. For an algebro-geometric interpretation of the fiber polytope $\Sigma(P \overset{\pi}{\to} Q)$ in terms of *Chow quotients* of toric varieties; see [Kapranov et al. 1991; Hu $\geq$ 1999].

As a consequence of Theorem 2.1, if one removes the top element $\hat{1}$ from $\omega_{\mathrm{coh}}(P \overset{\pi}{\to} Q)$, corresponding to the improper π-coherent subdivision $\mathcal{F} = \{P\}$, one obtains the face poset of a polytopal $(d' - d - 1)$-sphere, that is the boundary of $\Sigma(P \overset{\pi}{\to} Q)$. The generalized Baues problem asks roughly how close the whole Baues poset $\omega(P \overset{\pi}{\to} Q) - \hat{1}$ is topologically to this sphere. Before phrasing the problem precisely, we must first give the poset $\omega(P \overset{\pi}{\to} Q) - \hat{1}$ a topology. The standard way to do this is to consider its *order complex*, the abstract simplicial complex of chains in the poset [Björner 1995, (9.3)]. From here on, we will abuse notation and use the name of any poset also to refer to the topological space which is the geometric realization of its order complex.

We can now state the *Generalized Baues Problem*, in at least two forms, one stronger than the other. Both of these forms appear, implicitly or explicitly, either in the first mention of the problem by Billera and Sturmfels [1992, p. 545] or in the later formulation of [Billera et al. 1994, §3].

QUESTION 2.2 (WEAK GBP). *Is $\omega(P \overset{\pi}{\to} Q) - \hat{1}$ homotopy equivalent to a $(d' - d - 1)$-sphere ?*

QUESTION 2.3 (STRONG GBP). *Is the inclusion*

$$\omega_{\mathrm{coh}}(P \overset{\pi}{\to} Q) - \hat{1} \quad \hookrightarrow \quad \omega(P \overset{\pi}{\to} Q) - \hat{1}$$

a strong deformation retraction?

The strong GBP captures the sense we had from Figures 2 and 13 that the incoherent subdivisions were nothing more than "warts" attached to the spherical subcomplex indexed by the coherent subdivisions, and that these warts could be retracted onto this subcomplex. We should beware, however, that these pictures of small examples can be deceptive. In particular, we mention a vague metaconjecture that has several examples of empirical evidence; see [Athanasiadis 1999; de Loera et al. 1996; Athanasiadis et al. 1997, Remark 3.6]:

VAGUE METACONJECTURE 2.4. *Let $P_n \to Q_n$ be a "naturally occurring" infinite sequence of polytope surjections in which either*

$$\dim(Q_n) \to \infty \quad or \quad \dim(P_n) - \dim(Q_n) \to \infty$$

as n approaches infinity. Assume also that for some value of n there exist π-induced subdivisions of Q_n that are π-incoherent. Then as n approaches infinity, the fraction of the number of π-coherent subdivisions out of the total number of π-induced subdivisions approaches 0.

In other words, the warts take over eventually.

Besides the weak and strong versions, one can imagine other intermediate versions of the GBP. For example, one might ask whether the inclusion referred to in the strong GBP induces only a homotopy equivalence, rather than the stronger property of being a deformation retraction. We will resist naming these other versions, since they seem not to have been addressed in the literature.

Knowing that $\omega_{\mathrm{coh}}(P \overset{\pi}{\to} Q)$ is the poset of faces of a (polytopal) regular cell complex, the reader may be disappointed that we have not defined the entire Baues poset $\omega(P \overset{\pi}{\to} Q)$ to be the poset of faces in some regular cell complex, since it appears to be so in all of our small examples. For example, Figure 15 shows the order complex of $\omega(P \overset{\pi}{\to} Q) - \hat{1}$ for the example in Figure 13, which turns out to be the *barycentric subdivision* of the regular cell complex one would have liked to call "the Baues complex". Whenever such a regular cell complex exists, then of course, the order complex of $\omega(P \overset{\pi}{\to} Q) - \hat{1}$ will be its barycentric subdivision, and hence homeomorphic to the original regular cell complex. Unfortunately,

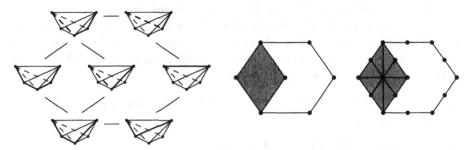

Figure 15. Left: Monotone paths as in Figure 13, defining P and Q. Middle: A regular cell complex that has $\omega(P \overset{\pi}{\to} Q) - \hat{1}$ as its poset of faces. Right: The order complex of $\omega(P \overset{\pi}{\to} Q) - \hat{1}$.

such a regular cell complex does not exist in general; relatively small examples show that lower intervals in $\omega(P \xrightarrow{\pi} Q) - \hat{1}$ need not be *homeomorphic* to spheres, which is the necessary condition for a poset to be the poset of faces of a regular cell complex [Björner 1995, (12.5)]. One way to obtain such an example is to add a seventh point a_0 to the point configuration $\mathcal{A}$ in Figure 2, in any location in the same plane. Then the unique proper subdivision of $\mathcal{A} \cup \{a_0\}$ which leaves the convex hull of $\mathcal{A}$ completely unrefined lies at the top of a lower interval that is not homeomorphic to a sphere.

3. Relations to Other Problems

Having stated the GBP, we can now explain how it relates to some of our previous questions, and to other problems in discrete geometry and topological combinatorics.

Connectivity questions. Questions 1.2, 1.3, and 1.4 are clearly related to the GBP, and appear at first glance to be weaker, in that they only ask for connectivity of a certain graph rather than homotopy sphericity of a complex. However, a positive answer to the strong GBP does not quite imply a positive answer to either of these questions. There are at least two subtleties associated with this conclusion, which we will now attempt to make precise.

For an element of a finite poset, let its *rank* be the length of the shortest saturated chain below it in the poset, so that minimal elements have rank 0. Let A_π, B_π denote the elements at rank 0 and rank 1 respectively in the Baues poset $\omega(P \xrightarrow{\pi} Q)$, and let G_π be the graph on the union $A_\pi \cup B_\pi$ obtained by restricting the Hasse diagram for $\omega(P \xrightarrow{\pi} Q)$ to this union of its bottom two ranks. Given a point set $\mathcal{A}$, let $G_\mathcal{A}$ denote its graph of triangulations and bistellar operations. Similarly, for a zonotope Z, let G_Z be its graph of cubical tilings and cube flips, and for a polytope P with a linear functional f, let $G_{P,f}$ be the graph of monotone paths and polygon moves.

The first subtlety we encounter is the relation between the graphs $G_\mathcal{A}$, G_Z, $G_{P,f}$ and the graph G_π. It is tempting to say that the barycentric subdivision of $G_\mathcal{A}$, G_Z, or $G_{P,f}$ is the same as G_π for the appropriate map π, since the vertex sets of each of these barycentric subdivisions forms a subset of the vertices of the appropriate G_π. However, it takes some work to show that these graphs coincide.

- In the case of point sets $\mathcal{A}$, a slight generalization of the *pulling construction* described by Lee [1991, § 2] can be used to show that every subdivision can be refined to a triangulation, so elements of rank 0 in the Baues poset coincide with the triangulations. Furthermore, the results and ideas of [Santos 1999] can be used to show that the elements of rank 1 coincide with the bistellar operations. Both of these assertions are easy when $\mathcal{A}$ is in general position, but otherwise become subtle.

- In the case of zonotopes Z, it is known that every zonotopal subdivision can be refined to a cubical tiling [Björner et al. 1993, Corollary 7.7.9], so elements of rank 0 in the Baues poset coincide with cubical tilings. For cube flips, one must first define these flips "correctly" for zonotopes in dimensions higher than 2, using the oriented matroid notion of mutations on the single element lifting associated to the tiling. Then a result of Santos [1997a, Theorem 4.14(ii)] combined with our previous assertion about bistellar flips can be used to prove that the elements of rank 1 in the Baues poset correspond to cube flips [Santos 1998].

- In the case of a polytope P and functional f, one can check directly that every cellular string can be refined to a monotone path, so elements of rank 0 in the Baues poset coincide with monotone paths. However, whenever the functional f is not generic in the sense that it is constant on some edge of P, one needs to be careful about how one defines polygon moves. Our previous naive definition will not suffice, as illustrated by Figure 16. Nevertheless, it is possible to correct this definition so that all elements of rank 1 in the Baues poset correspond to these corrected polygon moves [Santos 1998].

The second subtlety arises from the fact that even in cases where the GBP has a positive answer, connectivity of $\omega(P \xrightarrow{\pi} Q)$ does not necessarily imply connectivity of the graph G_π, since the 1-skeleton of $\omega(P \xrightarrow{\pi} Q)$ (or rather its order complex) contains some vertices corresponding to poset elements with ranks higher than $0, 1$. On the other hand, one would like to be able to apply the following easily verified lemma to $\omega(P \xrightarrow{\pi} Q)$:

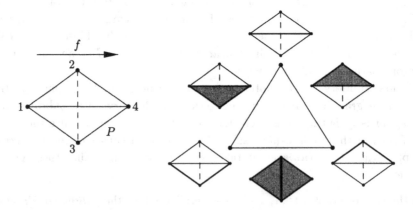

Figure 16. Left: A tetrahedron P with a nongeneric functional f for which vertices $2, 3$ have the same f-value. Right: The boundary complex of the fiber polytope $\Sigma(P \xrightarrow{\pi} Q)$, whose face poset coincides with the poset of all cellular strings (all are coherent). Note that the cellular string $(123, 234)$ labelling the bottom edge of this complex does not correspond to a polygon move as we had earlier defined it.

LEMMA 3.1. *Let X be a finite poset with a top element $\hat{1}$, and assume that X has the property that every strict principal order ideal $X_{<x} := \{x' \in X : x' < x\}$ either is connected, empty, or consists of two incomparable poset elements.*

Then the graph obtained by restricting X to its elements at rank 0 and 1 is connected.

Of course the GBP only implies the hypotheses of this lemma are satisfied with $X = \omega(P \xrightarrow{\pi} Q)$ for the strict principal order ideal $X_{<\hat{1}}$. But there is some hope that if one could prove the weak GBP in some case, then one can also prove homotopy sphericity for the rest of the order ideals $X_{<x}$, and hence can use the lemma. Under the genericity assumptions which were mentioned above for triangulations and monotone paths, one can check that these principal order ideals are Cartesian products of Baues posets for smaller polytopes, and hence their connectivity follows from positive answers to the GBP for these smaller polytopes. In particular, the positive answer for the strong GBP for monotone paths [Billera et al. 1994] (to be discussed in the next section) implies a positive answer to all of Question 1.4 under the assumption that the functional f is generic. Without such genericity assumptions, the structure of these principal order ideals may be more complicated. A specific study of these principal order ideals in the case of triangulations of a point set $\mathcal{A}$ was initiated by Santos [1999].

Flip deficiency. While we are discussing Questions 1.2, 1.3, and 1.4, it is appropriate to mention questions about the number of bistellar neighbors of a triangulation, the number of cube flip neighbors of a tiling, and the number of polygon-move neighbors of a monotone path. In the general setting of $\pi : P \to Q$, every tight π-coherent subdivision of Q represents a vertex of the $(d'-d)$-polytope $\Sigma(P \xrightarrow{\pi} Q)$, and therefore will have at least $d' - d$ neighboring tight π-coherent subdivisions lying along the edges of the polytope. On the other hand, π-induced subdivisions which are not π-coherent may have fewer neighbors, in which case we will say that the subdivision in question has *flip deficiency*. If the subdivision has no neighbors we say that it is *isolated*, which of course gives a negative answer to the GBP if $d' - d > 1$ in that case. Note that the example with all coherent cellular strings in Figure 16 shows that for monotone paths, we must either be careful to restrict ourselves to the case of a generic functional f, or else redefine what is meant by a "polygon-move" in talking about flip-deficiency.

Flip deficiency has been very well-explored for cubical tilings of zonotopes in the guise of counting *simplicial regions* of hyperplane arrangements or *mutations* in oriented matroids; see [Richter-Gebert 1992, Introduction § 3] for a nice summary. For triangulations, flip-deficiency has been explored only more recently; see [de Loera et al. 1999; Santos 1997b]. For monotone paths, the question of flip deficiency appears not to have been considered much at all.

A related question concerns the level of connectivity of the graphs G_A, G_Z, $G_{P,f}$ of triangulations and bistellar moves, tilings and cube flips, monotone paths and polygon moves respectively. For each of these graphs, the induced subgraph on the coherent elements is the 1-skeleton of $(d' - d)$-polytope and hence is $(d' - d)$-*vertex-connected* in the graph-theoretic sense by Balinski's Theorem [Ziegler 1995, § 3.5]. One can ask whether the entire graphs G_A, G_Z, $G_{P,f}$ share the same level of vertex-connectivity, which is stronger than saying that every vertex has at least $d' - d$ neighbors. Very recently, Azaola and Santos [Azaola and Santos 1999] proved the first nontrivial positive result in this direction, showing that for point sets A in $\mathbb{R}^d$ with $d + 4$ points (so $d' - d = 3$), the graph of triangulations and bistellar moves is 3-connected. The question has only been resolved negatively in some cases where flip-deficiency exists [de Loera et al. 1999; Santos 1997b].

Extension spaces, MacPhersonians and OM-Grassmannians. Let Z be a d-dimensional zonotope generated by a set of n vectors V. As mentioned in Section 1, one can associate to V its oriented matroid $\mathcal{M}$. The Bohne–Dress Theorem [Bohne 1992; Richter-Gebert and Ziegler 1994] then implies that the Baues poset $\omega(P \xrightarrow{\pi} Q)$ is isomorphic to the *extension poset* $\mathcal{E}(\mathcal{M}^*)$, consisting of all single-element extensions of the dual oriented matroid $\mathcal{M}^*$ ordered by weak maps. The following *Extension Space Conjecture* [Björner et al. 1993, § 7.2] appears not to be attributable to any single source:

CONJECTURE 3.2. *For a realizable oriented matroid $\mathcal{N}$, the order complex of the extension poset $\mathcal{E}(\mathcal{N}) - \hat{1}$ is homotopy equivalent to a* $(\mathrm{rank}(\mathcal{N}) - 1)$-*sphere.*

Hence the extension space conjecture is equivalent to the special case of the weak GBP dealing with zonotopal subdivisions. We will discuss positive cases of this conjecture, mostly taken from [Sturmfels and Ziegler 1993], in Section 4, but we mention that the results of Mnëv and Richter-Gebert [1993] show that one cannot remove the assumption that $\mathcal{N}$ is realizable. They cleverly construct nonrealizable oriented matroids $\mathcal{N}$ of rank 4 for which $\mathcal{E}(\mathcal{N}) - \hat{1}$ is disconnected!

The extension space $\mathcal{E}(\mathcal{N})$ is also closely related to certain combinatorial models of Grassmannians called *OM-Grassmannians* (see [Anderson 1999a] in this same volume, or [Richter-Gebert 1992, Introduction, § 4; Rambau 1996, § 1.2; Mnëv and Ziegler 1993] for fuller discussions). Briefly, given an oriented matroid $\mathcal{M}$, the *OM-Grassmannian* $\mathbb{G}_k(\mathcal{M})$ is the poset of rank k oriented matroids which are *strong images* [Björner et al. 1993, § 7.7] of $\mathcal{M}$, ordered by weak maps. If $\mathcal{M}$ has rank d, the order complex of $\mathbb{G}_k(\mathcal{M})$ is intended as a combinatorial model for the Grassmannian of k-planes in $\mathbb{R}^d$. In the special case where $\mathcal{M}$ is the *Boolean* or *free* oriented matroid on d elements, $\mathbb{G}_k(\mathcal{M})$ is called the *MacPhersonian* MacP(d, k), due to its occurrence in the work of Gelfand and MacPherson [1992; [MacPherson 1993]] on combinatorial formulas for characteristic classes. The following was conjectured by MacPherson when $\mathcal{M}$ is Boolean, and for all realizable $\mathcal{M}$ by Mnëv and Ziegler [Richter-Gebert 1992, Conjecture 4.2; Mnëv and Ziegler 1993, Conjecture 2.2].

CONJECTURE 3.3. *If $\mathfrak{M}$ is a realizable oriented matroid of rank d, then $\mathbb{G}_k(\mathfrak{M})$ is homotopy equivalent to the Grassmannian of k-planes in $\mathbb{R}^d$.*

Babson [1993] showed that Conjecture 3.3 is true for $k \leq 2$, and in the Boolean case, that MacP($d, 3$) is homotopy equivalent to the appropriate Grasmmannian. See [Mnëv and Ziegler 1993].

The relation to extension spaces and the Baues problem is that the extension poset $\mathcal{E}(\mathfrak{M})$ is a double cover of $\mathbb{G}_{d-1}(\mathfrak{M})$ in the sense that there is a two-to-one order-preserving map $\mathcal{E}(\mathfrak{M}) \to \mathbb{G}_{d-1}(\mathfrak{M})$. As a consequence, one can view the conjecture that $\mathbb{G}_{d-1}(\mathfrak{M})$ is homotopy equivalent to the Grassmannian of $(d-1)$-planes in $\mathbb{R}^d$ (or $(d-1)$-dimensional real projective space) as a projectivized version of the Extension Space Conjecture. This also implies that the positive results of Sturmfels and Ziegler [1993] on the Extension Space Conjecture 3.2 give some special cases of Conjecture 3.3.

4. Positive Results

In this section we review results which give a positive answer to the weak or strong GBP. The methods used tend to segregate into the three paradigms described below, where we have indicated the references whose proofs exemplify these paradigms:

Retraction: A proof of the strong GBP, by exhibiting an explicit homotopy retracting $\omega(P \overset{\pi}{\to} Q)$ onto $\omega_{\mathrm{coh}}(P \overset{\pi}{\to} Q)$. See [Billera et al. 1994, Theorem 2.3; Rambau and Ziegler 1996, Theorem 1.4; Athanasiadis et al. 1997, Theorem 1.2].

Homotopies: A proof of the weak GBP by a short chain of homotopy equivalences from $\omega(P \overset{\pi}{\to} Q)$ to some poset known to have spherical homotopy type. See [Björner 1992, Theorem 2; Edelman et al. 1997, Theorem 1.2].

Deletion-Contraction: An inductive proof of the weak GBP using (sometimes implicitly) the notion of *deletion-contraction* from matroid theory. See [Billera et al. 1994, Theorem 1.2; Sturmfels and Ziegler 1993, Theorem 1.2; Edelman and Reiner 1998, Theorem 3; Rambau and Santos 1997, Theorem 1.1].

Recall the general set-up: we consider a linear surjection of polytopes $\pi : P \to Q$ with P, Q being d', d-dimensional, respectively, and with P having n vertices. We divide our discussion of positive results into the following categories:

- $d = 1$ (monotone paths),
- $d' - d = 2$ (low codimension),
- P=cube (zonotopal tilings),
- $n - d' = 1$ or P=simplex (triangulations),
- cyclic polytopes.

The case $d = 1$: Monotone paths. The original paper of Billera, Kapranov and Sturmfels that posed the GBP [Billera et al. 1994] proves both the weak and strong GBP for monotone paths and cellular strings, under our usual genericity assumption that f is nonconstant along each edge of P. Rambau and Ziegler [1996] claim that the proofs in [Billera et al. 1994] can be adapted to remove this assumption. There are two proofs given in [Billera et al. 1994], one which follows the Retraction paradigm in proving the strong GBP (their Theorem 2.3) and one which implicityly uses the Deletion-Contraction paradigm (their Theorem 1.2) to prove the weak GBP.

This settles the original problem of Baues [1980, Conjecture 7.4], which is the special case in which the polytope P is a permutohedron and f is a generic functional. The weak GBP for cellular strings on zonotopes, as in Baues' special case, also follows from work of Björner [1992, Theorem 2] (motivated by the preprint version of [Billera et al. 1994]), which is a good example of the Homotopies paradigm. Björner observes that cellular strings on a zonotope Z are the same as what he calls the *essential chains* in the *poset of regions* [Edelman 1984] of the hyperplane arrangement which is the polar dual [Ziegler 1995, § 7.3] to Z. An essential chain in a bounded poset is a chain from the bottom element to the top element in which every step corresponds to a noncontractible (open) interval. Björner shows that the subposet of essential chains ordered by refinement is homotopy equivalent to the order complex of all chains in the proper part of the poset. For the poset of regions of a hyperplane arrangement, the homotopy type is known to be spherical by work of Edelman and Walker [1985]. Björner actually proves his result not just for zonotopes or realized oriented matroids, but for an arbitrary oriented matroid, where the notion of a cellular string and the poset of regions still make sense.

The case $d' - d \leq 2$: Low codimension. In very low codimension there is not much to say. If $d' - d = 0$ then $P = Q$ and the only π-induced subdivision of Q is the improper one. In the case $d' - d = 1$, there are exactly two proper π-induced subdivisions of Q, one coming from the "top" faces of P with respect to the projection π, the other coming from the "bottom" faces. Both of these subdivisions are coherent, and hence $\omega_{\mathrm{coh}}(P \xrightarrow{\pi} Q)$, $\omega(P \xrightarrow{\pi} Q)$ are both 0-spheres.

In the case $d' - d = 2$, the fiber polytope $\Sigma(P \xrightarrow{\pi} Q)$ is a polygon, and hence $\omega_{\mathrm{coh}}(P \xrightarrow{\pi} Q)$ is its boundary circle. Rambau and Ziegler [1996] use the Retraction paradigm to prove the strong GBP in this case.

The case $P =$ cube: zonotopal tilings. We saw in Section 2 that the case when P is a d'-cube corresponds to the case of zonotopal subdivisions and tilings of the zonotope $Z = Q = \pi(P)$. Furthermore, if $\mathcal{M}$ denotes the oriented matroid associated to the generating segments V of Z, then we saw that the Baues poset is the same as the poset of *single-element extensions* $\mathcal{E}(\mathcal{M}^*)$ for the dual oriented matroid $\mathcal{M}^*$, and the weak GBP is the same as the Extension Space Conjecture (Conjecture 3.2) for $\mathcal{M}^*$.

The extension space conjecture was investigated by Sturmfels and Ziegler [1993], who proved most of the strongest positive results at present. They showed that an inductively defined technical hypothesis called *strong Euclideanness* on the oriented matroid $\mathcal{M}$ implies that the extension space conjecture holds, using the Deletion-Contraction paradigm. They then showed that an oriented matroid on n elements with rank r is strongly Euclidean under various hypotheses: if $r \leq 3$, or $n - r \leq 2$, or when $\mathcal{M}$ is the *alternating* oriented matroid $C^{n,r}$ that comes from a cyclic arrangement of vectors [Björner et al. 1993, § 9.4]. Since oriented matroid duality exchanges r for $n - r$ and keeps n fixed, and since the alternating oriented matroids satisfy $(C^{n,r})^* \cong C^{n,n-r}$, their results imply the weak GBP when P is a d'-cube and $Q = Z$ is a d-dimensional zonotope under the following conditions:

- $d' - d \leq 3$, or
- $d \leq 2$, or
- Z is a cyclic zonotope.

It was also shown by Bailey [1997] that the hypothesis of strong Euclideanness holds for $\mathcal{M}^*$ when $\mathcal{M}$ is the oriented matroid associated to a d-dimensional zonotope having $d + 1$ generic generating segments, but with arbitrary multiple copies of each segment. Hence the weak GBP also holds for tilings of such zonotopes. We remark that for $d = 2$, the cubical tilings of these zonotopes (hexagons) were enumerated by MacMahon [1915–16, vol. 2, § X] in 1899.

Before closing our discussion of the Baues problem for tilings, we would like to mention an important result of Santos which shows that the GBP for zonotopal tilings is a special case of the GBP for triangulations. To any realized oriented matroid $\mathcal{M}$ one can associate a polytope $\Lambda(\mathcal{M})$ known as its *Lawrence polytope* [Bayer and Sturmfels 1990; Billera and Munson 1984; Santos 1997a, Chapter 4; Björner et al. 1993, § 9.3], using the technique of *Gale transforms*. This construction, due to Jim Lawrence (unpublished; see [Ziegler 1995, p. 183]) gives an encoding of all the information of the oriented matroid $\mathcal{M}$ into the face lattice of the polytope $\Lambda(\mathcal{M})$, and is useful for transferring matroid constructions and examples into the world of polytopes.

THEOREM 4.1 [Santos 1997a, Theorem 4.14; Huber et al. 1998]. *Let Z be a zonotope with associated oriented matroid $\mathcal{M}$. There is a natural bijection between the subdivisions of $\Lambda(\mathcal{M})$ and the zonotopal subdivisions of Z (=single-element liftings of $\mathcal{M}$) which induces an isomorphism between the associated Baues posets.*

Consequently, a negative answer to the GBP for zonotopal tilings produces a negative answer for triangulations. We remark that the Lawrence construction applies more generally to oriented matroids $\mathcal{M}$ which are not necessarily realizable, yielding a *matroid polytope* $\Lambda(\mathcal{M})$ [Björner et al. 1993, § 9.1] rather than a polytope. Santos' result also applies in this situation, where one defines the triangulation of a matroid polytope via his definition of a *triangulation of an oriented matroid* [Santos 1997a]. This definition unifies previous notions of such triangulations that had been proposed by Billera and Munson [1984] and Anderson [1999b].

The case $n - d' = 1$ or $P = $ simplex: triangulations. When $n - d' = 1$ the polytope P must be an n-dimensional simplex Δ^{n-1}. We saw that in this case the Baues poset is the poset of subdivisions of the point set $\mathcal{A}$, where $\mathcal{A}$ is the image under π of the vertices of Δ^{n-1}.

When the dimension d of $\mathcal{A}$'s ambient space is very small, as in the case of zonotopal tilings in $\mathbb{R}^1$, there is not much to say. For $d \leq 1$ every subdivision is coherent, and the secondary (or fiber) polytope $\Sigma(\mathcal{A})$ is a Cartesian product of simplices whose dimensions are given by the multiplicities of the interior points in $\mathcal{A}$, and one less than the multiplicities of the end points.

For $d = 2$, things start to get interesting. The fact that the graph of triangulations and bistellar operations is connected follows from work of Lawson [1977], who gave an algorithm which starts with any triangulation and moves it toward a particular coherent triangulation called the *Delaunay triangulation*. Joe [1989] observed that this procedure does *not* work in general for $d = 3$. However, Rajan [1994] and Edelsbrunner and Shah [1992] observed that a generalization of this flipping procedure works to move a particular coherent triangulation to the unique coherent triangulation which is induced by some chosen set of heights. In fact, this procedure amounts to nothing more than linear programming on the secondary polytope $\Sigma(\mathcal{A})$.

It is claimed at the end of [Billera et al. 1994] that one can positively answer the GBP for $\mathcal{A}$ in $\mathbb{R}^2$, and this was justified under the extra assumption that the points lie in general position by Edelman and Reiner [1998] using the Deletion-Contraction paradigm. The idea in their proof is to choose an extreme point v of $\mathcal{A}$, and use the fact that every subdivision of $\mathcal{A}$ gives rise to a lower-dimensional subdivision of the *vertex figure* of $\mathcal{A}$ at v. This gives an order-preserving map of subdivision posets which is shown to induce a homotopy equivalence by a technical argument, akin to the usual Quillen Fiber Lemma [Björner 1995, (10.5)]. The question of flip-deficiency for $\mathcal{A}$ in $\mathbb{R}^2$ was resolved by de Loera, Santos and Urrutia [1999]. They give a clever counting argument involving Euler's formula, showing that every triangulation has at least $|\mathcal{A}| - 3$ bistellar neighbors, so there is no flip deficiency.

For $\mathcal{A}$ in $\mathbb{R}^3$ and higher dimensions, our knowledge of bistellar connectivity and the GBP for triangulations is astoundingly limited. Very recent work of Azaola and Santos [1999] shows that in low codimension, $d' - d = 3$, the graph of triangulations and bistellar operations is 3-connected, so in particular, there is no flip deficiency in this case. De Loera, Santos and Urrutia [de Loera et al. 1999] used a similar counting argument as in the $d = 2$ case to show that for $\mathcal{A}$ in $\mathbb{R}^3$ in general position *and convex position* (i.e., no point of $\mathcal{A}$ is in the convex hull of the rest) there can be no flip deficiency. Both of these positive results are tight, in a sense, since an unpublished example of de Loera, Santos and Urrutia gives a triangulation of a configuration of 8 points in $\mathbb{R}^3$ with one point interior, having only 3 bistellar neighbors. They also exhibit in [de Loera et al. 1999], a triangulation of 9 points in $\mathbb{R}^3$ in general position with on point interior, having only 4 bistellar neighbors, and a triangulation of 10 points in convex general position in $\mathbb{R}^4$ having only 4 bistellar neighbors.

There are relatively few families of polytopes in higher dimensions whose triangulations have been well-studied, other than the cyclic polytopes which will be discussed in the next heading. We mention a few of these other families here.

Triangulations of the d-cube which use few maximal simplices are desirable for the purposes of fixed point algorithms [Todd 1976; Ziegler 1995, Problem 5.10]. Therefore one would be interested in algorithms which enumerate the triangulations, such as the program PUNTOS [de Loera 1995a], which enumerates all the triangulations lying in the same connected component of the graph of bistellar operations as the coherent triangulations. Unfortunately, de Loera [de Loera 1995b, Theorem 2.3.20; 1996] has shown that incoherent triangulations of the d-cube exist for $d \geq 4$ (including some with flip deficiency) so it is not known whether one can produce all triangulations of the cube by this method.

We momentarily digress to point out a (perhaps) surprising fact about the triangulations of a point set $\mathcal{A}$ which are extremal with respect to the number of maximal simplices — they need not be coherent! Such an example comes from work of Ohsugi and Hibi [1997], and was further analyzed by de Loera, Firla and Ziegler; see [Firla and Ziegler 1997]. This example is a point configuration $\mathcal{A}$ having 15 points in $\mathbb{R}^9$ lying in convex position (in fact, having all coordinates 0 or 1), for which the the maximal number of maximal simplices in a regular triangulation is smaller than for an arbitrary triangulation. This example also has the same property for triangulations with the minimal number of maximal simplices.

Cartesian products of simplices $\Delta^m \times \Delta^n$ were conjectured to have only coherent triangulations (see [Ziegler 1995, Problem 5.3]). This is true when m or n is equal to 1, as the secondary polytope in this case is known to be the *permutohedron* [Gel'fand et al. 1994, p. 243]. However, de Loera [1995b, Theorem 2.2.17; 1996] showed that there are incoherent triangulations whenever $m, n \geq 3$, and Sturmfels [1996, Theorem 10.15] showed that they exist when $m = 2$ and $n \geq 5$. A close study of the secondary polytope $\Sigma(\Delta^m \times \Delta^n)$ and its facets was initiated by Billera and Babson [1998], whose point of departure was the fact that a typical fiber of the map $\Delta^{(m+1)(n+1)-1} \to \Delta^m \times \Delta^n$ is a *transportation polytope*, i.e., the polytope of nonnegative $(m+1) \times (n+1)$ matrices with some prescribed row and column sums. The Ph.D. thesis of R. Hastings [1998] contains some interesting ways to view arbitrary triangulations of $\Delta^m \times \Delta^n$, and a few different ways to view incoherence for triangulations of point sets in general.

Another interesting family of polytopes are the (k, n)-*hypersimplices* $\Delta(k, n)$ defined in [Gelfand and MacPherson 1992] as the convex hull of all sums of k distinct standard basis vectors $e_{i_1} + \cdots + e_{i_k}$ in $\mathbb{R}^n$. Particular triangulations of the second hypersimplex $\Delta(2, n)$ were studied by de Loera, Thomas and Sturmfels [de Loera et al. 1995], and by Gelfand, Kapranov, and Zelevinsky (see [de Loera 1995b, § 2.5]). Stanley [1977] gave a triangulation of $\Delta(k, n)$ in general, which recovers a computation of its normalized volume due to Lagrange. In the special case $k = 2$ this triangulation coincides with the one given in [de Loera et al. 1995].

Cyclic polytopes $C(n, d)$. The cyclic polytope $C(n, d)$ is defined to be the convex hull of any n distinct points on the d-dimensional *moment curve*

$$\{(t, t^2, \ldots, t^d); t \in \mathbb{R}\}.$$

Cyclic polytopes play an important role in polytope theory because of the *Upper Bound Theorem* of McMullen [Ziegler 1995, § 8.4]: for any i, the cyclic polytope $C(n, d)$ achieves the maximum number of i-dimensional faces possible for a d-dimensional polytope with n vertices. Although the definition of $C(n, d)$ implicitly depends upon the parameters $t_1 < \cdots < t_n$ which are the x_1-coordinates of the points chosen on the moment curve, much of the combinatorial structure of $C(n, d)$ (including its face lattice, and its set of triangulations and subdivisions) does not depend upon this choice. Therefore we will omit the reference to these parameters except when necessary.

Note that the moment curve in $\mathbb{R}^{d'}$ maps to the moment curve in $\mathbb{R}^d$ under the natural surjection $\pi : \mathbb{R}^{d'} \to \mathbb{R}^d$ which forgets the last $d' - d$ coordinates. This equips the cyclic polytopes with natural surjections $\pi : C(n, d') \to C(n, d)$. Much has been said recently about the fiber polytopes and GBP for these natural maps, which include as special cases the study of triangulations of $C(n, d)$ when $d' = n - 1$, and the monotone paths on $C(n, d')$ with respect to the functional $f(x) = x_1$ when $d = 1$. The culmination of much of this work on the GBP was achieved very recently by Athanasiadis, Rambau, and Santos [Athanasiadis et al. 1998]. They use the deletion-contraction paradigm along with the "sliding" technique from [Rambau 1997b; Rambau and Santos 1997] to give a positive answer to the weak GBP for all of the maps $\pi : C(n, d') \to C(n, d)$. The following table gives a chronological summary of the progress on positive answers to the (weak) GBP for $\pi : C(n, d') \to C(n, d)$.

$d' = n - 1, \ d = 1$	(folklore)
$d' = n - 1, \ d = 2$	[Stasheff 1963]
$d = 1$	[Billera et al. 1994]
$d' - d \leq 2$	[Rambau and Ziegler 1996]
$d' = n - 1, \ d \leq 3$	[Edelman et al. 1997]
$d' = n - 1$	[Rambau and Santos 1997]
$d = 2, d' = n - 2$	[Athanasiadis et al. 1997]
$d = 2, n < 2d' + 2, d' \geq 9$	[Reiner 1998]
arbitrary n, d', d	[Athanasiadis et al. 1998]

In [Athanasiadis et al. 1997], the authors determine when the fiber polytope $\Sigma(C(n, d') \xrightarrow{\pi} C(n, d))$ is canonical in either of the following two ways:

- all π-induced subdivisions of $C(n, d)$ are π-coherent (this happens only when $d - d' \leq 2$, or $n - d' = 1$ and $d = 2$, excepting a few sporadic cases), or
- not all π-induced subdivisions are π-coherent, but the subset of π-coherent subdivisions does not depend upon the choice of parameters $t_1 < \cdots < t_n$ (this happens exactly if $d = 1$, $d' - d \geq 2$, and $n - d' \geq 2$).

The remaining results about cyclic polytopes deal exclusively with the case of triangulations of $C(n, d)$ and their combinatorics, that is, $n - d' = 1$. The philosophy here has been to try and generalize as many things as possible from the case $d = 2$, where the cyclic polytope $C(n, 2)$ is a convex polygon as in Figure 3. For $d = 2$ we know almost everything about the triangulations and subdivisions, as was described in Section 1. All these subdivisions of $C(n, 2)$ are coherent, so the poset of subdivisions is the face poset of the secondary polytope $\Sigma(\mathcal{A})$, the $(n - 3)$-dimensional associahedron. The 1-skeleton of the associahedron is the Hasse diagram for the Tamari poset (see Figure 3), and this poset turns out to be a lattice (oriented sideways in that figure).

In contrast to the $d = 2$ case, not every triangulation of $C(n, d)$ is coherent in general, starting with $C(9, 3)$, $C(9, 4)$, $C(9, 5)$; see [Athanasiadis et al. 1997]. It is also perhaps disappointing that triangulations of $C(n, d)$ can have flip-deficiency [Rambau and Santos 1997], and the higher Stasheff–Tamari posets are not lattices for $d \geq 4$ [Edelman et al. 1997]. On the bright side, Rambau and Santos [1997] prove that even though triangulations of $C(n, d)$ are not always coherent, they do enjoy the somewhat weaker property of being *lifting* triangulations; see [Santos 1997a, Definition 3.4; Björner et al. 1993, p. 410]. Rambau [1997b] also proves the interesting fact that triangulations of $C(n, d)$ are always *shellable* as simplicial complexes (see [Björner 1995, § 11.1] for the definition and significance of shellability).

Kapranov and Voevodsky [1991] suggested a generalization of the Tamari poset on triangulations of $C(n, 2)$ to a partial order on triangulations of $C(n, d)$, which they called the *higher Stasheff orders*, and which were studied by Edelman and Reiner [1996] under the name of *higher Stasheff-Tamari orders*. Actually, this latter paper defines two possible such orders which are related to each other, and it is not quite clear (though presumably true) that one of these orders is the same as that considered by Kapranov and Voevodsky. In [Edelman and Reiner 1996] it was proved for $d \leq 3$ that these two partial orders coincide and both are lattices, and also that for $d \leq 5$ the graph of bistellar operations on triangulations of $C(n, d)$ is connected. This last result was greatly improved by Rambau [1997b], who showed that the graph is connected for all d. In this paper, Rambau introduces the important "sliding" idea mentioned earlier: when one slides the n-th vertex on the moment curve down toward the $(n-1)$-st vertex, a subdivision of $C(n, d)$ induces a subdivision of $C(n-1, d)$. This map on subdivisions plays a crucial role in the positive answer to the weak GBP for triangulations of $C(n, d)$ in [Rambau and Santos 1997], and more generally, the weak GBP for $\pi : C(n, d') \to C(n, d)$ in [Athanasiadis et al. 1998]. Previously Edelman, Rambau and Reiner [Edelman et al. 1997] had used the lattice structure on the poset of triangulations and the Homotopies paradigm to positively answer the weak GBP for triangulations of $C(n, d)$ with $d \leq 3$. In that same paper, the authors show that for arbitrary d, both higher Stasheff–Tamari orders on the set of triangulations have proper parts which are homotopy equivalent to $(d - 4)$-spheres.

5. Negative Results

Recall from the previous section that we had a positive answer to the strong GBP by Billera, Kapranov, and Sturmfels [Billera et al. 1994] for $d = 1$, and by Rambau and Ziegler [1996] for $d' - d \leq 2$. Together these imply that a negative answer to the weak GBP would require a surjection $\pi : P \to Q$ with $d \geq 2$ and $d' - d \geq 3$ so that $d' \geq 5$. In the paper that includes their positive result, Rambau and Ziegler cleverly construct such a counterexample $\pi : P \to Q$ with the minimum possible dimensions $d' = 5, d = 2$. In fact, they show that the Baues poset $\omega(P \xrightarrow{\pi} Q)$ in this case is not homotopy equivalent to a 2-sphere by showing that it has an isolated element: a π-coherent subdivision of Q which lies below no other element of the poset! Their counterexample is also quite small and uncomplicated, in the sense that P has only 10 vertices, and the point configuration $\mathcal{A}$ in $\mathbb{R}^2$ which is the image of these 10 vertices under π is relatively simple: it consists of a triangle with three copies of each corner vertex, along with one interior point of the triangle. They also give a perturbed version of this same example in which the point configuration $\mathcal{A}$ lies in general position in $\mathbb{R}^2$, and the Baues poset is again disconnected (although it does not have any isolated points). These counterexamples can also be used to produce negative answers to the weak GBP for all d', d with $d \geq 2$ and $d' - d \geq 3$.

In light of this counterexample, attention has shifted to the motivating special cases of the GBP dealing with triangulations of point sets $\mathcal{A}$ and zonotopal tilings of a zonotope Z. Here no counterexamples have been found. The construction closest to a counterexample was provided by the previously mentioned work of Mnëv and Richter-Gebert [1993]. They produce (by two different methods) examples of rank 4 oriented matroids $\mathcal{M}$ whose extension posets $\mathcal{E}(\mathcal{M})$ contain isolated points. These examples do not give a counterexample to the Extension Space Conjecture or to the weak GBP because the oriented matroids in question are not *realizable*, that is they do not come from a zonotope Z. However, they do settle in the negative an earlier extension space conjecture which did not assume realizability of $\mathcal{M}$.

We should also view the instances of flip deficiency for triangulations found in [de Loera et al. 1999; Santos 1997b] as negative results, although they are far from settling the GBP. In particular, Santos' constructions [Santos 1997b] show that the ratio of the number of bistellar flips of a triangulation of $\mathcal{A}$ to the "expected lower bound" $|\mathcal{A}| - d - 1$ can approach zero.

We summarize the main open cases of the (weak) GBP here:

QUESTION 5.1. (i) *Is the poset of zonotopal subdivisions of a d-dimensional zonotope with d' generators homotopy equivalent to a $(d' - d - 1)$-dimensional sphere?*

(ii) *Is the poset of subdivisions of a point set $\mathcal{A}$ in $\mathbb{R}^d$ homotopy equivalent to a $(|\mathcal{A}| - d - 2)$-dimensional sphere?*

As was mentioned earlier, the work of Santos [1997a] shows that the first question is a special case of the second, and therefore a counterexample for the first would also settle the second, as well as the Extension Space Conjecture 3.2 and Conjecture 3.3.

6. Open Questions, Problems, Conjectures

The main open problems related to the GBP are Questions 1.2, 1.3, 5.1. In this section, we collect other problems and questions, some of which address more specifically the expected frontier between the cases of $\pi : P \to Q$ for which the GBP has positive and negative answer. In some cases, we go out on a limb by offering our predictions, but we warn the reader that many of these opinions are not based on very much data, and are only the opinion of this author.

We begin by conjecturing the frontier between good and bad behavior for triangulations, inspired by the positive and negative results contained in [Azaola and Santos 1999; de Loera et al. 1999].

CONJECTURE 6.1. *Let $\mathcal{A}$ be a point configuration in $\mathbb{R}^d$ with $d \leq 2$, or $d = 3$ and in convex position, or $|\mathcal{A}| - d \leq 4$. Then*

(a) *The strong GBP has positive answer for subdivisions of $\mathcal{A}$ (without the general position assumption needed in [Edelman and Reiner 1998]).*

(b) *Furthermore, the graph of triangulations and bistellar operations is $(|\mathcal{A}| - d - 1)$-vertex-connected, so in particular every triangulation of $\mathcal{A}$ has at least $|\mathcal{A}| - d - 1$ bistellar neighbors.*

(c) *On the other hand, there exists a point configuration in convex position in $\mathbb{R}^4$ and also one not in convex position in $\mathbb{R}^3$, each of which has an isolated triangulation which refines no other subdivision.*

In fact, it would be nice to have a simpler proof of the weak GBP for $\mathcal{A}$ in $\mathbb{R}^2$, even assuming general position, or perhaps a proof of the strong GBP via the Retraction paradigm.

For zonotopal tilings, one wonders whether there are realizable oriented matroids exhibiting the behavior of the counterexamples of Mněv and Richter-Gebert [1993].

QUESTION 6.2. *Does there exist a zonotope with a cubical tiling that refines no other zonotopal tiling?*

For monotone paths we know that the strong GBP has a positive answer, but an interesting question remains about connectivity via polygon moves. Note that as was mentioned in Section 3, one must be careful to define polygon moves correctly in the case where f is constant on some of the edges of P.

CONJECTURE 6.3. *For a d-dimensional polytope P, the graph of f-monotone paths and polygon moves is $(d-1)$-vertex-connected.*

In particular, every f-monotone path has at least $d-1$ neighbors in the graph $G_{P,f}$ of polygon moves.

Perhaps this can be proven by adapting the proof of the weak GBP for monotone paths given in [Billera et al. 1994, Theorem 1.2]?

One can also ask for bounds on the number of monotone paths. The existence of *neighborly* polytopes [Ziegler 1995, p. 16], in which every pair of vertices forms a boundary edge, shows that the number of f-monotone paths on P can grow exponentially in the number of vertices of P. For *coherent* monotone paths, the story is different. Let $r_d(n)$ denote the maximum number of coherent f-monotone paths for any linear functional f on a d-dimensional polytope P having n vertices. It is shown in [Athanasiadis et al. 1997, Remark 3.8] by a simple geometric argument that for d fixed, $r_d(n)$ grows no faster than $O(n^{3d-6})$. Motivated by McMullen's Upper Bound Theorem [Ziegler 1995, §8.4], one might expect that the linear functional $f(x) = x_1$ on the cyclic polytope $C(n, d)$ which induces the natural surjection $C(n, d) \to C(n, 1)$, also achieves this maximum value $r_d(n)$ for f-monotone paths. However, a counterexample is given in [Athanasiadis et al. 1997, Remark 3.8]. Nevertheless, the results of [Athanasiadis et al. 1997] show that the number of f-monotone paths for $f(x) = x_1$ on $C(n, d)$ grows like a polynomial in n of degree $d-2$, and the authors pose the following question [Athanasiadis et al. 1997, Question 3.10]:

QUESTION 6.4. *For fixed d, does $r_d(n)$ grow no faster than $O(n^{d-2})$?*

Athanasiadis [Athanasiadis 1998] has answered this question positively for $d \leq 4$.

In the special case where P is a d-dimensional zonotope having n generators, counting f-monotone paths turns out to be equivalent to counting the number of different possible linear orderings by linear functionals of a certain affine point configuration with n points in $\mathbb{R}^{d-1}$. Upper and lower bounds for the number of such functionals were addressed recently by Edelman [1998].

Although much is known about cyclic polytopes relating to the GBP, there remain several interesting open questions. One challenge is to count the number of triangulations of $C(n, d)$. Some data is given in [Athanasiadis et al. 1997, Table 4], compiled using PUNTOS [de Loera 1995a] and software of Rambau dedicated to this task. As said in the introduction, almost the only nontrivial known formula counting triangulations is the Catalan number

$$\frac{1}{n-1}\binom{2n-4}{n-2},$$

enumerating triangulations of the n-gon $C(n,2)$. We also have the following mostly trivial results:

- $C(n,1)$ has 2^{n-2} triangulations,
- $C(d+1,d)$ has 1 triangulation,
- $C(d+2,d)$ has 2 triangulations, and
- $C(d+3,d)$ has $d+3$ triangulations by the results of [Lee 1991].

Santos [1998] recently made the following conjecture for $C(d+4,d)$, based on the known data:

CONJECTURE 6.5. *Let a_d be the number of triangulations of $C(d+4,d)$. Then the second difference $a_d - 2a_{d-1} + a_{d-2}$ has the following form:*

$$a_{2k} - 2a_{2k-1} + a_{2k-2} = 2^k \qquad \text{(proved by Santos),}$$

$$a_{2k+1} - 2a_{2k} + a_{2k-1} = 2^{k+1} + k2^{k-1}.$$

He points out that this conjecture easily leads to simple closed form for a_d. Specifically, one would have

$$a_d = k_d 2^{d/2} - (d+4),$$

where

$$k_d = \begin{cases} d+8 & \text{if } d \text{ is even,} \\ (3d+23)/(2^{3/2}) \approx 1.06d + 8.13 & \text{if } d \text{ is odd.} \end{cases}$$

As was mentioned earlier, in [Athanasiadis et al. 1998] the authors give a positive answer to the GBP for all of the projections between cyclic polytopes. The special case of the projection $\pi : C(n,d') \to C(n,2)$, along with the methods used in [Athanasiadis et al. 1997, Theorem 1.2; Reiner 1998] inspire the following conjecture, which would also partly explain the importance of the interior point of Q present in the Rambau–Ziegler counterexample [Rambau and Ziegler 1996].

CONJECTURE 6.6. *The strong GBP has positive answer for $\pi : P \to Q$ if Q lies in $\mathbb{R}^2$ and all vertices of P project under π to the boundary of Q.*

The proof of the weak GBP for triangulations of cyclic polytopes given by Rambau and Santos [1997] uses the Deletion-Contraction paradigm. We next discuss some other conjectural approaches to this result, involving the relation of cyclic polytopes to *cyclic zonotopes* and *alternating matroids* [Björner et al. 1993, §9.4].

For any point configuration $\mathcal{A}$ in $\mathbb{R}^d$, the *dual point configuration* or *Gale transform* $\mathcal{A}^*$ lives in $\mathbb{R}^{|\mathcal{A}|-d-1}$ [Ziegler 1995, Lecture 6]. A single element extension of the oriented matroid $\mathcal{M}$ corresponding to $\mathcal{A}^*$ gives rise to a subdivision of $\mathcal{A}$ called a *lifting subdivision* [Björner et al. 1993, p. 410], and hence gives a map from the extension poset $\mathcal{E}(\mathcal{M})$ to the poset of subdivisions of $\mathcal{A}$. In the case where $\mathcal{A}$ is the set of vertices of a cyclic polytope $C(n,d)$, $\mathcal{A}^*$ is a cyclic arrangement of vectors [Ziegler 1995, Problem 6.13] so that $\mathcal{M}$ is the alternating matroid $C^{n,n-d-1}$. Therefore the results of [Sturmfels and Ziegler 1993] imply that $\mathcal{E}(C^{n,n-d-1})$ is homotopy equivalent to an $(n-d-2)$-sphere. Furthermore, [Rambau and Santos 1997] shows that every triangulation of $C(n,d)$ is a lifting triangulation (although it is not known whether this is true for all subdivisions), so that this map has a chance to be surjective.

CONJECTURE 6.7. *When $\mathcal{A}$ is the set of vertices of a cyclic polytope $C(n,d)$, the map described in the previous paragraph is a homotopy equivalence from the extension space $\mathcal{E}(C^{n,n-d-1})$ to the poset of subdivisions.*

A different approach relates the triangulations of cyclic polytopes to cyclic hyperplane arrangements and the work of Manin and Schechtman [1989], Ziegler [1993], and Kapranov and Voevodsky [1991] on *cyclic hyperplane arrangements and zonotopes* and the *higher Bruhat orders*.

Let $Z(n,d)$ be the d-dimensional *cyclic zonotope* with n generating segments in the directions $\{(1, t_i, t_i^2, \ldots, t_i^{d-1})\}_{i=1}^n$ for any n distinct values of the parameters $t_1 < \cdots < t_n$. The higher Bruhat orders $B(n,d)$ were defined in [Manin and Schechtman 1989], and may be thought of as a natural poset structure on the cubical tilings of $Z(n,d)$. For $d = 1$, $B(n,1)$ is the the *weak Bruhat order* [Björner et al. 1993, §2.3(b)] on the symmetric group. Ziegler [1993] observed that there were actually two natural and related (but different!) definitions for higher Bruhat orders, which he called $B(n,d)$ and $B_\subseteq(n,d)$. Among other things, he showed that the homotopy type of the second of these posets $B_\subseteq(n,d)$ is spherical. Rambau [1997a] later showed that $B(n,d)$ also has spherical homotopy type.

As was mentioned in Section 4, Kapranov and Voevodsky [1991] define a partial order on triangulations of $C(n,d)$, and Edelman and Reiner [1996] consider two such related partial orders $S_1(n,d)$ and $S_2(n,d)$, generalizing the Tamari poset on triangulations of $C(n,2)$. It is not quite clear, although presumably true, that the order $S_1(n,d)$ coincides with the order defined in [Kapranov and Voevodsky 1991]. In [Edelman et al. 1997], it is shown that both posets $S_1(n,d)$ and $S_2(n,d)$ have spherical homotopy type. Kapranov and Voevodsky also define an order-preserving map $B(n,d) \to S_1(n+2, d+1)$, and a similar map was given two definitions by Rambau in [1997b]. Rambau shows that his two definitions give the same map, but it is not clear that his map is the same as the one in [Kapranov and Voevodsky 1991]. For $d = 1$, this map coincides with a map from permutations to triangulations of an n-gon studied by Björner and Wachs [1997, §9], and by Tonks [Tonks 1997].

CONJECTURE 6.8. *The maps $B(n,d) \to S_1(n+2, d+1)$ defined by Kapranov–Voevodsky and Rambau are the same map f_{KVR}, and f_{KVR} induces a homotopy equivalence between the proper parts of these posets. (True for $d = 1$, by [Björner and Wachs 1997, §9].)*

The fact that these posets have homotopy equivalent proper parts already follows from the sphericity results previously mentioned. What does this have to do with the GBP? It is easy to see that for any zonotopal subdivision of $Z(n,d)$, the set of all cubical tilings which refine it forms an interval both in $B(n,d)$ and in $B_{\subseteq}(n,d)$. This gives a very natural order-preserving map from the Baues poset

$$\omega(I^n \to Z(n,d))$$

to the *poset of proper intervals* in $B(n,d)$ or $B_{\subseteq}(n,d)$. Similarly, for any subdivision of $C(n,d)$, the set of triangulations which refine it forms an interval both in $S_1(n,d)$ and in $S_2(n,d)$, giving an order-preserving map from the Baues poset $\omega(\Delta^{n-1} \to C(n,d))$ to the poset of proper intervals in $S_1(n,d)$ or $S_2(n,d)$.

CONJECTURE 6.9.   (a) *The image of the map from* $\omega(I^n \to Z(n,d))$ *to the poset of proper intervals in either* $B(n,d)$ *or* $B_{\subseteq}(n,d)$ *is exactly the set of noncontractible (open) intervals.* (True for $d = 1$, by [Björner and Wachs 1997, §9].)

(b) *The image of the map from* $\omega(\Delta^{n-1} \to C(n,d))$ *to the poset of proper intervals in either* $S_1(n,d)$ *or* $S_2(n,d)$ *is exactly the set of noncontractible intervals.* (True for $d \leq 3$, by [Edelman et al. 1997, Lemma 6.3].)

The previous conjecture would have two nice consequences:

(i) It would completely describe the homotopy type of all intervals (and hence compute the *Möbius function*) in both higher Bruhat orders $B(n,d), B_{\subseteq}(n,d)$ and in both higher Stasheff–Tamari orders $S_1(n,d), S_2(n,d)$. The intervals which are the images of the above maps are always isomorphic to Cartesian products of posets $B(n',d), B_{\subseteq}(n',d)$ or $S_1(n',d), S_2(n',d)$ for smaller values $n' < n$, and hence by the known sphericity results, are also homotopy spherical.

(ii) It would imply that $\omega(I^n \to Z(n,d))$ is homotopy equivalent to the suspension of the proper part of $B(n,d)$ or $B_{\subseteq}(n,d)$, and similarly $\omega(\Delta^{n-1} \to C(n,d))$ is homotopy equivalent to the suspension of the proper part of $S_1(n,d)$ or $S_2(n,d)$. This follows from the fact observed by Walker [Walker 1988] that the poset of proper intervals in a bounded poset P is homeomorphic to the suspension of the proper part of P, and the fact that the poset of proper noncontractible intervals in P is a deformation retract of the poset of all proper intervals in P [Edelman et al. 1997, Lemma 6.5].

Conjectures 6.8 and 6.9 fit into a diagram of conjectural homotopy equivalences (among spaces which are all known to be homotopy equivalent to an $(n-d-1)$-sphere) connecting the Baues posets for triangulations of $C(n,d)$ and zonotopal tilings of $Z(n,d)$ to each other and to the higher Bruhat and higher Stasheff–Tamari orders:

$$\text{Susp}(B(n,d) - \{\hat{0}, \hat{1}\}) \xrightarrow{\text{Susp}(f_{KVR})} \text{Susp}(S_1(n+2, d+1) - \{\hat{0}, \hat{1}\})$$

$$\omega(I^n \to Z(n,d)) \qquad\qquad\qquad \omega(\Delta^{n+1} \to C(n+2, d+1)).$$

In addition to the previous specific conjectures, we would also like to describe some more general problems related to the GBP.

The first of these relates to the concept of an *iterated fiber polytope* introduced by Billera and Sturmfels [Billera and Sturmfels 1994]. Given a tower $P \xrightarrow{\pi} Q \xrightarrow{\rho} R$ of linear surjections of polytopes, it was shown in [Billera and Sturmfels 1992] that the map π induces a surjection of the fiber polytopes

$$\pi : \Sigma(P \xrightarrow{\rho \circ \pi} R) \quad \longrightarrow \quad \Sigma(Q \xrightarrow{\rho} R).$$

They called the fiber polytope of this surjection the *iterated fiber polytope*

$$\Sigma(P \xrightarrow{\pi} Q \xrightarrow{\rho} R).$$

It is also clear how one can iterate this construction further, to define higher iterated fiber polytopes associated to longer towers of surjections.

QUESTION 6.10. *Study the iterated fiber polytopes for subsequences of the tower of natural surjections*

$$\Delta^{n-1} = C(n, n-1) \to C(n, n-2) \to \cdots \to C(n, 2) \to C(n, 1)$$

between cyclic polytopes. Are there any cases (like those classified in [Athanasiadis et al. 1997]) *where the structure of the iterated fiber polytope does not depend upon the choice of points on the moment curve defining $C(n,d)$?*

Another line of inquiry is suggested by the first part of Conjecture 6.9. Ziegler [1993] considers the higher Bruhat orders $B(n,d), B_{\subseteq}(n,d)$ as posets of *uniform extensions* of the affine oriented matroid corresponding to a cyclic hyperplane arrangement of hyperplanes. More generally, he introduces these *uniform extension posets* $\mathcal{U}(\mathcal{M}, g), \mathcal{U}_{\subseteq}(\mathcal{M}, g)$ for *any* affine oriented matroid $(\mathcal{M}, g)$. The map considered in Conjecture 6.9 can be generalized to a map from the extension poset $\mathcal{E}(\mathcal{M})$ to the set of proper intervals in $\mathcal{U}_{\subseteq}(\mathcal{M}, g)$: send a single element extension of $\mathcal{M}$ to the set of uniform extensions which lie below it in $\mathcal{E}(\mathcal{M})$.

QUESTION 6.11. *Study the map from the extension space $\mathcal{E}(\mathcal{M})$ to the poset of proper intervals in $\mathcal{U}_{\subseteq}(\mathcal{M}, g)$. Are there any nice classes of affine oriented matroids $(\mathcal{M}, g)$ where the image of the map is exactly the set of noncontractible proper intervals?*

As was the case in Conjecture 6.9, whenever the above question has a positive answer, the map in question induces a homotopy equivalence between $\mathcal{E}(\mathcal{M})$ and the suspension of the proper part of $\mathcal{U}(\mathcal{M}, g)$. The examples of Mnëv and Richter-Gebert [1993] show that $\mathcal{E}(\mathcal{M})$ does not always have spherical homotopy type, but it is still possible that such a homotopy equivalence may exist even in cases where sphericity fails.

Our last question relates to Stembridge's $q = -1$ *phenomenon* occurring in the context of cubical tilings of zonotopes; see [Stembridge 1994]. A zonotope Z is a centrally-symmetric polytope, and hence the antipodal map induces a natural involution ω on its set of cubical tilings. Say that a tiling of Z is *centrally symmetric* if it is fixed by this involution ω. Consider also the graph G_Z of cubical tilings and cube flips on Z. It can be shown that this graph will always be bipartite. We say that the $q = -1$ *phenomenon holds for* Z if the number of centrally symmetric tilings of Z is the same as the difference in cardinality of the two sides of the bipartition of G_Z.

Stembridge [1994] observed that known formulas counting symmetry classes of plane partitions implied the $q = -1$ phenomenon for zonotopal hexagons in the plane (with multiple copies of the three line segments which generate the hexagon as a zonotope). Further examples involving certain zonotopal octagons were found by Elnitsky [1997] and Bailey [1997]. However, one can check that the phenomenon does not hold for all zonotopes Z, as there are already examples of zonotopal octagons for which it fails.

QUESTION 6.12. *For which zonotopes Z does the $q = -1$ phenomenon hold?*

Acknowledgments

The author thanks Laura Anderson, Christos Athanasiadis, Lou Billera, Paul Edelman, Silvio Levy, Jesus de Loera, Jörg Rambau, Jürgen Richter-Gebert, Paco Santos, Colin Springer, Jim Stasheff, Günter Ziegler, and an anonymous referee. They provided many helpful conversations, insights, edits, and in some cases gave permission to use their figures or include their conjectures in this paper.

Note Added in Proof

The recent paper [Athanasiadis et al. 1999] resolves Conjecture 6.3 affirmatively for simple polytopes and for 3-dimensional polytopes, but negatively in general. Specifically, it is shown that for $d \geq 3$, the graph of f-monotone paths in a d-polytope with respect to a generic functional f is at least 2-connected, but there exist examples for each $d \geq 3$ in which the graph contains a vertex of degree 2.

References

[Adams 1956] J. F. Adams, "On the cobar construction", *Proc. Nat. Acad. Sci. U.S.A.* **42** (1956), 409–412.

[Anderson 1999a] L. Anderson, "Matroid bundles", in *New perspectives in algebraic combinatorics*, edited by L. Billera et al., Math. Sci. Res. Inst. Publications **37**, Cambridge University Press, New York, 1999.

[Anderson 1999b] L. Anderson, "Topology of combinatorial differential manifolds", *Topology* **38**:1 (1999), 197–221.

[Athanasiadis 1998] C. Athanasiadis, 1998. Personal communication.

[Athanasiadis 1999] C. A. Athanasiadis, "Piles of cubes, monotone path polytopes, and hyperplane arrangements", *Discrete Comput. Geom.* **21**:1 (1999), 117–130.

[Athanasiadis et al. 1997] C. A. Athanasiadis, J. de Loera, V. Reiner, and F. Santos, "Fiber polytopes for the projections between cyclic polytopes", preprint, 1997. See http://xxx.lanl.gov/abs/math.CO/9712257. To appear in *European J. Combin.*

[Athanasiadis et al. 1998] C. Athanasiadis, J. Rambau, and F. Santos, 1998. Personal communication.

[Athanasiadis et al. 1999] C. Athanasiadis, P. H. Edelman, and V. Reiner, "Monotone paths on polytopes", preprint, 1999. See www.math.umn.edu/~reiner/Papers/.

[Azaola and Santos 1999] M. Azaola and F. Santos, "The graph of triangulations of $d+4$ points is 3-connected", preprint, 1999. See http://matsun1.matesco.unican.es/CAG/people/santos/Articulos/index.html.

[Babson 1993] E. Babson, *A combinatorial flag space*, Ph.D. thesis, Mass. Inst. Technology, Cambridge, 1993.

[Babson and Billera 1998] E. K. Babson and L. J. Billera, "The geometry of products of minors", *Discrete Comput. Geom.* **20**:2 (1998), 231–249.

[Bailey 1997] G. D. Bailey, *Tilings of zonotopes: discriminantal arrangements, oriented matroids, and enumeration*, Ph.D. thesis, Univ. of Minnesota, Minneapolis, 1997.

[Baues 1980] H. J. Baues, *Geometry of loop spaces and the cobar construction*, Mem. Amer. Math. Soc. **230**, Amer. Math. Soc., 1980.

[Bayer and Brandt 1997] M. M. Bayer and K. A. Brandt, "Discriminantal arrangements, fiber polytopes and formality", *J. Algebraic Combin.* **6**:3 (1997), 229–246.

[Bayer and Sturmfels 1990] M. Bayer and B. Sturmfels, "Lawrence polytopes", *Canad. J. Math.* **42**:1 (1990), 62–79.

[Billera and Munson 1984] L. Billera and J. Munson, "Triangulations of oriented matroids and convex polytopes", *SIAM J. Algebraic Discrete Methods* **5** (1984), 515–525.

[Billera and Sturmfels 1992] L. J. Billera and B. Sturmfels, "Fiber polytopes", *Ann. of Math.* (2) **135**:3 (1992), 527–549.

[Billera and Sturmfels 1994] L. J. Billera and B. Sturmfels, "Iterated fiber polytopes", *Mathematika* **41**:2 (1994), 348–363.

[Billera et al. 1990] L. J. Billera, P. Filliman, and B. Sturmfels, "Constructions and complexity of secondary polytopes", *Adv. Math.* **83**:2 (1990), 155–179.

[Billera et al. 1993] L. J. Billera, I. M. Gel'fand, and B. Sturmfels, "Duality and minors of secondary polyhedra", *J. Combin. Theory Ser. B* **57**:2 (1993), 258–268.

[Billera et al. 1994] L. J. Billera, M. M. Kapranov, and B. Sturmfels, "Cellular strings on polytopes", *Proc. Amer. Math. Soc.* **122**:2 (1994), 549–555.

[Björner 1992] A. Björner, "Essential chains and homotopy type of posets", *Proc. Amer. Math. Soc.* **116**:4 (1992), 1179–1181.

[Björner 1995] A. Björner, "Topological methods", pp. 1819–1872 in *Handbook of combinatorics*, v. 2, edited by R. Graham et al., North-Holland, Amsterdam, 1995.

[Björner and Wachs 1997] A. Björner and M. L. Wachs, "Shellable nonpure complexes and posets, II", *Trans. Amer. Math. Soc.* **349**:10 (1997), 3945–3975.

[Björner et al. 1993] A. Björner, M. Las Vergnas, B. Sturmfels, N. White, and G. M. Ziegler, *Oriented matroids*, Encyclopedia of Mathematics and its Applications **46**, Cambridge University Press, New York, 1993.

[Bohne 1992] J. Bohne, *Eine kombinatorische Analyse zonotopaler Raumaufteilungen*, Dissertation, Universität Bielefeld, 1992. Also available as Preprint 92–041, Sonderforschungsbereich 343 "Diskrete Strukturen in der Mathematik", 1992.

[Connelly and Henderson 1980] R. Connelly and D. W. Henderson, "A convex 3-complex not simplicially isomorphic to a strictly convex complex", *Math. Proc. Cambridge Philos. Soc.* **88**:2 (1980), 299–306.

[Destainville et al. 1997] N. Destainville, R. Mosseri, and F. Bailly, "Configurational entropy of codimension-one tilings and directed membranes", *J. Statist. Phys.* **87**:3-4 (1997), 697–754.

[Edelman 1984] P. H. Edelman, "A partial order on the regions of $\mathbb{R}^n$ dissected by hyperplanes", *Trans. Amer. Math. Soc.* **283**:2 (1984), 617–631.

[Edelman 1998] P. H. Edelman, "Ordering points by linear functionals", preprint, 1998.

[Edelman and Reiner 1996] P. H. Edelman and V. Reiner, "The higher Stasheff–Tamari posets", *Mathematika* **43**:1 (1996), 127–154.

[Edelman and Reiner 1998] P. H. Edelman and V. Reiner, "Visibility complexes and the Baues problem for triangulations in the plane", *Discrete Comput. Geom.* **20**:1 (1998), 35–59.

[Edelman and Walker 1985] P. H. Edelman and J. W. Walker, "The homotopy type of hyperplane posets", *Proc. Amer. Math. Soc.* **94**:2 (1985), 221–225.

[Edelman et al. 1997] P. H. Edelman, J. Rambau, and V. Reiner, "On subdivision posets of cyclic polytopes", preprint 1997-030, MSRI, 1997. See http://www.msri.org/publications/preprints/online/1997-030.html.

[Edelsbrunner and Shah 1992] H. Edelsbrunner and N. Shah, "Incremental flipping works for regular triangulations", pp. 43–52 in *Proceedings of the Eighth Annual ACM Symposium on Computational Geometry* (Berlin, 1992), ACM Press, New York, 1992.

[Elnitsky 1997] S. Elnitsky, "Rhombic tilings of polygons and classes of reduced words in Coxeter groups", *J. Combin. Theory Ser. A* **77**:2 (1997), 193–221.

[Firla and Ziegler 1997] R. T. Firla and G. M. Ziegler, "Hilbert bases, unimodular triangulations, and binary covers of rational polyhedral cones", preprint, 1997. To appear in *Discrete Comput. Geom.*

[Friedman and Tamari 1967] H. Friedman and D. Tamari, "Problèmes d'associativité: une structure de treillis finis induite par une loi demi-associative", *J. Combinatorial Theory* **2** (1967), 215–242. Erratum in **4** (1968), 201.

[Gelfand and MacPherson 1992] I. M. Gelfand and R. D. MacPherson, "A combinatorial formula for the Pontrjagin classes", *Bull. Amer. Math. Soc. (N.S.)* **26**:2 (1992), 304–309.

[Gel'fand et al. 1994] I. M. Gel'fand, M. M. Kapranov, and A. V. Zelevinsky, *Discriminants, resultants, and multidimensional determinants*, Birkhäuser, Boston, MA, 1994.

[Goulden and Jackson 1983] I. P. Goulden and D. M. Jackson, *Combinatorial enumeration*, Wiley, New York, 1983.

[Haiman 1984] M. Haiman, "Constructing the associahedron", unpublished manuscript, Mass. Inst. Technology, Cambridge, MA, 1984.

[Hastings 1998] R. Hastings, *Triangulations of point configurations and polytopes*, Ph.D. thesis, Cornell University, Ithaca, NY, 1998.

[Hu ≥ 1999] Y. Hu, "A geometric invariant theory of toric varieties". In preparation.

[Huang and Tamari 1972] S. Huang and D. Tamari, "Problems of associativity: A simple proof for the lattice property of systems ordered by a semi-associative law", *J. Combinatorial Theory Ser. A* **13** (1972), 7–13.

[Huber et al. 1998] B. Huber, J. Rambau, and F. Santos, "The Cayley trick, lifting subdivisions and the Bohne–Dress Theorem on zonotopal tilings", preprint, 1998. See http://matsun1.matesco.unican.es/CAG/people/santos/Articulos/index.html.

[Huguet and Tamari 1978] D. Huguet and D. Tamari, "La structure polyédrale des complexes de parenthésages", *J. Combin. Inform. System Sci.* **3**:2 (1978), 69–81.

[Joe 1989] B. Joe, "Three-dimensional triangulations from local transformations", *SIAM J. Sci. Statist. Comput.* **10**:4 (1989), 718–741.

[Kapranov 1993] M. M. Kapranov, "The permutoassociahedron, Mac Lane's coherence theorem and asymptotic zones for the KZ equation", *J. Pure Appl. Algebra* **85**:2 (1993), 119–142.

[Kapranov and Saito 1997] M. M. Kapranov and M. Saito, "Hidden Stasheff polytopes in algebraic *K*-theory and in the space of Morse functions", preprint, 1997.

[Kapranov and Voevodsky 1991] M. M. Kapranov and V. A. Voevodsky, "Combinatorial-geometric aspects of polycategory theory: pasting schemes and higher Bruhat orders (list of results)", *Cahiers Topologie Géom. Différentielle Catégoriques* **32**:1 (1991), 11–27.

[Kapranov et al. 1991] M. M. Kapranov, B. Sturmfels, and A. V. Zelevinsky, "Quotients of toric varieties", *Math. Ann.* **290**:4 (1991), 643–655.

[Lawson 1977] C. L. Lawson, "Software for C^1-interpolation", in *Mathematical Software III*, edited by J. Rice, Academic Press, New York, 1977.

[Lee 1989] C. W. Lee, "The associahedron and triangulations of the *n*-gon", *European J. Combin.* **10**:6 (1989), 551–560.

[Lee 1991] C. W. Lee, "Regular triangulations of convex polytopes", pp. 443–456 in *Applied geometry and discrete mathematics: The Victor Klee Festschrift*, edited

by P. Gritzmann and B. Sturmfels, DIMACS Series in Discrete Mathematics and Theoretical Computer Science **4**, Amer. Math. Soc., Providence, RI, 1991.

[de Loera 1995a] J. A. de Loera, "PUNTOS", 1995. See ftp://geom.umn.edu/priv/ deloera/PUNTOS.tar. A Maple program for computing triangulations of point configurations.

[de Loera 1995b] J. A. de Loera, *Triangulations of polytopes and computational algebra*, Ph.D. thesis, Cornell University, Ithaca, NY, 1995.

[de Loera 1996] J. A. de Loera, "Nonregular triangulations of products of simplices", *Discrete Comput. Geom.* **15**:3 (1996), 253–264.

[de Loera et al. 1995] J. A. de Loera, B. Sturmfels, and R. R. Thomas, "Gröbner bases and triangulations of the second hypersimplex", *Combinatorica* **15**:3 (1995), 409–424.

[de Loera et al. 1996] J. A. de Loera, S. Hoşten, F. Santos, and B. Sturmfels, "The polytope of all triangulations of a point configuration", *Doc. Math.* **1** (1996), 103–119.

[de Loera et al. 1999] J. A. de Loera, F. Santos, and J. Urrutia, "The number of geometric bistellar neighbors of a triangulation", *Discrete Comput. Geom.* **21**:1 (1999), 131–142.

[MacMahon 1915–16] P. A. MacMahon, *Combinatory analysis* (2 v.), Cambridge University Press, 1915–16. Reprinted by Chelsea, New York, 1960.

[MacPherson 1993] R. D. MacPherson, "Combinatorial differential manifolds: a symposium in honor of John Milnor's sixtieth birthday", pp. 203–221 in *Topological methods in modern mathematics* (Stony Brook, NY, 1991), edited by L. R. Goldberg and A. Phillips, Publish or Perish, Houston, 1993.

[Manin and Schechtman 1989] Y. I. Manin and V. V. Schechtman, "Arrangements of hyperplanes, higher braid groups and higher Bruhat orders", pp. 289–308 in *Algebraic number theory: in honor of K. Iwasawa*, Advanced studies in pure mathematics **17**, Academic Press, and Tokyo, Kinokuniya, Boston, 1989.

[Milgram 1966] R. J. Milgram, "Iterated loop spaces", *Ann. of Math.* (2) **84** (1966), 386–403.

[Mnëv and Richter-Gebert 1993] N. E. Mnëv and J. Richter-Gebert, "Two constructions of oriented matroids with disconnected extension space", *Discrete Comput. Geom.* **10**:3 (1993), 271–285.

[Mnëv and Ziegler 1993] N. E. Mnëv and G. M. Ziegler, "Combinatorial models for the finite-dimensional Grassmannians", *Discrete Comput. Geom.* **10**:3 (1993), 241–250.

[Mosseri and Bailly 1993] R. Mosseri and F. Bailly, "Configurational entropy in octagonal tiling models", *Internat. J. Modern Phys. B* **7**:6-7 (1993), 1427–1436.

[MRI n.d.] Hexa-grid, a toy puzzle produced by the Mathematical Research Institute in The Netherlands (MRI). Information: Prof. D. Siersma, Budapestlaan 6, 3584 CD Utrecht, The Netherlands, mri@math.ruu.nl.

[Nabutovsky 1996] A. Nabutovsky, "Geometry of the space of triangulations of a compact manifold", *Comm. Math. Phys.* **181**:2 (1996), 303–330.

[Ohsugi and Hibi 1997] H. Ohsugi and T. Hibi, "A normal (0, 1)-polytope none of whose regular triangulations is unimodular", preprint, 1997. To appear in *Discrete Comput. Geom.*

[Pachner 1991] U. Pachner, "P.L. homeomorphic manifolds are equivalent by elementary shellings", *European J. Combin.* **12**:2 (1991), 129–145.

[Pallo 1987] J. Pallo, "On the rotation distance in the lattice of binary trees", *Inform. Process. Lett.* **25**:6 (1987), 369–373.

[Pallo 1988] J. Pallo, "Some properties of the rotation lattice of binary trees", *Comput. J.* **31**:6 (1988), 564–565.

[Pallo 1990] J. M. Pallo, "A distance metric on binary trees using lattice-theoretic measures", *Inform. Process. Lett.* **34**:3 (1990), 113–116.

[Pallo 1993] J. M. Pallo, "An algorithm to compute the Möbius function of the rotation lattice of binary trees", *RAIRO Inform. Théor. Appl.* **27**:4 (1993), 341–348.

[Rajan 1994] V. T. Rajan, "Optimality of the Delaunay triangulation in $\mathbb{R}^{d}$", *Discrete Comput. Geom.* **12**:2 (1994), 189–202.

[Rambau 1996] J. Rambau, *Polyhedral subdivisions and projections of polytopes*, Ph.D. thesis, Technische Universität Berlin, Aachen, 1996.

[Rambau 1997a] J. Rambau, "A suspension lemma for bounded posets", *J. Combin. Theory Ser. A* **80**:2 (1997), 374–379.

[Rambau 1997b] J. Rambau, "Triangulations of cyclic polytopes and higher Bruhat orders", *Mathematika* **44**:1 (1997), 162–194.

[Rambau and Santos 1997] J. Rambau and F. Santos, "The generalized Baues problem for cyclic polytopes", preprint, 1997. See http://matsun1.matesco.unican.es/CAG/people/santos/Articulos/index.html. To appear in *European J. Combin.*

[Rambau and Ziegler 1996] J. Rambau and G. M. Ziegler, "Projections of polytopes and the generalized Baues conjecture", *Discrete Comput. Geom.* **16**:3 (1996), 215–237.

[Reiner 1998] V. Reiner, "On some instances of the generalized Baues problem", preprint, 1998. See http://www.math.umn.edu/~reiner/Papers/papers.html.

[Richter-Gebert 1992] J. Richter-Gebert, *New construction methods for oriented matroids*, Ph.D. thesis, Royal Institute of Technology, Stockholm, 1992.

[Richter-Gebert and Ziegler 1994] J. Richter-Gebert and G. M. Ziegler, "Zonotopal tilings and the Bohne–Dress Theorem", pp. 211–232 in *Jerusalem Combinatorics '93* (Jerusalem, 1993), edited by H. Barcelo and G. Kalai, Contemporary Mathematics **178**, Amer. Math. Soc., Providence, 1994.

[Santos 1997a] F. Santos, "Triangulations of oriented matroids", preprint, 1997. See http://matsun1.matesco.unican.es/CAG/people/santos/Articulos/index.html.

[Santos 1997b] F. Santos, "Triangulations with very few geometric bistellar neighbors", preprint, 1997. See http://matsun1.matesco.unican.es/CAG/people/santos/Articulos/index.html. To appear in *Discrete Comput. Geom.*

[Santos 1998] F. Santos, January 1998. Personal communication.

[Santos 1999] F. Santos, "On the refinements of a polyhedral subdivision", preprint, 1999. See http://matsun1.matesco.unican.es/CAG/people/santos/index.html.

[Schönhardt 1928] E. Schönhardt, "Über die Zeulegung von Dreieckspolyedern in Tetraeder", *Math. Annalen* **98** (1928), 309–312.

[Sleator et al. 1988] D. D. Sleator, R. E. Tarjan, and W. P. Thurston, "Rotation distance, triangulations, and hyperbolic geometry", *J. Amer. Math. Soc.* **1**:3 (1988), 647–681.

[Stanley 1977] R. P. Stanley, "Eulerian partitions of a unit hypercube", pp. 49 in *Higher Combinatorics*, edited by M. Aigner, NATO Advanced Study Institute Series. Ser. C **31**, D. Reidel, Dordrecht, Holland, 1977. Appendix to article by D. Foata, "Distributions eulériennes".

[Stanley 1997] R. P. Stanley, *Enumerative combinatorics*, vol. 1, Cambridge Studies in Advanced Mathematics **49**, Cambridge University Press, Cambridge, 1997. Corrected reprint of the 1986 original.

[Stanley 1999] R. P. Stanley, *Enumerative combinatorics*, vol. 2, Cambridge Studies in Advanced Mathematics **62**, Cambridge University Press, Cambridge, 1999.

[Stanton and White 1986] D. Stanton and D. White, *Constructive combinatorics*, Undergraduate Texts in Mathematics, Springer, New York, 1986.

[Stasheff 1963] J. D. Stasheff, "Homotopy associativity of H-spaces, I", *Trans. Amer. Math. Soc.* **108** (1963), 275–292.

[Stembridge 1994] J. R. Stembridge, "Some hidden relations involving the ten symmetry classes of plane partitions", *J. Combin. Theory Ser. A* **68**:2 (1994), 372–409.

[Sturmfels 1991] B. Sturmfels, "Fiber polytopes: a brief overview", pp. 117–124 in *Special differential equations*, edited by M. Yoshida, Kyushu University, Fukuoka, 1991.

[Sturmfels 1996] B. Sturmfels, *Gröbner bases and convex polytopes*, University Lecture Series **8**, Amer. Math. Soc., Providence, RI, 1996.

[Sturmfels and Ziegler 1993] B. Sturmfels and G. Ziegler, "Extension spaces of oriented matroids", *Discrete Comput. Geometry* **10** (1993), 23–45.

[Tamari 1951] D. Tamari, *Monoïdes préordonnés et chaînes de Mal'cev*, Ph.D. thesis, Paris, 1951.

[Tamari 1962] D. Tamari, "The algebra of bracketings and their enumeration", *Nieuw Arch. Wisk.* (3) **10** (1962), 131–146.

[Todd 1976] M. J. Todd, *The computation of fixed points and applications*, Lecture Notes in Economics and Mathematical Systems **124**, Springer, Berlin, 1976.

[Tonks 1997] A. Tonks, "Relating the associahedron and the permutohedron", pp. 33–36 in *Operads: Proceedings of Renaissance Conferences* (Hartford, CT and Luminy, France, 1995)), edited by J.-L. Loday et al., Contemporary mathematics **202**, Amer. Math. Soc., Providence, RI, 1997.

[Walker 1988] J. W. Walker, "Canonical homeomorphisms of posets", *European J. Combin.* **9**:2 (1988), 97–107.

[Ziegler 1993] G. M. Ziegler, "Higher Bruhat orders and cyclic hyperplane arrangements", *Topology* **32**:2 (1993), 259–279.

[Ziegler 1995] G. M. Ziegler, *Lectures on polytopes*, Graduate Texts in Mathematics **152**, Springer, New York, 1995.

[Ziegler 1996] G. M. Ziegler, "Oriented matroids today", *Dynamic surveys in combinatorics* **4** (1996). See http://www.combinatorics.org/Surveys/index.html. Frequent updates.

VICTOR REINER
SCHOOL OF MATHEMATICS
UNIVERSITY OF MINNESOTA
MINNEAPOLIS, MN 55455
UNITED STATES
  reiner@math.umn.edu

Littlewood–Richardson Semigroups

ANDREI ZELEVINSKY

ABSTRACT. We discuss the problem of finding an explicit description of the semigroup LR_r of triples of partitions of length at most r such that the corresponding Littlewood–Richardson coefficient is non-zero. After discussing the history of the problem and previously known results, we suggest a new approach based on the "polyhedral" combinatorial expressions for the Littlewood–Richardson coefficients.

This article is based on my talk at the workshop on Representation Theory and Symmetric Functions, MSRI, April 14, 1997. I thank the organizers (Sergey Fomin, Curtis Greene, Phil Hanlon and Sheila Sundaram) for bringing together a group of outstanding combinatorialists and for giving me a chance to bring to their attention some of the problems that I find very exciting and beautiful.

In preparing the note for this volume (October 1998), I made a few small changes in the original version [Zelevinsky 1997], and added in the end a brief (and undoubtedly incomplete) account of some exciting progress achieved since April 1997. I am grateful to the referee for helpful suggestions.

For $r \geq 1$, let

$$P_r = \{(\lambda_1, \ldots, \lambda_r) \in \mathbb{Z}^r : \lambda_1 \geq \cdots \geq \lambda_r \geq 0\}$$

be the semigroup of partitions of length at most r. Our main object of study will be the set

$$LR_r = \{(\lambda, \mu, \nu) : \lambda, \mu, \nu \in P_r, \ c_{\mu\nu}^\lambda > 0\},$$

where $c_{\mu\nu}^\lambda$ is the Littlewood–Richardson coefficient. Recall that P_r is the set of highest weights of polynomial irreducible representations of $GL_r(\mathbb{C})$; if V_λ is the irreducible representation of $GL_r(\mathbb{C})$ with highest weight λ then $c_{\mu\nu}^\lambda$ is the multiplicity of V_λ in $V_\mu \otimes V_\nu$. Equivalently, the $c_{\mu\nu}^\lambda$ are the structure constants of the algebra of symmetric polynomials in r variables with respect to the basis of Schur polynomials. We call LR_r the *Littlewood–Richardson semigroup* of order r; this name is justified by the following result:

This work is supported in part by NSF grant DMS-9625511; research at MSRI is supported in part by NSF grant DMS-9022140.

Theorem 1. LR_r *is a finitely generated subsemigroup of the additive semigroup* $P_r^3 \subset \mathbb{Z}^{3r}$.

This is a special case of a much more general result well known to the experts in invariant theory. A short proof (valid for any reductive group instead of $GL_r(\mathbb{C})$) can be found in [Èlashvili 1992]; A. Elashvili attributes this proof to M. Brion and F. Knop. The semigroup property also follows at once from "polyhedral" expressions for $c_{\mu\nu}^\lambda$ that will be discussed later (see Theorem 5 and below).

Problem A. Describe LR_r explicitly.

I have been interested in this problem for several years. For example, in [Berenstein and Zelevinsky 1992] we determined the set $\{\lambda : (\lambda, \delta, \delta) \in LR_r\}$, where $\delta = (r-1, \ldots, 1, 0)$; this proves a special case of Kostant's conjecture that describes, for any semisimple Lie algebra, the irreducible components of the tensor square of the irreducible representation whose highest weight is the half-sum of positive roots. Practically nothing is known about the list of indecomposable generators of LR_r for general r. We will discuss the "dual" approach, namely we would like to describe the facets of the polyhedral convex cone $LR_r^{\mathbb{R}} \subset \mathbb{R}^{3r}$ generated by LR_r. Remarkable progress in this direction was recently made by A. Klyachko [1996]. Before discussing his results, we note that $c_{\mu\nu}^\lambda$ is given by the classical Littlewood–Richardson rule (see [Macdonald 1995], for example), which in principle makes Problem A purely combinatorial. In particular, the Littlewood–Richardson rule (or just the definition) readily implies the following properties of LR_r.

Homogeneity. $|\lambda| = |\mu| + |\nu|$ for $(\lambda, \mu, \nu) \in LR_r$, where $|\lambda| = \lambda_1 + \cdots + \lambda_r$.

Stability. $LR_{r+1} \cap \mathbb{Z}^{3r} = LR_r$, where

$$\mathbb{Z}^{3r} = \{(\lambda, \mu, \nu) \in \mathbb{Z}^{3(r+1)} : \lambda_{r+1} = \mu_{r+1} = \nu_{r+1} = 0\}.$$

Even stronger, we have $LR_{r+1} \cap \mathbb{Z}^{3r+2} = LR_r$, where

$$\mathbb{Z}^{3r+2} = \{(\lambda, \mu, \nu) \in \mathbb{Z}^{3(r+1)} : \lambda_{r+1} = 0\}.$$

Littlewood–Richardson semigroups appear naturally in several other contexts:

1. Hall algebra, extensions of abelian p-groups: see [Macdonald 1995].

2. Schubert calculus on Grassmannians: see [Fulton 1997].

3. Polynomial matrices and their invariant factors: see [Thompson 1989].

4. Eigenvalues of sums of Hermitian matrices.

We discuss the last item in more detail. For a Hermitian matrix A of order r, let $\lambda(A)$ denote the sequence of eigenvalues of A arranged in a weakly decreasing order (recall that A is Hermitian if $A^* = A$, and such a matrix always has real eigenvalues). Let HE_r denote the set of triples $(\lambda, \mu, \nu) \in \mathbb{R}^{3r}$ such that $\lambda = \lambda(A+B), \mu = \lambda(A)$, and $\nu = \lambda(B)$ for some Hermitian matrices A and B of order r. The following counterpart of Theorem 1 for HE_r is highly non-trivial.

Theorem 2. HE$_r$ *is a polyhedral convex cone in* $\mathbb{R}^{3r}$.

This theorem was announced by several authors (see below) but apparently the first complete proof was given by A. Klyachko [1996].

Problem B. Describe HE$_r$ explicitly.

Problems A and B are closely related to each other. They have a long history. Problem B was probably first posed by I. M. Gelfand in the late 40's (eigenvalues of the sum of two Hermitian matrices were studied already by H. Weyl in 1912, but I believe that I. M. Gelfand was the first who suggested studying the cone HE$_r$ as a whole rather than concentrate on individual eigenvalues). A solution was announced by V. B. Lidskii [1950], but the details of the proof were never published. F. A. Berezin and I. M. Gelfand [1956] discussed the relationships between Problems A and B; in particular, they suggested the remarkable equality

$$\mathrm{HE}_r \cap \mathbb{Z}_{\geq 0}^{3r} = \mathrm{LR}_r, \tag{1}$$

where $\mathbb{Z}_{\geq 0}$ stands for the set of nonnegative integers. A. Horn [1962] solved Problem B for $r \leq 4$ and conjectured a general answer. To formulate his conjecture we need some terminology. Let $[1, r]$ denote the set $\{1, 2, \ldots, r\}$. For a subset $I = \{i_1 < i_2 < \cdots < i_s\} \subset [1, r]$, we denote by $\rho(I) \subset P_s$ the partition

$$\rho(I) = (i_s - s, \ldots, i_2 - 2, i_1 - 1).$$

A triple (I, J, K) of subsets of $[1, r]$ will be called *HE-consistent* if I, J, K have the same cardinality s and $(\rho(I), \rho(J), \rho(K)) \in \mathrm{HE}_s$. For $\lambda \in \mathbb{R}^r$ and $I \subset [1, r]$, we will write $|\lambda|_I = \sum_{i \in I} \lambda_i$; in particular, $|\lambda|_{[1,r]} = |\lambda| = \lambda_1 + \cdots + \lambda_r$.

Horn's Conjecture. Let λ, μ, and ν be vectors in $\mathbb{R}^r$ with weakly decreasing components. Then $(\lambda, \mu, \nu) \in \mathrm{HE}_r$ if and only if $|\lambda| = |\mu| + |\nu|$ and $|\lambda|_I \leq |\mu|_J + |\nu|_K$ for all HE-consistent triples (I, J, K) of subsets of $[1, r]$.

The proofs of Horn's Conjecture and equality (1) were announced by B. V. Lidskii [1982]; unfortunately, as in the case of the paper by V. B. Lidskii [1950] mentioned earlier, the detailed proofs never appeared.

We now discuss the results in [Klyachko 1996]. First the author proves Theorem 2; moreover, he gives the following description of a set of defining linear inequalities for HE$_r$, which is very close (but not totally equivalent) to Horn's Conjecture. Modifying the definition of HE-consistent triples, we will call a triple (I, J, K) of subsets of $[1, r]$ *LR-consistent* if I, J, K have the same cardinality s and $(\rho(I), \rho(J), \rho(K)) \in \mathrm{LR}_s$.

Theorem 3 [Klyachko 1996]. *Horn's conjecture becomes true if HE-consistency in the formulation is replaced by LR-consistency.*

The fact that any $(\lambda, \mu, \nu) \in \mathrm{HE}_r$ satisfies the inequalities $|\lambda|_I \leq |\mu|_J + |\nu|_K$ for all LR-consistent triples (I, J, K) was proved independendly in [Helmke and

Rosenthal 1995]. A. Klyachko proves that these inequalities are necessary and sufficient. In fact, he makes a stronger statement which was reproduced in [Zelevinsky 1997]: he claims that all these inequalities are independent, i.e., they correspond to facets of the polyhedral convex cone HE_r. It was recently discovered by C. Woodward and P. Belkale that the last statement is false! As reported in [Fulton 1999], P. Belkale has recently shown that all the inequalities for which $c^{\rho(I)}_{\rho(J),\rho(K)} > 1$ are redundant.

A. Klyachko also proves the following weaker version of (1). Let $\mathrm{LR}_r^{\mathbb{Q}}$ be the set of all linear combinations of triples in LR_r with positive rational coefficients; equivalently, $\mathrm{LR}_r^{\mathbb{Q}} = \cup_{N \geq 1} \frac{1}{N} \mathrm{LR}_r$.

Theorem 4 [Klyachko 1996]. $\mathrm{HE}_r \cap \mathbb{Q}_{\geq 0}^{3r} = \mathrm{LR}_r^{\mathbb{Q}}$.

Theorems 3 and 4 appear in [Klyachko 1996] as a by-product of the study of stability criteria for toric vector bundles on the projective plane P^2. In view of these theorems, the equality (1) and Horn's Conjecture would follow from the affirmative answer to the following problem:

Saturation Problem. Is it true that $\mathrm{LR}_r^{\mathbb{Q}} \cap \mathbb{Z}_{\geq 0}^{3r} = \mathrm{LR}_r$?

In other words, does the fact that $c^{N\lambda}_{N\mu,N\nu} > 0$ for some $N \geq 1$ imply that $c^{\lambda}_{\mu\nu} > 0$? This is true and easy to check for $r \leq 4$. On the other hand, an obvious analogue of the problem for type B has negative answer (as pointed out to me by M. Brion, counterexamples can be found in [Èlashvili 1992]).

Examples. We list here the linear inequalities corresponding to LR-consistent triples for $r \leq 3$; combined with the conditions $\lambda_1 \geq \cdots \geq \lambda_r$, $\mu_1 \geq \cdots \geq \mu_r$, $\nu_1 \geq \cdots \geq \nu_r$, and $|\lambda| = |\mu| + |\nu|$, they provide a description of the cone HE_r.

- $r = 1$: No inequalities.

- $r = 2$: $\lambda_1 \leq \mu_1 + \nu_1$, $\lambda_2 \leq \min(\mu_1 + \nu_2, \mu_2 + \nu_1)$.

- $r = 3$: $\lambda_1 \leq \mu_1 + \nu_1$,

$$\lambda_2 \leq \min(\mu_1 + \nu_2, \mu_2 + \nu_1),$$
$$\lambda_3 \leq \min(\mu_1 + \nu_3, \mu_2 + \nu_2, \mu_3 + \nu_1),$$
$$\lambda_1 + \lambda_2 \leq \mu_1 + \mu_2 + \nu_1 + \nu_2,$$
$$\lambda_1 + \lambda_3 \leq \min(\mu_1 + \mu_2 + \nu_1 + \nu_3, \mu_1 + \mu_3 + \nu_1 + \nu_2),$$
$$\lambda_2 + \lambda_3 \leq \min(\mu_1 + \mu_2 + \nu_2 + \nu_3, \mu_1 + \mu_3 + \nu_1 + \nu_3, \mu_2 + \mu_3 + \nu_1 + \nu_2).$$

For instance, the inequality $\lambda_2 + \lambda_3 \leq \mu_1 + \mu_3 + \nu_1 + \nu_3$ corresponds to the triple $(I, J, K) = (\{2, 3\}, \{1, 3\}, \{1, 3\})$, which is LR-consistent because the triple of partitions $(\rho(I), \rho(J), \rho(K)) = ((1, 1), (1, 0), (1, 0))$ obviously belongs to LR_2.

Assuming the affirmative answer in the Saturation Problem, Theorem 3 provides a recursive procedure for describing the semigroup LR_r. Although quite

elegant, this procedure is not very explicit from combinatorial point of view. Thus, we would like to formulate the following problem:

Problem C. Find a non-recursive description of LR_r.

Equivalently, Problem C asks for a non-recursive description of LR-consistent triples. We would like to suggest an elementary combinatorial approach to this problem based on the "polyhedral" expressions for the coefficients $c_{\mu\nu}^\lambda$ given in [Berenstein and Zelevinsky 1992]. To present such an expression, it will be convenient to modify Littlewood–Richardson coefficients as follows. We will consider triples $(\bar\lambda, \bar\mu, \bar\nu)$ of dominant integral weights for the group SL_r. Let $V_{\bar\lambda}$ be the irreducible SL_r-module with highest weight $\bar\lambda$, and let $c_{\bar\lambda\bar\mu\bar\nu}$ denote the dimension of the space of SL_r-invariants in the triple tensor product $V_{\bar\lambda} \otimes V_{\bar\mu} \otimes V_{\bar\nu}$. The relationship between the $c_{\bar\lambda\bar\mu\bar\nu}$ and the Littlewood–Richardson coefficients is as follows. We will write each of the weights $\bar\lambda, \bar\mu$ and $\bar\nu$ as a nonnegative integer linear combination of fundamental weights $\omega_1, \omega_2, \ldots, \omega_{r-1}$ (in the standard numeration):

$$
\begin{aligned}
\bar\lambda &= l_1\omega_1 + \cdots + l_{r-1}\omega_{r-1}, \\
\bar\mu &= m_1\omega_1 + \cdots + m_{r-1}\omega_{r-1}, \\
\bar\nu &= n_1\omega_1 + \cdots + n_{r-1}\omega_{r-1}.
\end{aligned}
\tag{2}
$$

The definitions readily imply that if $\lambda, \mu, \nu \in P_r$ are such that $|\lambda| = |\mu| + |\nu|$ then $c_{\mu\nu}^\lambda = c_{\bar\lambda\bar\mu\bar\nu}$, where the coordinates l_s, m_s and n_s in (2) are given by

$$
l_s = \lambda_{r-s} - \lambda_{r-s+1}, \quad m_s = \mu_s - \mu_{s+1}, \quad n_s = \nu_s - \nu_{s+1}. \tag{3}
$$

Thus, the knowledge of LR_r is equivalent to the knowledge of the semigroup

$$
\overline{LR}_r = \{(\bar\lambda, \bar\mu, \bar\nu) \in \mathbb{Z}_{\geq 0}^{3(r-1)} : c_{\bar\lambda\bar\mu\bar\nu} > 0\}.
$$

Passing from LR_r to $\overline{LR}_r$ has two important advantages. First, the coefficients $c_{\bar\lambda\bar\mu\bar\nu}$ are more symmetric than the original Littlewood–Richardson coefficients: they are invariant under the 12-element group generated by all permutations of three weights $\bar\lambda, \bar\mu$ and $\bar\nu$, together with the transformation replacing each of these weights with its dual (i.e., sending (l_s, m_s, n_s) to $(l_{r-s}, m_{r-s}, n_{r-s})$). Second, the dimension of the ambient space reduces by 2, from $3r-1$ to $3(r-1)$. On the other hand, $\overline{LR}_r$ has at least one potential disadvantage: the condition $|\lambda| = |\mu| + |\nu|$ is replaced by a more complicated condition that $\sum_s s(l_s + m_s + n_s)$ is divisible by r (in more invariant terms, this means that $\bar\lambda + \bar\mu + \bar\nu$ must be a radical weight, i.e., belongs to the root lattice). To illustrate both phenomena, one can compare the description of LR_2 given above with the following description of $\overline{LR}_2$ which is equivalent to the classical Clebsch–Gordan rule: $\overline{LR}_2$ consists of triples of nonnegative integers (l_1, m_1, n_1) satisfying the triangle inequality and such that $l_1 + m_1 + n_1$ is even.

We now give a combinatorial expression for $c_{\bar{\lambda}\bar{\mu}\bar{\nu}}$ (this is one of several such expressions found in [Berenstein and Zelevinsky 1992]). Consider a triangle in $\mathbb{R}^2$, and subdivide it into small triangles by dividing each side into r equal parts and joining the points of the subdivison by the line segments parallel to the sides of our triangle. Let Y_r denote the set of all vertices of the small triangles, with the exception of the three vertices of the original triangle. Introducing barycentric coordinates, we identify Y_r with the set of integer triples (i, j, k) such that $0 \leq i, j, k < r$ and $i + j + k = r$. Let $\mathbb{Z}^{Y_r}$ be the set of integer families (y_{ijk}) indexed by Y_r; we think of $y \in \mathbb{Z}^{Y_r}$ as an integer "matrix" with Y_r as the set of "matrix positions." To every $y \in \mathbb{Z}^{Y_r}$ we associate the partial line sums

$$l_{ts}(y) = \sum_{j=t}^{s} y_{r-s,j,s-j},$$

$$m_{ts}(y) = \sum_{k=t}^{s} y_{s-k,r-s,k}, \qquad (4)$$

$$n_{ts}(y) = \sum_{i=t}^{s} y_{i,s-i,r-s},$$

where $0 \leq t \leq s \leq r$. We call these linear forms on $\mathbb{R}^{Y_r}$ *tails*, and we say that $y \in \mathbb{R}^{Y_r}$ is *tail-positive* if all tails of y are ≥ 0. We also say that a linear form on $\mathbb{R}^{Y_r}$ is *tail-positive* if it is a nonnegative linear combination of tails.

Theorem 5 [Berenstein and Zelevinsky 1992]. *For any triple* $(\bar{\lambda}, \bar{\mu}, \bar{\nu})$ *as in (2), the coefficient* $c_{\bar{\lambda}\bar{\mu}\bar{\nu}}$ *is equal to the number of tail-positive* $y \in \mathbb{Z}^{Y_r}$ *with prescribed values of line sums*

$$l_{0s}(y) = l_s, \quad m_{0s}(y) = m_s, \quad n_{0s}(y) = n_s, \qquad (5)$$

where $1 \leq s \leq r - 1$.

In other words, let $T_r \subset \mathbb{Z}^{Y_r}$ denote the semigroup of tail-positive elements, and let $\sigma : \mathbb{Z}^{Y_r} \to \mathbb{Z}^{3(r-1)}$ be the projection given by (5). Then Theorem 5 says that

$$\sigma(T_r) = \overline{\mathrm{LR}}_r. \qquad (6)$$

In particular, this implies at once that $\overline{\mathrm{LR}}_r$ (and hence LR_r) is a semigroup. Furthermore, Theorem 5 implies the following description of the convex cone $\overline{\mathrm{LR}}_r^{\mathbb{R}}$ generated by $\overline{\mathrm{LR}}_r$.

Corollary 6. *A linear form f on* $\mathbb{R}^{3(r-1)}$ *takes nonnegative values on* $\overline{\mathrm{LR}}_r^{\mathbb{R}}$ *if and only if the form $f \circ \sigma$ on* $\mathbb{R}^{Y_r}$ *is tail-positive.*

Returning to the Littlewood–Richardson semigroup LR_r, we have the projection $\partial : \mathrm{LR}_r \to \overline{\mathrm{LR}}_r$ given by (3). This projection extends by linearity to a projection $\partial : \mathbb{R}^{3r-1} \to \mathbb{R}^{3(r-1)}$, where $\mathbb{R}^{3r-1}$ is the subspace of triples $(\lambda, \mu, \nu) \in \mathbb{R}^{3r}$ satisfying $|\lambda| = |\mu| + |\nu|$. It is clear that the cone $\mathrm{LR}_r^{\mathbb{R}} \subset \mathbb{R}^{3r-1}$ is given by

the linear inequalities $f \circ \partial \geq 0$ for all linear forms f as in Corollary 6. This suggests the following strategy for determining the set of LR-consistent triples. Take a triple of subsets (I, J, K) in $[1, r]$ of the same cardinality s, consider the corresponding linear form $|\mu|_J + |\nu|_K - |\lambda|_I$ on $\mathbb{R}^{3r-1}$, write this form as $f \circ \partial$, and compute the form $f \circ \sigma$ on $\mathbb{R}^{Y_r}$. A straightforward calculation gives

$$(f \circ \sigma)(y) = \sum_{(i,j,k) \in Y_r} \left(\#(I_{>i}) - \#(J_{>r-j}) - \#(K_{>r-k}) \right) y_{ijk}, \qquad (7)$$

where $\#(I_{>i})$ stands for the number of elements of I which are $> i$. Taking into account Theorem 3, we obtain the following new criterion for LR-consistency.

Theorem 7. *A triple of subsets (I, J, K) of the same cardinality s in $[1, r]$ is LR-consistent if and only if $|\rho(I)| = |\rho(J)| + |\rho(K)|$ and the form in (7) is tail-positive.*

In particular, since every tail-positive linear form is obviously a nonnegative linear combination of the y_{ijk}, we obtain the following necessary condition for LR-consistency.

Corollary 8. *If a triple of subsets (I, J, K) in $[1, r]$ is LR-consistent then*

$$\#(I_{>i}) \geq \#(J_{>r-j}) + \#(K_{>r-k}) \qquad (8)$$

for all $(i, j, k) \in Y_r$.

It would be interesting to deduce this corollary directly from the Littlewood–Richardson rule. One can show that (8) is not sufficient for LR-consistency. In fact, Theorem 7 can be used to produce other necessary conditions for LR-consistency. One can hope to solve Problem C by generating a system of necessary and sufficient conditions for LR-consistency using this method.

Added in October 1998: Since April 1997, important progress has been achieved in the problems discussed above. Here is a very brief and incomplete discussion of some of these developments.

First, a nice self-contained exposition of Klyachko's results was given in the seminar talk [Fulton 1999]. One can also find there an account of some new developments in related areas, and an expanded list of references.

A beautiful affirmative solution to the Saturation Problem has been announced in [Knutson and Tao 1998]. The proof is entirely combinatorial, and it basically follows the "polyhedral" approach discussed above. The main new ingredient is a geometric reformulation of the Littlewood–Richardson rule in terms of certain planar configurations of line segments (the honeycomb model).

Several interesting analogues and generalizations of the polyhedral cones HE_r and LR_r were introduced and studied in [Brion 1998; Berenstein and Sjamaar 1998; Agnihotri and Woodward 1997]. It would be interesting to use a geometric approach developed in [Brion 1998] for a solution of Problem C above.

Let me conclude with the following remark. The Littlewood–Richardson coefficients and the corresponding semigroups LR_r have an obvious generalization for the tensor products of any given number (instead of just two) of polynomial irreducible representations of $GL_r(\mathbb{C})$. Let $c^\lambda_{\mu^{(1)},\dots,\mu^{(p)}}$ denote the multiplicity of V_λ in $V_{\mu^{(1)}} \otimes V_{\mu^{(p)}}$, and define

$$LR_r^{(p)} = \{(\lambda, \mu^{(1)}, \dots, \mu^{(p)}) : \lambda, \mu^{(k)} \in P_r, \ c^\lambda_{\mu^{(1)},\dots,\mu^{(p)}} > 0\} \ .$$

It might look surprising but the study of "multiple LR-semigroups" $LR_r^{(p)}$ can be reduced to that of the ordinary ones using the following fact:

Proposition 9. *We have*

$$c^\lambda_{\mu^{(1)},\dots,\mu^{(p)}} = c^{\tilde{\lambda}}_{\tilde{\mu},\tilde{\nu}} \ , \tag{9}$$

where partitions $\tilde{\lambda}, \tilde{\mu}, \tilde{\nu} \in P_{pr}$ *are given by*

$$\tilde{\nu}_{(j-1)r+i} = \delta_{j1}\lambda_i, \quad \tilde{\mu}_{(j-1)r+i} = \sum_{k=j+1}^{p} \mu_1^{(k)}, \quad \tilde{\lambda}_{(j-1)r+i} = \mu_i^{(j)} + \sum_{k=j+1}^{p} \mu_1^{(k)} \tag{10}$$

for $1 \le j \le p$ *and* $1 \le i \le r$.

Notice that the partition $\tilde{\nu}$ in Proposition 9 has the same Young diagram as λ, and that $\tilde{\lambda} - \tilde{\mu}$ is a skew diagram whose connected components are translates of $\mu^{(1)}, \dots, \mu^{(p)}$. The equality (9) is then well known; see [Macdonald 1995]. Formulas (10) define a linear embedding $\varphi : P_r^{p+1} \to P_{pr}^3$ such that $\varphi(LR_r^{(p)}) = \varphi(P_r^{p+1}) \cap LR_{pr}$. It follows in particular that the saturation property for ordinary Littlewood–Richardson semigroups implies the saturation property for the semigroups $LR_r^{(p)}$.

References

[Agnihotri and Woodward 1997] S. Agnihotri and C. Woodward, "Eigenvalues of products of unitary matrices and quantum Schubert calculus", preprint, 1997. Available at http://xxx.lanl.gov/abs/math.AG/9712013.

[Berenstein and Sjamaar 1998] A. Berenstein and R. Sjamaar, "Projections of coadjoint orbits and the Hilbert–Mumford criterion", preprint, 1998. Available at http://xxx.lanl.gov/abs/math.SG/9810125.

[Berenstein and Zelevinsky 1992] A. D. Berenstein and A. V. Zelevinsky, "Triple multiplicities for sl(r + 1) and the spectrum of the exterior algebra of the adjoint representation", *J. Algebraic Combin.* **1**:1 (1992), 7–22.

[Berezin and Gel'fand 1956] F. A. Berezin and I. M. Gel'fand, "Some remarks on the theory of spherical functions on symmetric Riemannian manifolds", *Trudy Moskov. Mat. Obshch.* **5** (1956), 311–351. In Russian.

[Brion 1998] M. Brion, "On the general faces of the moment polytope", Technical Report 438, Institut Fourier, Grenoble, 1998.

[Èlashvili 1992] A. G. Èlashvili, "Invariant algebras", pp. 57–64 in *Lie groups, their discrete subgroups, and invariant theory*, edited by E. B. Vinberg, Advances in Soviet Math. **8**, Amer. Math. Soc., Providence, RI, 1992.

[Fulton 1997] W. Fulton, *Young tableaux*, London Math. Soc. Student Texts **35**, Cambridge University Press, Cambridge, 1997.

[Fulton 1999] W. Fulton, "Eigenvalues of sums of Hermitian matrices (after A. Klyachko)", pp. 255–269 in *Séminaire Bourbaki* 1997/98 (exposés 835–849), Astérisque **252**, Soc. math. France, Paris, 1999.

[Helmke and Rosenthal 1995] U. Helmke and J. Rosenthal, "Eigenvalue inequalities and Schubert calculus", *Math. Nachr.* **171** (1995), 207–225.

[Horn 1962] A. Horn, "Eigenvalues of sums of Hermitian matrices", *Pacific J. Math.* **12** (1962), 225–241.

[Klyachko 1996] A. A. Klyachko, "Stable bundles, representation theory and Hermitian operators", Report no. 1, 1996/97, Institut Mittag-Leffler, 1996. Earlier version, University of Marne-La-Vallée, 1994.

[Knutson and Tao 1998] A. Knutson and T. Tao, "The honeycomb model of the Berenstein-Zelevinsky polytope I. Klyachko's saturation conjecture", preprint, 1998. Available at http://xxx.lanl.gov/abs/math.RT/9807160.

[Lidskii 1950] V. B. Lidskii, "On the characteristic numbers of the sum and product of symmetric matrices", *Doklady Akad. Nauk SSSR (N.S.)* **75** (1950), 769–772. In Russian.

[Lidskii 1982] B. V. Lidskii, "Spectral polyhedron of a sum of two Hermitian matrices", *Funktsional. Anal. i Prilozhen.* **16**:2 (1982), 76–77. In Russian; translation in *Funct. Anal. Appl.* **16** (1982), 139–140.

[Macdonald 1995] I. G. Macdonald, *Symmetric functions and Hall polynomials*, 2nd ed., Oxford Univ. Press, 1995.

[Thompson 1989] R. C. Thompson, "Divisibility relations satisfied by the invariant factors of a matrix product", pp. 471–491 in *The Gohberg anniversary collection* (Calgary, 1988), vol. 1, edited by H. Dym et al., Operator Theory, Advances and Applications **40**, Birkhäuser, Basel, 1989.

[Zelevinsky 1997] A. Zelevinsky, "Littlewood–Richardson semigroups", preprint 1997-044, Mathematical Sciences Research Institute, Berkeley, 1997. Available at http://www.msri.org/publications/preprints/online/1997-044.html.

ANDREI ZELEVINSKY
DEPARTMENT OF MATHEMATICS
NORTHEASTERN UNIVERSITY
BOSTON, MA 02115
UNITED STATES
andrei@neu.edu